An Introduction to the
History of Mathematics

AN INTRODUCTION TO THE HISTORY OF MATHEMATICS

Revised Edition

Howard Eves

PROFESSOR OF MATHEMATICS, UNIVERSITY OF MAINE

HOLT, RINEHART AND WINSTON

NEW YORK   CHICAGO   SAN FRANCISCO   TORONTO   LONDON

TO MIMSE

in fond recollection of countless such things as eating
ice cream together in the middle of a lake during the rain

Preface to the
Revised Edition

The very kind and continued reception given to the first edition of this book has encouraged the production of a second edition. Though this second edition is fundamentally the same as the first edition, all those parts of the book that make contact with present time are brought to date. Also, a number of minor historical and typographical corrections have been made, certain passages have been improved, the final chapter has been considerably expanded, and the bibliographies at the ends of the chapters have been brought to date and amplified.

One of the pleasantest experiences connected with the first edition was the extensive correspondence received from high school teachers and from high school students. Many of the teachers have instituted in their schools a course in the history of elementary mathematics based upon this book, and a surprising number of high school students have used the book and its problem studies as a source for project material in various courses and in state mathematics and science fairs. It is hoped that the new amplified bibliographies may better serve these folks.

It was particularly pleasing to learn that the problem studies have been so widely used to introduce young students to "junior" mathematical research. Research is perhaps the noblest endeavor of the mathematician, and any young student showing an interest in the subject should, as early as possible, be given some training in research. Of the various ways of accomplishing this, the problem-study approach seems generally to be much the best. Some of the research papers produced by high school students from problem studies of this book approach the caliber of a college Masters thesis.

It is a delight also to express here deep-felt thanks to the many college instructors and college students who have used the book, and in particular to all those who took the trouble to write kind words about it— such letters are veritable vitamin pills to an author.

And finally, it is pleasant to express gratitude to Professor Carl B. Boyer of Brooklyn College whose scholarly comments have done so much to improve the historical accuracy of this second edition.

H. E.

Stillwater, Maine
December 1963

Preface to the First Edition

In 1948, while serving on the mathematics staff at Oregon State College, I was asked to develop an undergraduate course in the history of elementary mathematics. Wishing to make the students' participation in the course something more than the usual carrying out of reading assignments capped with a term paper, I conceived the innovation of accompanying Problem Studies. This injection of some genuine mathematics into the course was carried out with much conviction and not a little trepidation. The result was gratifying; the course flourished in an astonishing and exciting way. It was about this time that my friend Dr. C. V. Newsom urged that the course be written in textbook form.

It is a pleasure to acknowledge here the assistance I received from a number of sources. The task of writing the book was considerably lightened by the understanding of the administrative officers of both Oregon State College and Champlain College of the State University of New York; I appreciate this understanding more than words can tell. In a work of this kind one leans heavily upon the existing scholarly productions in the field; my debt to the historians of mathematics is, of course, immeasurable. I thank the house of John Wiley and Sons and the University of Chicago Press for so kindly allowing me to quote from some of their publications. I wish to acknowledge gratefully the enthusiasm of my students, and the work done by Miss Nathalia St. Denis, who painstakingly typed the manuscript for me. I am also grateful for the valuable advice and the numerous suggestions of Dr. M. A. Steiner, who gave so unstintingly of his time by critically reading the entire manuscript and by doing yeoman work on the index.

But my greatest debt is due to Dr. C. V. Newsom. His encouragement launched the project, and all along the way I was able to benefit from consultations with him. It pleases me to add this preface to that long and growing list of prefaces in which Dr. Newsom's generous assistance is so gratefully acknowledged.

H. E.

Plattsburgh, New York
July 1953

Contents

PART 1
Before the Seventeenth Century

PART 2

Seventeenth Century and Later European Mathematics

11 The Calculus and Related Concepts----- 318

11-1. Introduction; 11-2. Zeno's Paradoxes; 11-3. Eudoxus' Method of Exhaustion; 11-4. Archimedes' Method of Equilibrium; 11-5. The Beginnings of Integration in Western Europe; 11-6. Cavalieri's Method of Indivisibles; 11-7. The Beginning of Differentiation; 11-8. Wallis and Barrow; 11-9. Newton; 11-10. Leibniz.

Problem Studies --------------------- 345

11.1. The Method of Exhaustion; 11.2. The Method of Equilibrium; 11.3. Some Archimedean Problems; 11.4. The Method of Indivisibles; 11.5. The Prismoidal Formula; 11.6. Differentiation; 11.7. The Binomial Theorem; 11.8. An Upper Bound for the Roots of a Polynomial; 11.9. Approximate Solution of Equations; 11.10. Algebra of Classes. BIBLIOGRAPHY

12 Transition to the Twentieth Century ---- 352

12-1. Introduction; 12-2. The Bernoulli Family; 12-3. DeMoivre, Taylor, Maclaurin; 12-4. Euler, Clairaut, d'Alembert; 12-5. Lambert, Lagrange, Monge; 12-6. Three Significant Events of the Nineteenth Century; 12-7. The Evolution of Some Basic Concepts; 12-8. Laplace, Legendre, Gauss; 12-9. Geometry in the Nineteenth Century; 12-10. Cauchy, Weierstrass, Riemann, Cantor.

An Introduction to the
History of Mathematics

Introduction

This book is an attempt to write an introduction to the history of mathematics that can serve as a textbook for a one-semester undergraduate course which meets three hours a week. Consequently, the treatment is restricted principally to "elementary" mathematics, that is, mathematics through the beginnings of calculus. It is the author's conviction that the history of a subject cannot be appreciated properly without at least a fair acquaintance with the subject itself. The historical material in this book is presented roughly in chronological order, and the reader will find that a knowledge of simple arithmetic and of high school algebra, geometry, and trigonometry is in general sufficient for a proper understanding of the first nine chapters. A knowledge of the rudiments of plane analytic geometry is needed for Chapter 10, and a knowledge of the basic concepts of the calculus is required for Chapter 11. Any concepts or developments of a more advanced nature appearing in the book are, it is hoped, sufficiently explained at the points where they are introduced. A certain amount of mathematical maturity is desirable, and whether nine, ten, or all twelve chapters are to be covered will depend upon the students' previous preparation.

An important innovation in the present treatment is the inclusion of problems. At the end of each chapter will be found a set of Problem Studies, each Problem Study containing a number of associated problems and questions. It is felt that by discussing a number of these Problem Studies in class, and working others as home assignments, the course will become more concrete and meaningful for the student, and the student's grasp of a number of historically important concepts will become crystallized. No better appreciation and understanding of number systems can be gained than by actually working with these systems. And rather than just tell a student that the ancient Greeks solved quadratic equations geometrically, let him solve some by the Greek method; in so doing he will not only thoroughly understand the Greek method, but he will achieve a deeper appreciation of Greek mathematical accomplishment. Thus it is hoped that the student will learn much of his history, as well as some interesting things in mathematics, from these Problem Studies. Some of the Problem Studies concern themselves with historically important problems and procedures, others furnish valuable material for the future teacher of either high school or college mathematics, and still others are purely recreational. Of course there are many more Problem Studies than can be covered in any one semester, and they are of varying degrees of difficulty. This will permit an instructor to select problems according to his students' abilities and to vary his assignments from year to year. At the end of the book is a collection of suggestions for the solution of many of the Problem Studies.

There is often some difficulty experienced in pronouncing the Hindu and Arabian names. Some of this difficulty can be circumvented by observing the following accepted equivalents.

Hindu Names

a like *u* in *but,* ā as in *father*

e as in *they*

i as in *pin,* ī as in *pique*

o as in *so*

u as in *put,* ū as in *rule*

c like *ch* in *church*

ś like English *sh*

If the penult is long, it is accented; if it is short, the antepenult is accented

Arabian Names

a as in *ask,* â as in *father*

e as in *bed*

i as in *pin,* î as in *pique*

o as in *obey*

u as in *put,* û as in *rule*

ḏ like *th* in *that,* ṯ like *th* in *thin*

q like *c* or *k* in *cook*

The accent is on the last syllable containing a long vowel or a vowel followed by two consonants. Otherwise the accent falls on the first syllable.

Some assistance in pronouncing the Greek names correctly is given in the Chronological Table, where the accented syllable of each such name is indicated.

The history of mathematics, even that of elementary mathematics, is so vast that only an introduction to the subject is possible in a one-semester course. The interested student will want to consult further literature. Accordingly, to each chapter has been appended a first bibliography dealing with the material of that chapter. An additional general bibliography, given immediately after the final chapter, applies to every chapter. It must be realized that the bibliography makes no pretense to completeness and is intended merely to serve as a start in any search for further material. Few periodical references have been furnished; important references of this sort are very numerous and will soon be encountered by an inquiring student. All references given are accessible and in English.

BEFORE
THE
SEVENTEENTH
CENTURY

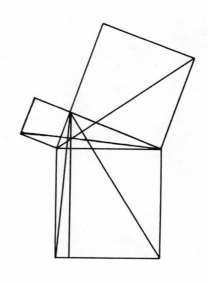

1

Numeral Systems

1-1 Primitive Counting

The concept of number and the process of counting developed so long before the time of recorded history that the manner of this development is largely conjectural. It is not difficult, though, to imagine how it probably came about. It seems fair to argue that mankind, even in most primitive times, had some number sense, at least to the extent of recognizing *more* and *less* when some objects were added to or taken from a small group, for studies have shown that some animals possess such a sense. With the gradual evolution of society, simple counting became imperative. A tribe had to know how many members it had and how many enemies, and a man found it necessary to know if his flock of sheep was decreasing in size. Probably the earliest way of keeping a count was by some simple tally method, employing the principle of one-to-one correspondence. In keeping a count on sheep, for example, one finger per sheep could be turned under. Counts could also be maintained by making collections of pebbles or sticks, by making scratches in the dirt or on a stone, by cutting notches in a piece of wood, or by tying knots in a string. Then, perhaps later, an assortment of vocal sounds was developed as a word tally against the number of objects in a small group. And still later, with the refinement of writing, an assortment of symbols was evolved to stand for these numbers. Such an imagined development is supported by reports of anthropologists in their studies of present-day primitive peoples.

In the earlier stages of the vocal period of counting, different sounds (words) were used, for example, for *two sheep* and *two men*. The abstraction of the common property of *two*, represented by some sound considered independently of any concrete association, probably was a long time in arriving. Our present number words in all likelihood originally referred to sets of certain concrete objects, but these connections, except for that perhaps relating five and hand, are now lost to us.

1-2 Number Bases

When it became necessary to make more extensive counts the counting process had to be systematized. This was done by arranging the numbers into convenient basic groups, the size of the groups being largely determined by the matching process employed. Somewhat simply put, the method was like this. Some number b was selected as a base (also called *radix* or *scale*) for counting, and names were assigned to the numbers $1, 2, \cdots, b$. Names for numbers larger than b were then essentially given by combinations of the number names already selected.

Since man's fingers furnished such a convenient matching device, it is not surprising that 10 was ultimately chosen far more often than not for the number base b. Consider, for example, our present number words, which are formed on 10 as a base. We have the special names *one, two, $\cdots$, ten* for the numbers $1, 2, \cdots, 10$. When we come to 11 we say *eleven*, which, the philologists tell us, derives from *ein lifon*, meaning *one left over*, or one over ten. Similarly *twelve* is from *twe lif* (two over ten). Then we have *thirteen* (three and ten), *fourteen* (four and ten), up through *nineteen* (nine and ten). Then comes *twenty* (*twe-tig*, or two tens), *twenty-one* (two tens and one), and so on. The word *hundred*, we are told, comes originally from a term meaning *ten times* (ten).

There is evidence that 2, 3, and 4 have served as primitive number bases. For example, there are natives of Queensland who count "one, two, two and one, two twos, much," and some African pygmies count "*a, oa, ua, oa-oa, oa-oa-a,* and *oa-oa-oa*" for 1, 2, 3, 4, 5, and 6. A certain tribe of Tierra del Fuego has its first few number names based on 3, and some South American tribes similarly use 4.

As might be expected, the quinary scale, or number system based on 5, was the first scale to be used extensively. To this day some South American tribes count by hands—"one, two, three, four, hand, hand and one," and so on. The Yukaghirs of Siberia use a mixed scale by counting "one, two, three, three and one, five, two threes, one more, two fours, ten with one missing, ten." German peasant calendars used a quinary scale as late as 1800.

There is also evidence that 12 may have been used as a base in prehistoric times, chiefly in relation to measurements. Such a base may have been suggested by the approximate number of lunations in a year, or perhaps because 12 has so many integral fractional parts. At any rate, we have 12 as the number of inches in a foot, ounces in the ancient pound, pence in a shilling, lines in an inch, hours about the clock, months in a year, and the words dozen and gross are used as higher units.

The vigesimal scale, or number system based on 20, has been widely used, and recalls man's barefoot days. This scale was used by American Indian peoples, and is best known in the well-developed Mayan number system. Celtic traces of a base 20 are found in the French *quatre-vingt* instead of *huitante*, and *quatre-vingt-dix* instead of *nonante*. Traces are also found in Gaelic, Danish, and Welsh. The Greenlanders use "one man" for 20, "two men" for 40, and so on. In English we have the frequently used word score.

The sexagesimal scale, or number system based on 60, was used by the ancient Babylonians, and is still used when measuring time and angles in minutes and seconds.

1-3   Written Number Systems

In addition to spoken numbers, *finger numbers* were at one time widely used. The expression of numbers by various positions of the fingers and hands had the advantage of transcending language differences, but, like the vocal numbers, lacked permanence and was not very suitable for performing calculations. We have already mentioned the use of marks and notches as early ways of recording numbers. In such devices we probably have man's first attempt at writing. At any rate, various written number systems gradually evolved from these primitive efforts to make permanent number records. A written number is called a *numeral,* and we now turn our attention to a simple classification of early numeral systems.

1-4   Simple Grouping Systems

Perhaps about the earliest type of numeral system that was developed is that which has been called a *simple grouping system*. In such a system some number b is selected for number base and symbols are adopted for 1, b, b^2, b^3, and so on. Then any number is expressed by using these symbols *additively,*

each symbol being repeated the required number of times. The following
illustrations will clarify the underlying principle.

A very early example of a simple grouping system is that furnished by
the Egyptian hieroglyphics, employed as far back as 3400 B.C. and chiefly
used by the Egyptians when making inscriptions on stone. Although the
hieroglyphics were sometimes used on other writing media than stone, the
Egyptians early developed two considerably more rapid writing forms for
work on papyrus, wood, and pottery. The earlier of these forms was a
running script, known as the hieratic, derived from the hieroglyphic and
used by the priesthood. From the hieratic there later evolved the demotic
writing, which was adopted for general use. The hieratic and demotic
numeral systems are not of the simple grouping type.

The Egyptian hieroglyphic numeral system is based on the scale of 10.
The symbols adopted for 1 and the first few powers of 10 are

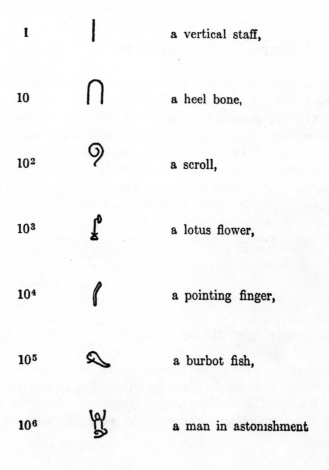

1		a vertical staff,
10		a heel bone,
10^2		a scroll,
10^3		a lotus flower,
10^4		a pointing finger,
10^5		a burbot fish,
10^6		a man in astonishment

Any number is now expressed by using these symbols additively, each symbol being repeated the required number of times. Thus

$$13015 = 1(10^4) + 3(10^3) + 1(10) + 5 = \text{𝌆 𝌆 𝌆 ⋂ ⦀}$$

We have written this number from left to right although it was more customary for the Egyptians to write from right to left.

The early Babylonians, lacking papyrus and having little access to suitable stone, resorted principally to clay as a writing medium. The inscription was pressed into a clay tablet by a stylus, resulting in the formation of wedge-shaped (cuneiform) characters. The tablet was then baked to a hardness which resulted in a permanent record. On cuneiform tablets dating from 2000 B.C. to 200 B.C. numbers less than 60 are expressed by a simple grouping system to base 10, and it is interesting that the writing is often simplified by using a subtractive symbol. The subtractive symbol and the symbols for 1 and 10 are

$$\triangledown^\triangleright, \; \triangledown, \; \langle$$

respectively. As examples of written numbers employing these symbols, we have

$$25 = 2(10) + 5 = \quad \langle\langle \;\; \triangledown\triangledown\triangledown$$

and

$$38 = 40 - 2 = \quad \langle\langle\langle\langle \; \triangledown \; \triangledown\triangledown$$

The method employed by the Babylonians for writing larger numbers will be considered below in Section 1–7.

The Attic, or Herodianic, Greek numerals were developed some time prior to the third century B.C. and constitute a simple grouping system to base 10 formed from initial letters of number names. In addition to the symbols I, Δ, H, X, M for 1, 10, 10^2, 10^3, 10^4, there is a special symbol for 5. This special symbol is an old form of Π, the initial letter of the Greek *pente* (five), and Δ is the initial letter of the Greek *deka* (ten). The other symbols can be similarly explained. The symbol for 5 was frequently used both alone and in combination with other symbols in order to shorten number representations. As an example in this numeral system we have

$$2857 = \text{X X 𝈍 H H H 𝈎 𝈍 I I}$$

in which one can note the special symbol for 5 appearing once alone and twice in combination with other symbols.

As a final example of a simple grouping system, again to base 10, we have the familiar Roman numerals. Here the basic symbols I, X, C, M for 1, 10, 10^2, 10^3 are augmented by V, L, D for 5, 50, and 500. The subtractive principle, where a symbol for a smaller unit placed before a symbol for a larger unit means the difference of the two units, was only sparingly used in ancient and medieval times. The fuller use of this principle was introduced in modern times. As an example in this system we have

$$1944 = MDCCCCXXXXIIII,$$

or, in more modern times, when the subtractive principle became common,

$$1944 = MCMXLIV$$

There has been no lack of imagination in the attempts to account for the origins of the Roman number symbols. Among the more plausible explanations, acceptable to many authorities on Latin history and epigraphy, is that I, II, III, IIII were derived from the raised fingers of the hand. The symbol X may be a compound of two V's, or may have been suggested by crossed hands or thumbs, or may have originated from the common practice when counting by strokes of crossing groups of ten. There is some evidence that the original symbols for 50, 100, and 1000 may have been the Greek aspirates X (*chi*), Θ (*theta*), and Φ (*phi*). Older forms for *chi* were

$$\downarrow, \perp, \perp, \mathsf{L},$$

all of which were used for 50 in early inscriptions. The symbol Θ for 100 probably later developed into the somewhat similar symbol C, under the influence of the fact that C is the initial letter of the Latin word *centum* (hundred). A commonly used early symbol for 1000 is C|Ɔ, which could be a variant of Φ. The symbol for 1000 became an M, under the influence of the fact that M is the initial letter of the Latin word *mille* (thousand). Five hundred, being half of 1000, was represented by |Ɔ, which later became a D. The symbols C|Ɔ and |Ɔ for 1000 and 500 are found as late as 1715.

1-5 Multiplicative Grouping Systems

There are instances where a simple grouping system developed into what may be called a *multiplicative grouping system*. In such a system, after a base b has been selected, symbols are adopted for $1, 2, \cdots, b - 1$, and a second set of symbols for $b, b^2, b^3, \cdots$. The symbols of the two sets are employed *multiplicatively* to show how many units of the higher groups are needed. Thus, if we should designate the first nine numbers by the usual symbols, but designate 10, 100, and 1000 by a, b, c, say, then in a multiplicative grouping system we would write

$$5625 = 5c6b2a5.$$

The traditional Chinese-Japanese numeral system is a multiplicative grouping system to base 10. Writing vertically, the symbols of the two basic groups and of the number 5625 are

1	一		10	十	Example: 5625
2	二		10^2	百	五
3	三		10^3	千	千
4	四				六
5	五				百
6	六				二
7	七				十
8	八				五
9	九				

1-6   Ciphered Numeral Systems

In a *ciphered numeral system*, after a base b has been selected, sets of symbols are adopted for 1, 2, $\cdots$, $b - 1$; b, $2b$, $\cdots$, $(b - 1)b$; b^2, $2b^2$, $\cdots$, $(b - 1)b^2$; and so on. Although many symbols must be memorized in such a system, the representation of numbers is compact.

The so-called Ionic, or alphabetic, Greek numeral system is of the ciphered type and can be traced as far back as about 450 B.C. It is a system based on 10 and employing 27 characters—the 24 letters of the Greek alphabet together with the symbols for the obsolete *digamma*, *koppa*, and *sampi*. Although the capital letters were used, the small letters being substituted much later, we shall here illustrate the system with the small letters. The following equivalents had to be memorized:

1	α	alpha	10	ι	iota	100	ρ	rho
2	β	beta	20	κ	kappa	200	σ	sigma
3	γ	gamma	30	λ	lambda	300	τ	tau
4	δ	delta	40	μ	mu	400	υ	upsilon
5	ϵ	epsilon	50	ν	nu	500	ϕ	phi
6	obsolete	digamma	60	ξ	xi	600	χ	chi
7	ζ	zeta	70	o	omicron	700	ψ	psi
8	η	eta	80	π	pi	800	ω	omega
9	θ	theta	90	obsolete	koppa	900	obsolete	sampi

As examples of the use of these symbols, we have

$$12 = \iota\beta, \qquad 21 = \kappa\alpha, \qquad 247 = \sigma\mu\zeta.$$

Accompanying bars or accents were used for larger numbers.

Other ciphered numeral systems are the Egyptian hieratic and demotic, Coptic, Hindu Brahmi, Hebrew, Syrian, and early Arabic. The last three, like the Ionic Greek, are *alphabetic* ciphered numeral systems.

1-7   Positional Numeral Systems

Our own numeral system is an example of a *positional numeral system* with base 10. For such a system, after the base b has been selected, basic symbols are adopted for 0, 1, 2, $\cdots$, $b - 1$. Thus there are b basic symbols, frequently called digits in our common system. Now any number N can be written uniquely in the form

$$N = a_n b^n + a_{n-1} b^{n-1} + \cdots + a_2 b^2 + a_1 b + a_0,$$

where $0 \leq a_i < b$, $i = 0, 1, \cdots, n$. We then represent the number N to base b by the sequence of basic symbols

$$a_n a_{n-1} \cdots a_2 a_1 a_0.$$

Thus a basic symbol in any given numeral represents a multiple of some power of the base, the power depending on the position in which the basic symbol occurs. In our own *Hindu-Arabic* numeral system, for example, the 2 in 206 stands for $2(10^2)$, or 200; whereas in 27, the 2 stands for $2(10)$, or 20. It is to be noted that for complete clarity some symbol for zero is needed to indicate any possible missing powers of the base. A positional numeral system is a logical, though not necessarily historical, outgrowth of a multiplicative grouping system.

The ancient Babylonians evolved, sometime between 2000 and 3000 B.C., a sexagesimal system employing the principle of position. The numeral system, however, is really a mixed one in that, although numbers exceeding 60 are written according to the positional principle, numbers within the basic 60 group are written by a simple grouping system to base 10 as explained in Section 1–4. As an illustration we have

$$524{,}551 = 2(60^3) + 25(60^2) + 42(60) + 31 = \quad \text{𒌋𒌋𒌍 𒌋𒌋𒌋𒐖𒐕𒌋𒌋𒌍𒌍 𒁹}$$

This positional numeral system suffered, until after 300 B.C., from the lack of a zero symbol. The resulting ambiguity on existing clay tablets can often be resolved only by a careful study of the context.

Very interesting is the Mayan numeral system, of remote but unknown date of origin, uncovered by the early sixteenth-century Spanish expeditions into Yucatán. This system is essentially a vigesimal one, except that the second number group is $(18)(20) = 360$ instead of $20^2 = 400$. The higher groups are of the form $(18)(20^n)$. The explanation of this discrepancy probably lies in the fact that the official Mayan year consisted of 360 days. The symbol for zero given in the table below, or some variant of this symbol, is consistently used. The numbers within the basic 20 group are written very simply by dots and dashes (pebbles and sticks) according to the following simple grouping scheme, the dot representing 1 and the dash 5.

1	•	**6**	·⎯	**11**	⎯⎯ (with •)	**16**	(• over ≡)
2	• •	**7**	·· ⎯	**12**	⎯⎯ (with ••)	**17**	(•• over ≡)
3	• • •	**8**	··· ⎯	**13**	⎯⎯ (with •••)	**18**	(••• over ≡)
4	• • • •	**9**	···· ⎯	**14**	⎯⎯ (with ••••)	**19**	(•••• over ≡)
5	⎯⎯	**10**	≡	**15**	≡≡≡	**0**	⬭

An example of a larger number, written in the vertical Mayan manner, is shown below.

$$43{,}487 = 6(18)(20^2) + 0(18)(20) + 14(20) + 7 =$$

1-8 Early Computing

Many of the computing patterns used today in elementary arithmetic, such as those for performing long multiplications and divisions, were developed as late as the fifteenth century. Two reasons are usually advanced to account for this tardy development, namely the mental difficulties and the physical difficulties encountered in such work.

The first of these, the mental difficulties, must be somewhat discounted. The impression that the ancient numeral systems are not amenable to even the simplest calculations is largely based on lack of familiarity with these systems. It is clear that addition and subtraction in a simple grouping system require only ability to count the number symbols of each kind and then to convert to higher units. No memorization of number combinations is needed. In a ciphered numeral system, if sufficient addition and multiplication tables have been memorized, the work can proceed much as we do it today. The French mathematician, Paul Tannery, attained considerable skill in multiplication with the Greek Ionic numeral system and even concluded that that system has some advantages over our present one.

The physical difficulties encountered, however, were quite real. Without a plentiful and convenient supply of some suitable writing medium, any very extended development of the arithmetic processes was bound to be hampered. It must be remembered that our common machine-made pulp paper is less than a hundred years old. The older rag paper was made by hand, was consequently expensive and scarce, and, even at that, was not introduced into Europe until the twelfth century, although it is likely that the Chinese knew how to make it a thousand years before.

An early paperlike writing material, called *papyrus*, was invented by the ancient Egyptians, and by 650 B.C. had been introduced into Greece. It was made from a water reed called *papu*. The stems of the reed were cut into long thin strips and laid side by side to form a sheet. Another layer of strips was laid on top and the whole soaked with water, after which the sheet was pressed out and dried in the sun. Probably because of a natural gum in the plant, the layers stuck together. After the sheets were dry they were readied for writing by laboriously smoothing them with a round hard object. Papyrus was too valuable to be used in any quantity as mere scratch paper.

Another early writing medium was parchment, made from the skins of animals, usually sheep and lambs. Naturally this was scarce and hard to get. Even more valuable was vellum, a parchment made from the skin of calves. In fact, so costly was parchment that the custom arose in the Middle Ages of washing the ink off old parchment manuscripts and using them over again. Such manuscripts are called *palimpsests* (*palin*, again; *psao*, rub smooth). In some instances, after the passage of years, the original writing of a palimpsest reappeared beneath the later treatment. Some interesting restorations have been made in this manner.

Small boards bearing a thin coat of wax, along with a stylus, formed a writing medium for the Romans of about two thousand years ago. Before and during the Roman Empire, sand trays were frequently used for simple counting and for the drawing of geometrical figures. And of course stone and clay were used very early for making written records.

The way around these mental and physical difficulties was the invention of the *abacus* (Greek *abax*, sand tray), which can be called the earliest mechanical computing device used by man. It appeared in many forms in parts of the ancient and medieval world. Let us describe a rudimentary form of abacus and illustrate its use in the addition and subtraction of some

Roman numbers. Draw four vertical parallel lines and label them from left to right by M, C, X, and I, and obtain a collection of convenient counters, like checkers or pennies. A counter will represent 1, 10, 100, or 1000 units according as it is placed on the I, X, C, or M line. To reduce the number of counters which may subsequently appear on a line we agree to replace any five counters on a line by one counter in the space just to the left of that line. Any number, less than 10,000, may then be represented on our frame of lines by placing not more than four counters on any line, and not more than one counter in the space just to the left of that line.

Let us now add

$$MDCCLXIX \text{ and } MXXXVII.$$

Represent the first of the two numbers by counters on the frame, as illustrated at the left in Figure 1. We now proceed to add the second number,

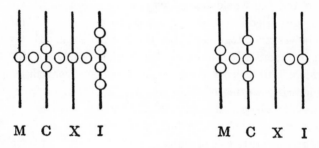

M   C   X   I   M   C   X   I

Fig. 1

working from right to left. To add the VII, put another counter between the X and I lines and two more counters on the I line. The I line now has six counters on it. We remove five of them and put instead another counter between the X and I lines. Of the three counters now between the X and I lines we "carry over" two of them as a single counter on the X line. We now add the XXX by putting three more counters on the X line. Since we now have a total of five counters on the X line they are replaced by a single counter between the C and X lines, and the two counters now found there are "carried over" as a single counter on the C line. We finally add the M by putting another counter on the M line. The final appearance of our frame is illustrated at the right in Figure 1 and the sum can be read off as $MMDCCCVI$. We have obtained the sum of the two numbers by simple mechanical operations and without requiring any scratch paper.

Subtraction is similarly carried out, except that now, instead of "carrying over" *to* the left, we may find it necessary to "borrow" *from* the left.

The Hindu-Arabic positional numeral system represents a number very simply by recording in order the number of counters belonging to the various lines of the abacus. The symbol 0 stands for a line with no counters on it. Our present addition and subtraction patterns, along with the concepts of "carrying over" and of "borrowing" may have originated in the processes for carrying out these operations on the abacus. Since, with the Hindu-Arabic numeral system, we are working with symbols instead of the actual counters, it becomes necessary either to commit the simple number combinations to memory or to have recourse to an elementary addition table.

1-9   The Hindu-Arabic Numeral System

The Hindu-Arabic numeral system is named after the Hindus, who may have invented it, and after the Arabs, who transmitted it to western Europe. The earliest preserved examples of our present number symbols are found on some stone columns erected in India about 250 B.C. by King Ásoka. Other early examples in India, if correctly interpreted, are found among records cut about 100 B.C. on the walls of a cave in a hill near Poona and in some inscriptions of about 200 A.D. carved in the caves at Nasik. These early specimens contain no zero and do not employ positional notation. Positional value, however, and also a zero, must have been introduced in India sometime before 800 A.D., for the Persian mathematician al-Khowârizmî describes such a completed Hindu system in a book of 825 A.D.

How and when the new numeral symbols first entered Europe is not settled. In all likelihood they were carried by traders and travelers of the Mediterranean coast. They are found in a tenth-century Spanish manuscript and may have been introduced into Spain by the Arabs who invaded the peninsula in 711 A.D., to remain there for several hundred years. The completed system was more widely disseminated by a twelfth-century Latin translation of al-Khowârizmî's treatise and by subsequent European works on the subject.

The next four hundred years saw the battle between the abacists and the algorists, as the advocates of the new system were called, and by 1500 our present rules in computing won supremacy. In another hundred years the abacists were almost forgotten, and by the eighteenth century no trace of an abacus was found in western Europe. Its reappearance, as a curiosity, was due to the French geometer Poncelet, who brought back a specimen to France after his release as a Russian prisoner of war following the Napoleonic Russian campaign.

There was considerable variation found in the number symbols until these symbols became stabilized by the development of printing. Our word *zero* probably comes from the Latinized form *zephirum* of the Arabic *sifr*, which in turn is a translation of the Hindu *sunya*, meaning void or empty. The Arabic *sifr* was introduced into Germany, in the thirteenth century by Nemorarius, as *cifra*, from which we have obtained our present word *cipher*.

1-10   Arbitrary Bases

We recall that to represent a number in a positional numeral system with base b we need basic symbols for the integers zero up through $b - 1$. Even though the base $b = 10$ is such an important part of our culture, the choice of 10 is really quite arbitrary, and other bases have great practical and theoretical importance. If $b \leqq 10$ we may use our ordinary digit symbols. Thus, for example, we may consider 3012 as a number expressed to base 4 with the basic symbols 0, 1, 2, 3. To make clear that the number is considered as expressed to base 4 we shall write it as $(3012)_4$. When no subscript is written it will be understood in this treatment that the number is expressed to the ordinary base 10. If $b > 10$ we must augment our digit symbols by some new basic symbols, for we always need b basic symbols. Thus, if $b = 12$, we may take 0, 1, 2, 3, 4, 5, 6, 7, 8, 9, t, e for our basic symbols, where t and e are symbols for *ten* and *eleven*. Thus, as an example, we might have $(3t1e)_{12}$.

It is easy to convert a number from a given base to the ordinary base 10. Thus we have

$$(3012)_4 = 3(4^3) + 0(4^2) + 1(4) + 2 = 198$$

and

$$(3t1e)_{12} = 3(12^3) + 10(12^2) + 1(12) + 11 = 6647.$$

If we have a number expressed in the ordinary scale we may express it to base b as follows. Letting N be the number, we have to determine the integers a_n, a_{n-1}, $\cdots$, a_0 in the expression

$$N = a_n b^n + a_{n-1} b^{n-1} + \cdots + a_2 b^2 + a_1 b + a_0,$$

where $0 \leqq a_i < b$. Dividing the above equation by b we have

$$N/b = a_n b^{n-1} + a_{n-1} b^{n-2} + \cdots + a_2 b + a_1 + a_0/b = N' + a_0/b.$$

That is, the remainder a_0 of this division is the last digit in the desired representation. Dividing N' by b we obtain

$$N'/b = a_n b^{n-2} + a_{n-1} b^{n-3} + \cdots + a_2 + a_1/b,$$

and the remainder of this division is the next to the last digit in the desired representation. Proceeding in this way we obtain all the digits a_0, a_1, $\cdots$, a_n. This procedure can be systematized quite conveniently as shown below. Suppose, for example, we wish to express 198 to the base 4. We find

$$
\begin{array}{r|l}
4 & 198 \\
4 & 49 \quad \text{remainder 2} \\
4 & 12 \quad \text{remainder 1} \\
4 & 3 \quad \text{remainder 0} \\
  & 0 \quad \text{remainder 3}
\end{array}
$$

The desired representation is $(3012)_4$. Again, suppose we wish to express 6647 to the base 12, where, again, t and e are employed to represent *ten* and *eleven*, respectively. We find

$$
\begin{array}{r|l}
12 & 6647 \\
12 & 553 \quad \text{remainder } e \\
12 & 46 \quad \text{remainder 1} \\
12 & 3 \quad \text{remainder } t \\
   & 0 \quad \text{remainder 3}
\end{array}
$$

The desired representation is $(3t1e)_{12}$.

One is apt to forget, when adding or multiplying numbers in our ordinary system, that the actual work is accomplished mentally and that the number symbols are used merely to retain a record of the mental results.

Our success and efficiency in carrying out such arithmetic operations depend on how well we have in mind the addition and multiplication tables, the learning of which absorbed so much of our time in the primary grades. With corresponding tables constructed for a given base b we can similarly perform additions and multiplications within the new system, without at any time reverting to the ordinary system.

Let us illustrate with base 4. We first construct the following addition and multiplication tables for base 4.

Addition

	0	1	2	3
0	0	1	2	3
1	1	2	3	10
2	2	3	10	11
3	3	10	11	12

Multiplication

	0	1	2	3
0	0	0	0	0
1	0	1	2	3
2	0	2	10	12
3	0	3	12	21

The addition of 2 and 3 therefore, by reference to the table, is 11, and the multiplication of 2 and 3 is 12. Using these tables, exactly as we are accustomed to use the corresponding tables for base 10, we can now perform additions and multiplications. As an example, for the multiplication of $(3012)_4$ by $(233)_4$ we have, omitting the subscript 4,

$$
\begin{array}{r}
3012 \\
233 \\
\hline
21102 \\
21102 \\
12030 \\
\hline
2101122
\end{array}
$$

Considerable familiarity with the tables will be needed in order to perform the inverse operations of subtraction and division. This, of course, is also true for the base 10; that is the secret of much of the difficulty encountered in teaching the inverse operations in the elementary grades.

Problem Studies

1.1 Written Numbers

Write 574 and 475 in (a) Egyptian hieroglyphs, (b) Roman numerals, (c) Attic Greek numerals, (d) Babylonian cuneiform, (e) traditional Chinese-Japanese, (f) alphabetic Greek, (g) Mayan numerals.

1.2 Alphabetic Greek Numeral System

(a) How many different symbols must one memorize in order to write numbers less than 1000 in alphabetic Greek? In Egyptian hieroglyphs? In Babylonian cuneiform?

(b) In the alphabetic Greek numeral system the numbers 1000, 2000, $\cdots$, 9000 were often represented by priming the symbols for 1, 2, $\cdots$, 9. Thus 1000 might appear as α'. The number 10,000, or myriad, was denoted by M. The multiplication principle was used for multiples of 10,000. Thus 20,000, 300,000, and 4,000,000 appeared as βM, λM, and υM. Write, in alphabetic Greek, the numbers 5,780, 72,803, 450,082, 3,257,888.

(c) Make an addition table up through 10 + 10 and a multiplication table up through 10 × 10 for the alphabetic Greek numeral system, using the symbols a, b, c to stand for the obsolete digamma, koppa, and sampi.

1.3 Old and Hypothetical Numeral Systems

(a) As an alternative to the cuneiform, or wedge-shaped, numeral symbols, the ancient Babylonians sometimes used *circular* numeral symbols, so named because they were formed by *circular-shaped* imprints, in clay tablets, made with a *round*-ended stylus. Here the symbols for 1 and 10 are) and $\cdot$. Write, with circular Babylonian numerals, the numbers 5,780, 72,803, 450,082, 3,257,888.

(b) Very interesting is the *Chinese scientific* (or *rod*) *numeral system*,

Fig. 2

probably two thousand or more years old. The system is essentially positional, with base 10. Figure 2 shows how the digits 1, 2, 3, 4, 5, 6, 7, 8, 9 are represented when they appear in an odd (units, hundreds, and so forth) position. But when they appear in an even (tens, thousands, and so forth) position, they are represented as shown in Figure 3. In this system a circle, O, was used for zero in the Sung Dynasty (960–1126) and later. Write, with rod numerals, the numbers 5,780, 72,803, 450,082, 3,257,888.

Fig. 3

(c) In a simple grouping system to base 5 let 1, 5, 5^2, 5^3 be represented by /, *,), (. Express the numbers 360, 252, 78, 33 in this system.

(d) In a positional numeral system to base 5 let 0, 1, 2, 3, 4 be represented by #, /, *,), (. Express the numbers 360, 252, 78, 33 in this system.

1.4   Finger Numbers

(a) Finger numbers were widely used for many centuries and from this use there developed finger processes for some simple computations. One of these processes, by giving the product of two numbers each between 5 and 10, served to reduce the memory work connected with the multiplication tables. For example, to multiply 7 by 9, raise $7 - 5 = 2$ fingers on one hand and $9 - 5 = 4$ fingers on the other hand. Now add the raised fingers, $2 + 4 = 6$, for the tens digit of the product, and multiply the closed fingers, $3 \times 1 = 3$, for the units digit of the product, giving the result 63. This process is still used by some European peasants. Prove that the method gives correct results.

(b) Locate illustrations of finger numbers and then explain the ninth-century riddle that is sometimes attributed to Alcuin (ca. 775): I saw a man holding eight in his hand, and from the eight he took seven, and six remained.

(c) Explain the following, found in Juvenal's tenth satire: "Happy is he indeed who has postponed the hour of his death so long and finally numbers his years upon his right hand."

1.5   Radix Fractions

Fractional numbers can be expressed, in the ordinary scale, by digits following a decimal point. The same notation is also used for other bases. Thus, just as the expression .3012 stands for

$$3/10 + 0/10^2 + 1/10^3 + 2/10^4,$$

the expression $(.3012)_b$ stands for

$$3/b + 0/b^2 + 1/b^3 + 2/b^4.$$

An expression like $(.3012)_b$ is called a *radix fraction for base b*. A radix fraction for base 10 is commonly called a *decimal fraction*.

(a) Show how to convert a radix fraction for base b into a decimal fraction.

(b) Show how to convert a decimal fraction into a radix fraction for base b.

(c) Express $(.3012)_4$ and $(.3t1e)_{12}$ as decimal fractions.

(d) Express .4402 as a radix fraction, first for base 7, and then for base 12.

1.6   Arithmetic in Other Scales

(a) Construct addition and multiplication tables for bases 7 and 12.

(b) Add and then multiply $(3406)_7$ and $(251)_7$, first using the tables of part (a) and then by converting to base 10. Similarly add and then multiply $(3t04e)_{12}$ and $(51tt)_{12}$.

(c) We may apply the tables for base 12 to simple mensuration problems involving feet and inches. For example, if we take one foot as a unit, then 3 feet 7 inches becomes $(3.7)_{12}$. To find, to the nearest square inch, the area of a rectangle 3 feet 7 inches long by 2 feet 4 inches wide we may multiply $(3.7)_{12}$ by $(2.4)_{12}$ and then convert the result to square feet and square inches. Complete this example.

1.7   Problems in Scales of Notation

(a) Express $(3012)_5$ and $(.3012)_5$ to base 8.

(b) For what base is $3 \times 3 = 10$? For what base is $3 \times 3 = 11$? For what base is $3 \times 3 = 12$?

(c) Can 27 represent an even number in some scale? Can 37? Can 72 represent an odd number in some scale? Can 82?

(d) Find b such that $79 = (142)_b$. Find b such that $72 = (2200)_b$.

(e) A three digit number in the scale of 7 has its digits reversed when expressed in the scale of 9. Find the three digits.

(f) What is the smallest base for which 301 represents a square integer?

(g) If $b > 2$ show that $(121)_b$ is a square integer. If $b > 4$ show that $(40,001)_b$ is divisible by $(221)_b$.

1.8 Some Recreational Aspects of the Binary Scale

The positional number system with base 2 has applications in various branches of mathematics. Also there are many games and puzzles, like the well-known game of *Nim* and the puzzle of the *Chinese rings*, having solutions which depend on this system. Following are two easy puzzles of this sort.

(a) Show how to weigh, on a simple equal arm balance, any weight w of a whole number of pounds using a set of weights of 1 pound, 2 pounds, 2^2 pounds, 2^3 pounds, and so forth, there being only one weight of each kind.

(b) Consider the following four cards containing numbers from 1 through 15.

1 9	2 10	4 12	8 12
3 11	3 11	5 13	9 13
5 13	6 14	6 14	10 14
7 15	7 15	7 15	11 15

On the first card are all those numbers whose last digit in the binary system is 1; the second contains all those numbers whose second digit from the end is 1; the third contains all those whose third digit from the end is 1; the fourth contains all those whose fourth digit from the end is 1. Now someone is asked to think of a number N from 1 through 15 and to tell on which cards N can be found. It is then easy to announce the number N by merely adding the top left numbers on the cards where it appears. Make a similar set of six cards for detecting any number from 1 through 63. It has been noted that if the numbers are written on cards weighing 1, 2, 4, $\cdots$ units, then an automaton in the form of a postal scale could express the number N.

1.9 Some Number Tricks

Many simple number tricks, where one is to "guess a selected number," have explanations depending on our own positional scale. Expose the following tricks of this kind.

(a) Someone is asked to think of a two digit number. He is then requested to multiply the tens digit by 5, add 7, double, add the units digit of the original number, and announce the final result. From this result the conjurer secretly subtracts 14 and obtains the original number.

(b) Someone is asked to think of a three digit number. He is then requested to multiply the hundreds digit by 2, add 3, multiply by 5, add 7, add the tens digit, multiply by 2, add 3, multiply by 5, add the units digit, and announce the result. From this result the conjurer secretly subtracts 235 and obtains the original number.

(c) Someone is asked to think of a three digit number whose first and third digits are different. He is then requested to find the difference between this number and that obtained by reversing the three digits. Upon disclosing only the last digit of this difference the conjurer announces the entire difference. How does the trickster do this?

Bibliography

ANDREWS, F. E., *New Numbers*. New York: Harcourt, Brace & World, Inc., 1935.

BALL, W. W. R., AND H. S. M. COXETER, *Mathematical Recreations and Essays*, 11th ed. New York: The Macmillan Company, 1939.

CAJORI, FLORIAN, *A History of Mathematical Notations*, 2 vols. Chicago: The Open Court Publishing Company, 1928–1929.

CONANT, L. L., *The Number Concept, Its Origin and Development*. New York: The Macmillan Company, 1923.

CUNNINGTON, SUSAN, *The Story of Arithmetic, a Short History of Its Origin and Development*. London: Swan Sonnenschein & Company, Ltd., 1904.

DANTZIG, TOBIAS, *Number, the Language of Science*. New York: The Macmillan Company, 1946.

FREITAG, H. T., AND A. H. FREITAG, *The Number Story*. Washington: National Council of Teachers of Mathematics, 1960.

HILL, G. F., *The Development of Arabic Numerals in Europe*. New York: Oxford University Press, 1915.

KRAITCHIK, MAURICE, *Mathematical Recreations*. New York: W. W. Norton & Company, Inc., 1942.

LARSEN, H. D., *Arithmetic for Colleges*. New York: The Macmillan Company, 1950.

MORLEY, S. G., *An Introduction to the Study of the Maya Hieroglyphs*. Washington: Government Printing Office, 1915.

ORE, OYSTEIN, *Number Theory and Its History*. New York: McGraw-Hill Book Company, Inc., 1948.

RINGENBERG, L. A., *A Portrait of 2*. Washington: National Council of Teachers of Mathematics, 1956.

SMELTZER, DONALD, *Man and Number*. New York: Emerson Books, Inc., 1958.

SMITH, D. E., *Number Stories of Long Ago*. Washington: National Council of Teachers of Mathematics, 1958.

————, AND JEKUTHIEL GINSBURG, *Numbers and Numerals*. Washington: National Council of Teachers of Mathematics, 1958.

————, AND L. C. KARPINSKI, *The Hindu-Arabic Numerals*. Boston: Ginn & Company, 1911.

SWAIN, R. L., *Understanding Arithmetic*. New York: Holt, Rinehart and Winston, Inc., 1957.

TERRY, G. S., *The Dozen System*. London: Longmans, Green & Co., Ltd., 1941.

————, *Duodecimal Arithmetic*. London: Longmans, Green & Co., Ltd., 1938.

2

Babylonian and Egyptian Mathematics

2-1 The Ancient Orient

Early mathematics required a practical basis for its development, and such a basis arose with the evolution of more advanced forms of society. It was along some of the great rivers of Africa and Asia that the new forms of society made their appearance, the Nile in Africa, the Tigris and Euphrates in western Asia, the Indus and then the Ganges in south-central Asia, and the Hwang Ho and then the Yangtze in eastern Asia. With marsh drainage, flood control, and irrigation it was possible to convert the lands along such rivers into rich agricultural regions. Extensive projects of this sort not only knit together previously separated localities, but the engineering, financing, and administration of both the projects and the purposes for which they were created required the development of considerable technical knowledge and its concomitant mathematics. Thus early mathematics can be said to have originated in certain areas of the ancient orient primarily as a practical science to assist in agricultural and engineering pursuits. These pursuits required the computation of a usable calendar, the development of systems of weights and measures to serve in the harvesting, storing, and apportioning of foods, the creation of surveying methods for canal and reservoir construction and for parceling land, and the evolution of financial and commercial practices for raising and collecting taxes and for purposes of trade.

As we have seen, the initial emphasis of mathematics was on practical arithmetic and mensuration. A special craft came into being for the cultivation, application, and instruction of this practical science. In such a situation, however, tendencies toward abstraction were bound to develop, and, to some

29

extent, the science was then studied for its own sake. It was in this way that algebra evolved ultimately from arithmetic and the beginnings of theoretical geometry grew out of mensuration.

It should be noted, however, that one cannot find in all ancient oriental mathematics even a single instance of what we today call a demonstration. In place of an argument there is merely a description of a process. One is instructed, "Do thus and so." Moreover, except possibly for a few specimens, these instructions are not even given in the form of general rules, but are simply applied to sequences of specific cases. Thus, if the solution of quadratic equations is to be explained, we do not find a derivation of the process used, nor do we find the process described in general terms, but instead we are offered a large number of specific quadratics, and we are told step by step how to solve each of these specific instances. Unsatisfactory as the "Do thus and so" procedure may seem to us, it should not seem strange, for it is very much the procedure we ourselves frequently use in teaching portions of grade school and high school mathematics.

There are difficulties in dating discoveries made in the ancient orient. One of these difficulties lies in the static nature of the social structure and the prolonged seclusion of certain areas. Another difficulty is due to the writing media upon which discoveries were preserved. The Babylonians used imperishable baked clay tablets and the Egyptians used stone and papyrus, the latter fortunately being long lasting because of the unusually dry climate of the region. But the early Chinese and Indians used very perishable media like bark and bamboo. Thus while a fair quantity of definite information is now known about the science and the mathematics of ancient Babylonia and Egypt, very little is known with any degree of certainty about these studies in ancient China and India. Accordingly this chapter, which is largely devoted to the mathematics of the pre-Hellenistic centuries, will be limited to Babylonia and Egypt.

BABYLONIA

2-2 Sources

Archeologists working in Mesopotamia have systematically unearthed, since before the middle of the nineteenth century, some half-million inscribed clay tablets. Over 50,000 tablets were excavated at the site of ancient Nippur alone. There are many excellent collections of these tablets, such as those in the great museums at Paris, Berlin, and London, and in the archeological exhibits at Yale, Columbia, and the University of Pennsylvania. Of the half-million tablets, about 300 have been identified as strictly mathematical tablets containing mathematical tables and lists of mathematical problems. We owe our knowledge of ancient Babylonian* mathematics to the scholarly deciphering and interpretation of many of these mathematical tablets.

The puzzle of the inscriptions was unlocked in 1847 by Rawlinson, who perfected a key earlier suggested by Grotefend. The tablets appear to bear upon all phases and interests of the life of the times and to range over many periods of Babylonian history. There are mathematical texts dating from the latest Sumerian period of perhaps 2100 B.C., a second and very large group from the succeeding First Babylonian Dynasty, of King Hammurabi's era, and on down to about 1600 B.C., and a third generous group running from about 600 B.C. to 300 A.D. covering the New Babylonian Empire of Nebuchadnezzar and the following Persian and Seleucidan eras. The lacuna between the second and third groups coincides with an especially turbulent period of Babylonian history. Most of our knowledge of the contents of these mathematical tablets does not predate 1935 and is largely due to the remarkable findings of Otto Neugebauer and F. Thureau-Dangin. Since the work of interpreting these tablets is still proceeding, new and perhaps equally remarkable discoveries are quite probable in the near future.

2-3   Commercial and Agrarian Mathematics

Even the oldest tablets show a high level of computational ability and make it clear that the sexagesimal positional system was already long established. There are thousands of texts of this early period dealing with farm deliveries and with arithmetical calculations based on these transactions. The tablets show that these ancient Sumerians were familiar with all kinds of legal and domestic contracts, like bills, receipts, promissory notes, accounts, both simple and compound interest, mortgages, deeds of sale, and guaranties: There are tablets which are records of business firms, and others that deal with systems of weights and measures.

Many arithmetical processes were carried out with the aid of various

*It should be understood that the descriptive term *Babylonian* is used merely for convenience, and that other peoples besides Babylonians—such as Sumerians, Persians, and others—are subsumed under the term.

tables. Of the 300 mathematical tablets, about 200 are table tablets. These table tablets show multiplication tables, tables of reciprocals, tables of squares and cubes, and even tables of exponentials. These latter were probably used, along with interpolation, for problems on compound interest. The reciprocal tables were used to reduce division to multiplication.

That the calendar used by the Babylonians was established ages earlier is evidenced by the facts that their year started with the vernal equinox and that the first month was named after Taurus. But the sun was in Taurus at this equinox around 4700 B.C. Thus it seems safe to say that the Babylonians had some kind of arithmetic as far back as the fourth or fifth millennium B.C.

2-4   Geometry

Babylonian geometry is intimately related to practical mensuration. From numerous concrete examples the Babylonians of 2000 to 1600 B.C. must have been familiar with the general rules for the area of a rectangle, the areas of right and isosceles triangles (and perhaps the general triangle), the area of a trapezoid having one side perpendicular to the parallel sides, the volume of a rectangular parallelepiped, and, more generally, the volume of a right prism with trapezoidal base. The circumference of a circle was taken as three times the diameter and the area as one-twelfth the square of the circumference (both correct for $\pi = 3$), and the volume of a right circular cylinder was then obtained by finding the product of the base and the altitude. The volume of a frustum of a cone or of a square pyramid is incorrectly given as the product of the altitude and half the sum of the bases. The Babylonians also knew that corresponding sides of two similar right triangles are proportional, that the perpendicular through the vertex of an isosceles triangle bisects the base, and that an angle inscribed in a semicircle is a right angle. The Pythagorean theorem was also known. In this connection see Section 2-6.

The chief feature of Babylonian geometry is its algebraic character. The more intricate problems which are clothed in geometric terminology are essentially nontrivial algebra problems. A typical example may be found in Problem Study 2.3. There are many problems concerning a transversal parallel to a side of a right triangle, which lead to quadratic equations; there are others which lead to systems of simultaneous equations, one instance giving ten equations in ten unknowns. There is a Yale tablet, possibly from 1600 B.C., in which a general cubic equation arises in a discussion of volumes of frustums of a pyramid, as the result of eliminating z from a system of equations of the type

$$z(x^2 + y^2) = A, \qquad z = ay + b, \qquad x = c.$$

We undoubtedly owe to the ancient Babylonians our present division of the circumference of a circle into 360 equal parts. Several explanations have been put forward to account for the choice of this number, but none is so plausible as the following, backed by the profound scholarship of Otto Neugebauer. In early Sumerian times there existed a large distance unit, a sort of *Babylonian mile*, equal to about seven of our miles. Since the Babylonian mile was used for measuring longer distances, it was natural that it should also become a time unit, namely the time required to travel a Babylonian mile. Later, sometime in the first millennium B.C., when Babylonian astronomy reached the stage in which systematic records of celestial phenomena were kept, the Babylonian time-mile was adopted for measuring spans of time. Since a complete day was found to be equal to 12 time-miles, and one complete day is equivalent to one revolution of the sky, a complete circuit was divided into 12 equal parts. But, for convenience, the Babylonian mile had been subdivided into 30 equal parts. We thus arrive at $(12)(30) = 360$ equal parts in a complete circuit.

2-5 Algebra

By 2000 B.C. Babylonian arithmetic had evolved into a well-developed rhetorical algebra. Not only were quadratic equations solved, both by the equivalent of substituting in a general formula and by completing the square, but some cubic (third degree) and biquadratic (fourth degree) equations were discussed. A tablet has been found giving a tabulation not only of the squares and the cubes of the integers from 1 to 30, but also of the combination $n^3 + n^2$ for this range. A number of problems are given which lead to cubics of the form $x^3 + x^2 = b$. These can be solved by using the $n^3 + n^2$ table. Problem Study 2.4 concerns itself with possible uses of this particular table.

There are some Yale tablets of about 1600 B.C. listing hundreds of unsolved problems involving simultaneous equations which lead to biquadratic equations for solution. As an example we have

$$xy = 600, \qquad 150(x - y) - (x + y)^2 = -1000.$$

As another illustration from the same tablets we have a pair of equations of the form

$$xy = a, \qquad bx^2/y + cy^2/x + d = 0,$$

which lead to an equation of the sixth degree in x, but which is quadratic in x^3.

Neugebauer has found two interesting series problems on a Louvre tablet of about 300 B.C. One of them states that

$$1 + 2 + 2^2 + \cdots + 2^9 = 2^9 + 2^9 - 1,$$

and the other one that

$$1^2 + 2^2 + 3^2 + \cdots + 10^2 = [1(1/3) + 10(2/3)]55 = 385.$$

One wonders if the Babylonians were familiar with the formulas

$$\sum_{i=0}^{n} r^i = (r^{n+1} - 1)/(r - 1)$$

and

$$\sum_{i=1}^{n} i^2 = [(2n + 1)/3] \sum_{i=1}^{n} i = n(n + 1)(2n + 1)/6.$$

The first of these was known to contemporary Greeks, and Archimedes found practically the equivalent of the second.

The Babylonians gave some interesting approximations to the square roots of nonsquare numbers, like 17/12 for $\sqrt{2}$ and 17/24 for $1/\sqrt{2}$. Perhaps the Babylonians used the approximation formula

$$(a^2 + h)^{1/2} \approx a + h/2a.$$

A very remarkable approximation for $\sqrt{2}$ is

$$1 + 24/60 + 51/60^2 + 10/60^3 = 1.414213,$$

found on a Yale tablet of about 1600 B.C.

There are astronomical tablets of the third century B.C. which make explicit use of the law of signs in multiplication.

In summary, we conclude that the ancient Babylonians were indefatigable table makers, computers of high skill, and definitely stronger in algebra than geometry. One is certainly struck by the depth and the diversity of the problems which they considered.

2-6   Plimpton 322

Perhaps the most remarkable of the Babylonian mathematical tablets yet analyzed is that known as *Plimpton 322*, meaning that it is the item with catalogue number 322 in the G. A. Plimpton collection at Columbia University. The tablet is written in Old Babylonian script, which dates it somewhere from 1900 to 1600 B.C., and it was first described by Neugebauer and Sachs in 1945.

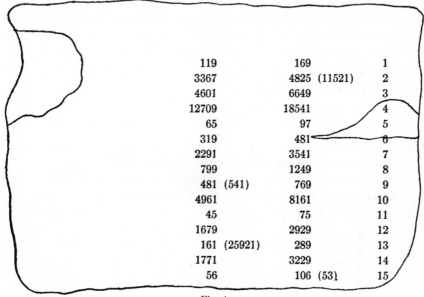

119		169		1
3367		4825	(11521)	2
4601		6649		3
12709		18541		4
65		97		5
319		481		6
2291		3541		7
799		1249		8
481	(541)	769		9
4961		8161		10
45		75		11
1679		2929		12
161	(25921)	289		13
1771		3229		14
56		106	(53)	15

Fig. 4

Figure 4 gives an idea of the shape of the tablet. Unfortunately a missing piece has been broken from the entire left edge and the tablet is further marred by a deep chip near the middle of the right edge and a flaked area in the top left corner. The tablet contains three essentially complete columns of figures which, for convenience, are reproduced on Figure 4 in our own decimal notation. There is a fourth but incomplete column of figures along the broken edge. We shall later reconstruct this column.

It is clear that the column on the extreme right merely serves to number the lines. The next two columns seem, at first glance, to be rather haphazard. One soon discovers, however, that corresponding numbers in these columns, with four unfortunate exceptions, constitute the hypotenuse and a leg of integral-sided right triangles. The four exceptions are noted in Figure 4 by placing the original readings in parentheses to the right of the corrected readings. It is hard to explain the exception in the second line, but the others can be accounted for. Thus, in the ninth line, 481 and 541 appear as (8,1) and (9,1) in the sexagesimal system. Clearly the occurrence of 9 instead of 8 could be a mere slip of the stylus when writing these numbers in cuneiform script. The number in line 13 is the square of the corrected value, and that in the last line is half of the corrected value.

Now a set of three positive integers, like (3,4,5), which can be the sides of a right triangle, is known as a *Pythagorean triple*. Again, if the triple contains no common integral factor other than unity, it is known as a *primitive Pythagorean triple*. Thus (3,4,5) is a primitive triple, whereas (6,8,10) is not. One of the achievements of the Greeks, a good thousand years after the date of the Plimpton tablet, was to show that all primitive Pythagorean triples (a,b,c) are given parametrically by

$$a = 2uv, \qquad b = u^2 - v^2, \qquad c = u^2 + v^2,$$

where u and v are relatively prime, of different parity, and $u > v$. Thus if $u = 2$ and $v = 1$, we obtain the primitive triple, $a = 4$, $b = 3$, $c = 5$.

Suppose we compute the other leg a of the integral-sided right triangles determined by the given hypotenuse c and leg b on the Plimpton tablet. We find the following Pythagorean triples:

a	b	c	u	v
120	119	169	12	5
3456	3367	4825	64	27
4800	4601	6649	75	32
13500	12709	18541	125	54
72	65	97	9	4
360	319	481	20	9
2700	2291	3541	54	25
960	799	1249	32	15
600	481	769	25	12
6480	4961	8161	81	40
60	45	75	2	1
2400	1679	2929	48	25
240	161	289	15	8
2700	1771	3229	50	27
90	56	106	9	5

It will be noticed that all of these triples, except the ones in lines 11 and 15, are primitive triples. For discussion we have also listed the values of the parameters u and v leading to these Pythagorean triples. The evidence seems good that the Babylonians of this remote period were acquainted with the general parametric representation of primitive Pythagorean triples as given above. This evidence is strengthened when we notice that u and v, and hence also a (since $a = 2uv$), are *regular* sexagesimal numbers (see Problem Study 2.1). It appears that the table on the tablet was constructed by deliberately choosing small regular numbers for the parameters u and v.

This choice of u and v must have been motivated by some subsequent process involving division, for regular numbers appear in tables of reciprocals and are used to reduce division to multiplication. An examination of the fourth, and partially destroyed, column gives the answer. For this column is found to contain the values of $(c/a)^2$ for the different triangles. To carry out the division, the side a, and hence the numbers u and v, had to be regular.

It is worth examining the column of values for $(c/a)^2$ a little more deeply. This column, of course, is a table giving the square of the secant of the angle B opposite side b of the right triangle. Because side a is regular, sec B has a finite sexagesimal expansion. Moreover it turns out, with the particular choice of triangles as given, that the values of sec B form a surprisingly regular sequence which decreases by almost exactly 1/60 as we pass from one line of the table to the next, and the corresponding angle decreases from 45° to 31°. We thus have a secant table for angles from 45° to 31°, formed by means of integral-sided right triangles, in which there is a regular jump in the function rather than in the corresponding angle. All this is truly remarkable. It seems highly probable that there were companion tables giving similar information for angles ranging from 0° to 15° and from 16° to 30°.

The analysis of Plimpton 322 shows the careful examination to which some of the Babylonian mathematical tablets must be subjected. Formerly such a tablet might have been summarily dismissed as being merely a business list or record.

EGYPT

2-7 Sources and Dates

The mathematics of ancient Egypt, contrary to much popular opinion, never reached the level attained by Babylonian mathematics. This may have been due to the more advanced economic development of Babylonia. Babylonia was located on a number of great caravan routes, while Egypt stood in semi-isolation. Nor did the relatively peaceful Nile demand such extensive engineering and administrational efforts as did the more erratic Tigris and Euphrates.

Nevertheless, until the recent deciphering of so many Babylonian mathematical tablets, Egypt was long the richest field for ancient historical research. The reasons for this lie in the veneration that the Egyptians had for their dead and in the unusually dry climate of the region. The former led to the erection of long-lasting tombs and temples with richly inscribed walls, and the latter played a leading role in the preservation of many papyri and objects that would otherwise have perished.

Following is a chronological list of some of the tangible items bearing on the mathematics of ancient Egypt. In addition to these items there are numerous wall inscriptions and minor papyri that contribute to our knowledge.

1. 3100 B.C. In a museum at Oxford is a royal Egyptian mace dating from this time. On the mace are several numbers in the millions and hundred thousands, written in Egyptian hieroglyphs, recording results of a successful military campaign.

2. 2900 B.C. The great pyramid of Gizeh was erected about 2900 B.C. and undoubtedly involved some mathematical and engineering problems. The structure covers 13 acres and contains over 2,000,000 stone blocks, averaging 2.5 tons in weight, very carefully fitted together. These stone blocks were brought from sandstone quarries located on the other side of the Nile. Some chamber roofs are made of 54-ton granite blocks, 27 feet long and 4 feet thick, hauled from a quarry 600 miles away, and set 200 feet above ground. It is reported that the sides of the square base involve a relative error of less than 1/14,000, and that the relative error in the right angles at the corners does not exceed 1/27,000. The engineering skill implied by these impressive statistics is considerably diminished when we realize that the task was accomplished by an army of 100,000 laborers working for a period of 30 years.

3. 1850 B.C. This is the approximate date of the Moscow papyrus, a mathematical text containing 25 problems which were already old when the manuscript was compiled. The Moscow papyrus was published with editorial commentary in 1930.

4. 1850 B.C. The oldest extant astronomical instrument, a combination plumb line and sight rod, dates from this time and is preserved in the Berlin Museum.

5. 1650 B.C. This is the approximate date of the Rhind (or Ahmes) papyrus, a mathematical text partaking of the nature of a practical handbook and containing 85 problems copied in hieratic writing by the scribe Ahmes from an earlier work. The papyrus was purchased in Egypt by the English Egyptologist A. Henry Rhind and then later acquired by the British Museum. This and the Moscow papyrus are our chief sources of information concerning ancient Egyptian mathematics. The Rhind papyrus was published in 1927.

6. 1500 B.C. The largest existing obelisk, erected before the Temple of the Sun at Thebes, was quarried about this time. It is 105 feet long with a square base 10 feet to the side and weighs about 430 tons.

7. 1500 B.C. The Berlin Museum possesses an Egyptian sundial dating from this period. It is the oldest sundial extant.

8. 1350 B.C. The Rollin papyrus of about 1350 B.C., now preserved in the Louvre, contains some elaborate bread accounts showing the practical use of large numbers at the time.

9. 1167 B.C. This is the date of the Harris papyrus, a document prepared by Rameses IV when he ascended the throne. It sets forth the great works of his father, Rameses III. The listing of the temple wealth of the time furnishes the best example of practical accounts that has come to us from ancient Egypt.

Egyptian sources of more recent dates than the above show no appreciable gain in either mathematical knowledge or mathematical technique. In fact, there are instances showing definite regression.

2-8   Arithmetic and Algebra

All of the 110 problems found in the Moscow and Rhind papyri are numerical, and many of them are very simple. Although most of the problems have a practical origin, there are some of a theoretical nature.

One consequence of the Egyptian numeral system is the additive character of the dependent arithmetic. Thus multiplication and division were usually performed by a succession of doubling operations depending on the fact that any number can be represented as a sum of powers of 2. As an example of multiplication let us find the product of 26 and 33. Since $26 = 16 +$

8 + 2, we have merely to add these multiples of 33. The work may be arranged as follows:

1	33
* 2	66
4	132
* 8	264
*16	528
	858

Addition of the proper multiples of 33, that is, those indicated by an asterisk, gives the answer 858. Again, to divide 753 by 26, say, we successively double the divisor 26 up to the point where the next doubling would exceed the dividend 753. The procedure is shown below.

1	26
2	52
* 4	104
* 8	208
*16	416
28	

Now, since

$$753 = 416 + 337$$
$$= 416 + 208 + 129$$
$$= 416 + 208 + 104 + 25,$$

we see, noting the starred items in the column above, that the quotient is 16 + 8 + 4 = 28, with a remainder of 25. This Egyptian process of multiplication and division not only eliminates the necessity of learning a multiplication table, but is so convenient on the abacus that it persisted as long as that instrument was in use, and even for some time beyond.

The Egyptians endeavored to avoid some of the computational difficulties encountered with fractions by representing all fractions, except 2/3, as the sum of so-called *unit fractions*, or fractions with unit numerators. This reduction was made possible by tables so representing fractions of the form 2/n, the only case necessary because of the dyadic nature of Egyptian multiplication. The problems of the Rhind papyrus are preceded by such a table for all odd n from 5 to 101. Thus we find 2/7 expressed as 1/4 + 1/28, 2/97 as 1/56 + 1/679 + 1/776, and 2/99 as 1/66 + 1/198, only one decomposition being offered for any particular case. The table is utilized in some of the problems of the papyrus.

Unit fractions were denoted in Egyptian hieroglyphs by placing an elliptical symbol above the denominator number. A special symbol was used also for the exceptional 2/3 and another symbol sometimes appeared for 1/2. These symbols are shown below in association with some modern numerals.

$$\underset{3}{\bigcirc} = 1/3, \quad \underset{4}{\bigcirc} = 1/4,$$

$$\underset{2}{\bigcirc} \text{ or } \sqsubset \ = 1/2,$$

$$\mathbb{\Phi} = 2/3.$$

There are interesting theories to explain how the Egyptians obtained their unit fraction decompositions (see Problem Study 2.7).

Many of the 110 problems in the Rhind and Moscow papyri show their practical origin by dealing with questions regarding the strength of bread and of beer, with feed mixtures for cattle and domestic fowl, and with the storage of grain. Many of these require nothing more than a simple linear equation, and are generally solved by the method later known in Europe as the *rule of false position*. Thus to solve

$$x + x/7 = 24$$

assume any convenient value for x, say $x = 7$. Then $x + x/7 = 8$, instead of 24. Since 8 must be multiplied by 3 to give the required 24, the correct x must be $3(7)$, or 21.

There are some theoretical problems involving arithmetic and geometric progressions. See, for example, Problem Study 2.8 (c) and Section 2-10. A papyrus of about 1950 B.c., found at Kahun, contains the following problem: "A given surface of 100 units of area shall be represented as the sum of two squares whose sides are to each other as 1 : 3/4." Here we have $x^2 + y^2 = 100$ and $x = 3y/4$. Elimination of x yields a pure quadratic in y. We may, however, solve the problem by false position. Thus, take $y = 4$. Then $x = 3$, and $x^2 + y^2 = 25$, instead of 100. We must therefore correct x and y by doubling the initial values, obtaining $x = 6$, $y = 8$.

There is some symbolism in Egyptian algebra. In the Rhind papyrus we find symbols for *plus* and *minus*. The first of these symbols represents a pair of legs walking from right to left, the normal direction for Egyptian writing, and the other a pair of legs walking from left to right, opposite to the direction of Egyptian writing. Symbols, or ideograms, were also employed for *equals* and for the unknown.

2-9   Geometry

Twenty-six of the 110 problems in the Moscow and Rhind papyri are geometric. Most of these problems stem from mensuration formulas needed for computing land areas and granary volumes. The area of a circle is taken as equal to that of the square on 8/9 of the diameter, and the volume of a right cylinder as the product of the area of the base by the length of the altitude. Recent investigations seem to show that the ancient Egyptians knew that the area of any triangle is given by half the product of base and altitude. Some of the problems seem to concern themselves with the cotangent of the dihedral angle between the base and.a face of a pyramid, and others show an acquaintance with the elementary theory of proportion. Contradicting repeated and apparently unfounded stories, no documentary evidence has been found showing that the Egyptians were aware of even a particular case of the Pythagorean theorem. In later Egyptian sources the incorrect formula $K = (a + c)(b + d)/4$ is used for finding the area of an arbitrary quadrilateral with successive sides of lengths a, b, c, d.

Very remarkable is the existence in the Moscow papyrus of a numerical example of the correct formula for the volume of a frustum of a square pyramid (see Problem Study 2.10(a)). No other unquestionably genuine example of this formula has been found in ancient oriental mathematics, and several conjectures have been formulated to explain how it might have been discovered. E. T. Bell aptly refers to this early Egyptian example as the "greatest Egyptian pyramid."

2-10   A Curious Problem in the Rhind Papyrus

Although little difficulty was encountered in deciphering and then interpreting most of the problems in the Rhind papyrus, there is one problem, Problem Number 79, for which the interpretation is not so certain. In this problem occurs the following curious set of data, here transcribed:

	Estate
Houses	7
Cats	49
Mice	343
Heads of wheat	2401
Hekat measures	16807
	19607

One easily recognizes the numbers as the first five powers of 7, along with their sum. Because of this it was at first thought that perhaps the writer was here introducing the symbolic terminology *houses*, *cats*, and so on, for *first power*, *second power*, and so on.

A more plausible and interesting explanation, however, was given by the historian Moritz Cantor in 1907. He saw in this problem an ancient forerunner of a problem which was popular in the Middle Ages, and which was given by Leonardo Fibonacci in 1202 in his *Liber abaci*. Among the many problems occurring in this work is the following: "There are seven old women on the road to Rome. Each woman has seven mules; each mule carries seven sacks; each sack contains seven loaves; with each loaf are seven knives; and each knife is in seven sheaths. Women, mules, sacks, loaves, knives, and sheaths, how many are there in all on the road to Rome?" As a later and more familiar version of the same problem we have the Old English children's rhyme:

> As I was going to St. Ives
> I met a man with seven wives;
> Every wife had seven sacks;
> Every sack had seven cats;
> Every cat had seven kits.
> Kits, cats, sacks, and wives,
> How many were going to St. Ives?

According to Cantor's interpretation, the original problem in the Rhind papyrus might then be formulated somewhat as follows: "An estate consisted of seven houses; each house had seven cats; each cat ate seven mice; each mouse ate seven heads of wheat; and each head of wheat was capable of yielding seven hekat measures of grain. Houses, cats, mice, heads of wheat, and hekat measures of grain, how many of these in all were in the estate?"

Here, then, is a problem which has been preserved as part of the puzzle lore of the world. It was apparently already old when Ahmes copied it, and older by three thousand years when Fibonacci incorporated a version of it in his *Liber abaci*. Seven hundred and fifty years later we are reading another variant of it to our children. One cannot help wondering if a surprise twist such as occurs in the Old English rhyme also occurred in the ancient Egyptian problem.

There are many puzzle problems popping up every now and then in our present-day magazines which have medieval counterparts. How much further back some of them go is now almost impossible to determine.*

*See D. E. Smith, "On the Origin of Certain Typical Problems," *American Mathematical Monthly*, 24 (February, 1917), 64–71.

Problem Studies

2.1   Regular Numbers

A number is said to be (*sexagesimally*) *regular* if its reciprocal has a finite sexagesimal expansion (that is, a finite expansion when expressed as a radix fraction for base 60). With the exception of a single tablet in the Yale collection, all Babylonian tables of reciprocals contain only reciprocals of regular numbers. A Louvre tablet of about 300 B.C. contains a regular number of seven sexagesimal places and its reciprocal of seventeen sexagesimal places.

(a) Show that a necessary and sufficient condition for n to be regular is that $n = 2^a 3^b 5^c$, where a, b, c are nonnegative integers.

(b) Express, by finite sexagesimal expansions, the numbers 1/2, 1/3, 1/5, 1/15, 1/360, 1/3600.

(c) Generalize part (a) to numbers having general base b.

(d) List all the sexagesimally regular numbers less than 100, and then list all the decimally regular numbers less than 100.

2.2   Compound Interest

There are tablets in the Berlin, Yale, and Louvre collections containing problems in compound interest, and there are some Istanbul tablets which appear originally to have had tables of a^n for $n = 1$ to 10 and $a = 9$, 16, 100, and 225. With such tables one can solve exponential equations of the type $a^x = b$.

(a) On a Louvre tablet of about 1700 B.C. occurs the problem: to find how long it would take for a certain sum of money to double itself at compound annual interest of 20 per cent. Solve this problem by modern methods.

(b) Find $(1.2)^3$ and $(1.2)^4$ and then, by linear interpolation, obtain x such that $(1.2)^x = 2$. The result so obtained agrees with a Babylonian solution of this problem.

2.3   Algebraic Geometry

(a) The algebraic character of Babylonian geometry problems is illustrated by the following, found on a Strassburg tablet of about 1800 B.C. "An area A, consisting of the sum of two squares is 1000. The side of one square is 10 less than 2/3 of the side of the other square. What are the sides of the squares?" Solve this problem.

(b) On a Louvre tablet of about 300 B.C. are four problems concerning rectangles of unit area and given semiperimeter. Let the sides and semiperimeter be x, y, and a. Then we have

$$xy = 1, \qquad x + y = a.$$

Solve this system by eliminating y and thus obtaining a quadratic in x.

(c) Solve the system of part (b) by using the identity

$$[(x - y)/2]^2 = [(x + y)/2]^2 - xy$$

This is essentially the method used on the Louvre tablet. It is interesting that the identity appeared contemporaneously as Proposition 5 of Book II of Euclid's *Elements*.

2.4 Cubics

(a) A Babylonian tablet has been discovered which gives the values of $n^3 + n^2$ for $n = 1$ to 30. Make such a table for $n = 1$ to 10.

(b) Find, by means of the above table, a root of the cubic equation $x^3 + 2x^2 - 3136 = 0$.

(c) A Babylonian problem of about 1800 B.C. seems to call for the solution of the simultaneous system $xyz + xy = 7/6$, $y = 2x/3$, $z = 12x$. Solve this system using the table of part (a).

(d) Otto Neugebauer believes that the Babylonians were quite capable of reducing the general cubic equation to the "normal form" $n^3 + n^2 = c$, although there is as yet no evidence that they actually did do this. Show how such a reduction might be made.

(e) In connection with the table of part (a), Neugebauer has noted that the Babylonians may well have observed the relation $\sum\limits_{i=1}^{n} i^3 = (\sum\limits_{i=1}^{n} i)^2$ for various values of n. Establish this relation by mathematical induction.

2.5 Square Root Approximations

It is known that the infinite series obtained by expanding $(a^2 + h)^{1/2}$ by the binomial theorem process converges to $(a^2 + h)^{1/2}$ if $-a^2 < h < a^2$.

(a) Establish the approximation formula

$$(a^2 + h)^{1/2} \approx a + h/2a, \qquad 0 < h < a^2.$$

(b) Take $a = 4/3$ and $h = 2/9$ in the approximation formula of part (a), and thus find a Babylonian rational approximation for $\sqrt{2}$. Find a rational approximation for $\sqrt{5}$ by taking $a = 2$, $h = 1$.

(c) Establish the better approximation formula

$$(a^2 + h)^{1/2} \approx a + h/2a - h^2/8a^3, \qquad 0 < h < a^2,$$

and approximate $\sqrt{2}$ and $\sqrt{5}$ by using the same values for a and h as in part (b).

2.6 Duplation and Mediation

The Egyptian process of multiplication later developed into a slightly improved method known as *duplation and mediation*, the purpose of which was mechanically to pick out the required multiples of one of the factors that have to be added in order to give the required product. Thus, taking the example in the text, suppose we wish to multiply 26 by 33. We may successively halve the 26 and double the 33, thus

26	33
13	66*
6	132
3	264*
1	528*
	858

In the doubling column we now add those multiples of 33 corresponding to the *odd* numbers in the halving column. Thus we add 66, 264, and 528 to obtain the required product 858. The process of duplation and mediation is utilized by high-speed electronic computing machines.

(a) Multiply 424 by 137 using duplation and mediation.

(b) Prove that the duplation and mediation method of multiplication gives correct results.

(c) Find, by the Egyptian method, the quotient and remainder when 1043 is divided by 28.

2.7 Unit Fractions

(a) Show that $z/pq = 1/pr + 1/qr$, where $r = (p + q)/z$. This method for finding possible decompositions of a fraction into two unit fractions is indicated on a papyrus written in Greek probably sometime between 500 and 800 A.D., and found at Akhmim, a city on the Nile River.

(b) Take $z = 2$, $p = 1$, $q = 7$, and thus obtain the unit fraction decomposition of 2/7 as given in the Rhind papyrus.

(c) Represent 2/99 as the sum of two unit fractions in at least five different ways.

2.8 Egyptian Algebra

The following problems are found in the Rhind papyrus.

(a) "If you are asked, what is 2/3 of 1/5, take the double and the sixfold; that is 2/3 of it. One must proceed likewise for any other fraction." Interpret this and prove the general statement.

(b) "A quantity, its 2/3, its 1/2, and its 1/7, added together, become 33. What is the quantity?" Solve this problem by the rule of false position.

(c) "Divide 100 loaves among 5 men in such a way that the shares received shall be in arithmetic progression and that one seventh of the sum of the largest three shares shall be equal to the sum of the smallest two." Solve this problem using modern methods.

2.9 Egyptian Geometry

(a) In the Rhind papyrus the area of a circle is repeatedly taken as equal to that of the square on 8/9 of the diameter. This leads to what value for π?

(b) Prove that of all triangles having two given sides, that in which these sides form a right angle is the maximum.

(c) Denote the lengths of the sides AB, BC, CD, DA of a quadrilateral $ABCD$ by a, b, c, d, and let K represent the area of the quadrilateral. Show that $K \leq (ad + bc)/2$, equality holding if and only if angles A and C are right angles.

(d) For the hypothesis of part (c) now show that $K \leq (a+c)(b+d)/4$, equality holding if and only if $ABCD$ is a rectangle. Thus the Egyptian formula for the area of a quadrilateral, cited in Section 2-9, gives too large an answer for all nonrectangular quadrilaterals.

2.10 The Greatest Egyptian Pyramid

(a) In the Moscow papyrus we find the following numerical example: "If you are told: A truncated pyramid of 6 for the vertical height by 4 on the base by 2 on the top. You are to square this 4, result 16. You are to double 4, result 8. You are to square 2, result 4. You are to add the 16, the 8, and the 4, result 28. You are to take one third of 6, result 2. You are to take 28 twice, result 56. See, it is 56. You will find it right." Show that this illustrates the general formula

$$V = h(a^2 + ab + b^2)/3,$$

giving the volume of a frustum of a square pyramid in terms of the height h and the sides a and b of the bases.

(b) If m and n are two positive numbers, $m \geqq n$, then we define the *arithmetic mean*, the *heronian mean*, and the *geometric mean* of m and n to be $A = (m + n)/2$, $R = (m + \sqrt{mn} + n)/3$, $G = \sqrt{mn}$. Show that $A \geqq R \geqq G$, the equality signs holding if and only if $m = n$.

(c) Assuming the familiar formula for the volume of any pyramid (volume equals one third the product of base and altitude), show that the volume of a frustum of the pyramid is given by the product of the height of the frustum and the heronian mean of the bases of the frustum.

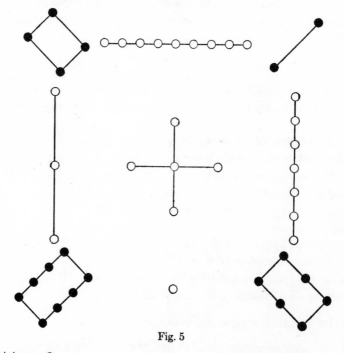

Fig. 5

2.11 Magic Squares

There are some Chinese mathematical works of which parts, at least, are claimed to date from very early times. This is difficult to verify because we lack original sources. As an added complication it was decreed, in 213 B.C. by the Emperor Shï Huang-ti, that all books in the country be burned.

Although the edict was most certainly not completely carried out, and many books that were burned were soon restored from memory, we are now in doubt as to the genuineness of anything claimed to be older than the unfortunate date.

One of the oldest of the Chinese mathematical classics is the *I-king*, or *Book on Permutations*. In this appears the configuration, shown in Figure 5, known as the *lo-shu*.

The lo-shu is the oldest known example of a magic square, and myth claims that it was first seen by the Emperor Yu, in about 2200 B.C., decorated upon the back of a divine tortoise along a bank of the Yellow River. It is a square array of numerals indicated by knots in strings, black knots for even numbers and white knots for odd numbers.

(a) An *nth order magic square* is a square array of n^2 distinct integers so arranged that the n numbers along any row, column, or main diagonal have the same sum, called the *magic constant* of the square. The magic square is said to be normal if the n^2 numbers are the first n^2 positive integers. Show that the magic constant of an *nth* order normal magic square is $n(n^2 + 1)/2$.

(b) De la Loubère, when envoy of Louis XIV to Siam in 1687–1688, learned a simple method for finding a normal magic square of any odd order. Let us illustrate the method by constructing one of the fifth order. Draw a

	18	25	2	9	
17	24	1	8	15	17
23	5	7	14	16	23
4	6	13	20	22	4
10	12	19	21	3	10
11	18	25	2	9	

square and divide it into 25 cells. Border the square with cells along the top and the right edge, and shade the added cell in the top right corner. Write 1 in the middle top cell of the original square. The general rule is then to proceed diagonally upward to the right with the successive integers. Exceptions to this general rule occur when such an operation takes us out of the original square or leads us into a cell already occupied. In the former situation we get back into the original square by shifting clear across the square, either from top to bottom or from right to left, as the case may be, and continue with the general rule. In the second situation we write the number in the cell immediately beneath the one last filled, and then continue with the general rule. The shaded cell is to be regarded as occupied.

Thus, in our illustration, the general rule would place 2 diagonally upward from 1 in the fourth cell bordered along the top. We must therefore shift the 2 to the fourth cell in the bottom row of the original square. When we come to 4, it first falls in the third cell up bordered along the right edge. It must, therefore, be written clear across to the left in the third cell up in the first column of the original square. The general rule would place 6 in the cell already occupied by 1. It is accordingly written in the cell just below that occupied by the last written number, 5. And so on.

Construct a normal magic square of order seven.

(c) Show that the central cell of a normal magic square of the third order must be occupied by 5.

(d) Show that in a normal magic square of the third order 1 can never occur in a corner cell.

2.12　The 3, 4, 5 Triangle

There are reports that ancient Egyptian surveyors laid out right angles by constructing 3, 4, 5 triangles with a rope divided into 12 equal parts by 11 knots. Since there is no documentary evidence to the effect that the Egyptians were aware of even a particular case of the Pythagorean theorem, the following purely academic problem arises:* Show, without using the Pythagorean theorem, its converse, or any of its consequences, that the 3, 4, 5 triangle is a right triangle. Try to solve this problem.

Bibliography

BALL, W. W. R., AND H. S. M. COXETER, *Mathematical Recreations and Essays*, 11th ed. New York: The Macmillan Company, 1939.

CHACE, A. B., L. S. BULL, H. P. MANNING, AND R. C. ARCHIBALD, eds., *The Rhind Mathematical Papyrus*, 2 vols. Buffalo, N. Y.: Mathematical Association of America, 1927–1929.

*See Victor Thébault, "A Note on the Pythagorean Theorem," *The Mathematics Teacher*, Vol. 43, Oct. 1950, p. 278.

COOLIDGE, J. L., *A History of Geometrical Methods.* New York: Oxford University Press, 1940.

DATTA, B., AND A. N. SINGH, *History of Hindu Mathematics.* Bombay: Asia Publishing House, 1962.

KRAITCHIK, MAURICE, *Mathematical Recreations.* New York: W. W. Norton & Company, Inc., 1942.

MIKAMI, YOSHIO, *The Development of Mathematics in China and Japan.* New York: Hafner Publishing Company, 1913. Reprinted by Chelsea Publishing Company, New York, 1961.

NEEDHAM, J., with the collaboration of WANG LING, *Science and Civilization in China,* Vol. 3. New York: Cambridge University Press, 1959.

NEUGEBAUER, OTTO, *The Exact Sciences in Antiquity,* 2d ed. Providence: Brown University Press, 1957.

————, AND A. J. SACHS, eds., *Mathematical Cuneiform Texts,* American Oriental Series, Vol. 29. New Haven: American Oriental Society, 1946.

ORE, OYSTEIN, *Number Theory and Its History.* New York: McGraw-Hill Book Company, Inc., 1948.

SANFORD, VERA, *The History and Significance of Certain Standard Problems in Algebra.* New York: Teachers College, Columbia University, 1927.

SMITH, D. E., AND YOSHIO MIKAMI, *A History of Japanese Mathematics.* Chicago: The Open Court Publishing Company, 1914.

VAN DER WAERDEN, B. L., *Science Awakening,* tr. by Arnold Dresden. Groningen, Netherlands: P. Noordhoff, N.V., 1954.

3

Pythagorean Mathematics

3-1 Birth of Demonstrative Mathematics

The last centuries of the second millennium B.C. witnessed many economic and political changes. Some civilizations disappeared, the power of Egypt and Babylonia waned, and new peoples, especially the Hebrews, Assyrians, Phoenicians, and Greeks, came to the fore. The Iron Age was ushered in and brought with it sweeping changes in warfare and in all pursuits requiring tools. The alphabet was invented and coins introduced. Trade was increasingly stimulated and geographical discoveries were made. The world was ready for a new type of civilization.

The new civilization made its appearance in the trading towns that sprang up along the coast of Asia Minor and later on the mainland of Greece, on Sicily, and on the Italian shore. The static outlook of the ancient orient became impossible and in a developing atmosphere of rationalism men began to ask why as well as how.

For the first time, in mathematics, as similarly in other fields, men began to ask fundamental questions such as, "*Why* are the base angles of an isosceles triangle equal?" and "*Why* does a diameter of a circle bisect the circle?" The empirical processes of the ancient orient, quite sufficient for the question how, no longer sufficed to answer these more scientific inquiries of why. Some attempt at demonstrative methods was bound to assert itself, and the deductive feature, which modern scholars regard as a fundamental characteristic of mathematics, came into being. It thus happened that mathematics, in the modern sense of the word, was born in this atmosphere of rationalism and in one of the new trading towns located on the west coast of Asia Minor. For tradition has it that demonstrative geometry began with Thales of Miletus, one of the "seven wise men" of antiquity, during the first half of the sixth century B.C.

Thales seems to have spent the early part of his life as a merchant, becoming wealthy enough to devote the latter part of his life to study and some travel. It is said that he resided for a time in Egypt, and there evoked admiration by calculating the height of a pyramid by means of shadows (see Problem Study 3.1). Returning to Miletus his many-sided genius won him a reputation as a statesman, counselor, engineer, businessman, philosopher, mathematician, and astronomer. Thales is the first known individual with whom mathematical discoveries are associated. In geometry he is credited with the following elementary results:

1. A circle is bisected by any diameter.
2. The base angles of an isosceles triangle are equal.
3. The vertical angles formed by two intersecting lines are equal.
4. Two triangles are congruent if they have two angles and one side in each respectively equal. [Thales probably used this result in his determination of the distance of a ship from shore (see Problem Study 3.1).]
5. An angle inscribed in a semicircle is a right angle. [This was recognized by the Babylonians some 1400 years earlier.]

The value of these results is not to be measured by the theorems themselves, but rather by the belief that Thales supported them by some logical reasoning instead of intuition and experiment.

As with other great men, many charming anecdotes are told about Thales, which, if not true, are at least apposite. Thus there was the occasion when he demonstrated how easy it is to get rich; foreseeing a heavy crop of olives coming, he obtained a monopoly on all the oil presses of the region and then later realized a fortune by renting them out. And there is the story of the recalcitrant mule which, when transporting salt, found that by rolling over in the stream he could dissolve the contents of his load and thus travel more lightly; Thales broke him of the troublesome habit by loading him with sponges. He answered Solon's query as to why he never married by having a runner appear next day with a fictitious message for Solon stating that Solon's favorite son had been suddenly killed in an accident; Thales then calmed the grief-stricken father, explained everything, and said, "I merely wanted to tell you why I never married." At another time, having fallen into a ditch while observing the stars, he was asked by an old woman how he could hope to see anything in the heavens when he couldn't even see what was at his own feet. Asked how we might lead more upright lives he advised, "By refraining from doing what we blame in others." When once asked what he would take for one of his discoveries he replied, "I will be sufficiently rewarded if, when telling it to others, you will not claim the discovery as your own, but will say it was mine." And when asked what was the strangest thing he had ever seen he answered, "An aged tyrant."

Recent research indicates that there is no evidence backing an often repeated story that Thales predicted a solar eclipse which took place in 585 B.C.

3-2   Pythagoras and the Pythagoreans

It must be borne in mind that there exist no primary sources covering early Greek mathematics, and that we rely chiefly on manuscripts and accounts dating from Arabian and Christian times. With great ingenuity and patience, scholars of classicism have reliably restored many of the original texts, such as those of Euclid, Apollonius, Archimedes, and others. From painstaking text criticism, from remarks by later commentators, and from many fragments and scattered notices by later authors and philosophers, some sort of consistent, although largely hypothetical, account of the history of early Greek mathematics is now known. A debt is owed, along these lines, to the profound and scholarly investigations of such men as Paul Tannery, T. L. Heath, H. G. Zeuthen, A. Rome, J. L. Heiberg, and E. Frank.

It is a misfortune that an apparently full history of Greek geometry and astronomy, covering the period prior to 335 B.C., written by Eudemus, a pupil of Aristotle, has been lost. A brief account of this history was given later (*ca.* 450 A.D.) by Proclus, in his *Commentary on Euclid*. This abstract, known as the *Eudemian Summary*, constitutes our most reliable source of information for the very early period of Greek mathematics now under discussion. The account of·the mathematical achievements of Thales, sketched in the preceding section, was furnished by the *Eudemian Summary*.

One of the early outstanding mathematicians mentioned in the *Eudemian Summary* is Pythagoras, who became enveloped by his followers in such a mythical haze that very little is known about him with any degree of certainty. It seems that he was born about 572 B.C. on the Aegean island of Samos. Being about fifty years younger than Thales and living so near to Thales' home city of Miletus, it may be that Pythagoras studied under the older man. He then appears to have sojourned in Egypt and may even have indulged in more extensive travel. Returning home he found Samos under the tyranny of Polycrates and Ionia under the dominion of the Persians, and accordingly he migrated to the Greek seaport of Crotona, located in southern Italy. There he founded the famous Pythagorean school, which, in addition to being an academy for the study of philosophy, mathematics, and natural science, developed into a closely knit brotherhood with secret rites and observances. In time the influence and aristocratic tendencies of the brotherhood became so great that the democratic forces of southern Italy destroyed the buildings of the school and caused the society to disperse. According to one report, Pythagoras fled to Metapontum where he died, maybe through murder, at an advanced age of 75 to 80. The brotherhood, although scattered, continued to exist for at least two centuries longer.

The Pythagorean philosophy rested on the assumption that whole number is the cause of the various qualities of matter. This led to an exhaltation and study of number properties, and arithmetic (considered as the theory of numbers), along with geometry, music, and spherics (astronomy) constituted the fundamental liberal arts of the Pythagorean program of study. This group of subjects became known, in the Middle Ages, as the *quadrivium*, to which was added the *trivium* of grammar, logic, and rhetoric. These seven liberal arts came to be looked upon as the necessary equipment of an educated person.

Because Pythagoras' teaching was entirely oral, and because of the custom of the brotherhood to refer all discoveries back to the revered founder, it is now difficult to know just which mathematical findings should be credited to Pythagoras himself, and which to other members of the fraternity.

3-3 Pythagorean Arithmetic

The ancient Greeks made a distinction between the study of the abstract relationships connecting numbers and the practical art of computing with numbers. The former was known as *arithmetic* and the latter as *logistic*. This classification persisted through the Middle Ages until about the close of the fifteenth century, when texts appeared treating both the theoretical and practical aspects of number work under the single name *arithmetic*. It is interesting that today *arithmetic* has its original significance in continental Europe, while in England and America the popular meaning of *arithmetic* is synonymous with that of ancient *logistic*, and in these two countries the descriptive term *number theory* is used to denote the abstract side of number study.

It is generally conceded that Pythagoras and his followers, in conjunction with the fraternity's philosophy, took the first steps in the development of number theory, and at the same time laid much of the basis of future number mysticism. Thus Iamblichus, an influential Neoplatonic philosopher of about 320 A.D., has ascribed to Pythagoras the discovery of *amicable*, or *friendly, numbers*. Two numbers are *amicable* if each is the sum of the proper divisors* of the other. For example, 284 and 220, constituting the pair ascribed to Pythagoras, are amicable, since the proper divisors of 220 are 1, 2, 4, 5, 10, 11, 20, 22, 44, 55, 110, and the sum of these is 284, while the proper divisors of 284 are 1, 2, 4, 71, 142, and the sum of these is 220. This pair of numbers attained a mystical aura, and superstition later maintained that two talismans bearing these numbers would seal perfect friendship between the wearers. The numbers came to play an important role in magic, sorcery, astrology, and the casting of horoscopes. Curiously enough, it seems that no new pair of amicable numbers was discovered until the great French number theorist Pierre de Fermat in 1636 announced 17,296 and 18,416 as another pair. Two years later the French mathematician and philosopher René Descartes gave a third pair. The Swiss mathematician Leonhard Euler undertook a systematic search for amicable numbers and, in 1747, gave a list of 30 pairs, which he later extended to more than 60. A second curiosity in the history of these numbers was the late discovery, by the sixteen-year-old Italian boy Nicolo Paganini in 1886, of the overlooked and relatively small pair of amicable numbers, 1184 and 1210. Over 400 pairs of amicable numbers are now known.

Other numbers having mystical connections essential to numerological speculations, and sometimes ascribed to the Pythagoreans, are the *perfect, deficient*, and *abundant numbers*. A number is *perfect* if it is the sum of its proper divisors, *deficient* if it exceeds the sum of its proper divisors, and *abundant* if it is less than the sum of its proper divisors. So God created the world in six days, a perfect number, since $6 = 1 + 2 + 3$. On the other hand, as Alcuin (735–804) observed, the whole human race descended from the eight souls of Noah's ark, and this second creation was imperfect, for 8, being greater than $1 + 2 + 4$, is deficient. Until 1952 there were only 12 known perfect numbers, all of them even numbers, of which the first three are 6, 28, and 496. The last proposition of the ninth book of Euclid's *Elements* (*ca.* 300 B.C.) proves that *if $2^n - 1$ is a prime number,† then $2^{n-1}(2^n - 1)$ is*

---

*The *proper divisors* of a positive integer N are all the positive integral divisors of N except N itself. Note that 1 is a proper divisor of N. A somewhat antiquated synonym for proper divisor is *aliquot part*.

† A *prime number* is a positive integer greater than 1 and having no positive integral divisors other than itself and unity. An integer greater than 1 that is not a prime number is called a *composite number*. Thus 7 is a prime number, whereas 12 is a composite number.

a perfect number. The perfect numbers given by Euclid's formula are even numbers, and Euler has shown that every even perfect number must be of this form. The existence or nonexistence of odd perfect numbers is one of the

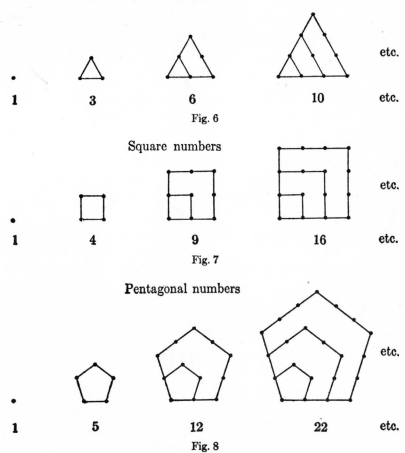

Triangular numbers

1   3   6   10   etc.

Fig. 6

Square numbers

1   4   9   16   etc.

Fig. 7

Pentagonal numbers

1   5   12   22   etc.

Fig. 8

celebrated unsolved problems in number theory. There certainly is no number of this type less than 2,000,000.

In 1952, with the aid of the SWAC digital computer, five more perfect numbers were discovered, corresponding to $n = 521$, 607, 1279, 2203, and 2281 in Euclid's formula. In 1957, the Swedish machine BESK found another, corresponding to $n = 3217$, and in 1961, an IBM 7090 found two more, for $n = 4253$ and 4423. There are no other perfect numbers for $n < 5000$. The perfect number for $n = 4423$ contains 2663 digits.

While not all historians of mathematics feel that amicable and perfect numbers can be ascribed to the Pythagoreans, there seems to be universal agreement that the *figurate numbers* did originate with the earliest members of the society. These numbers, considered as the number of dots in certain geometrical configurations, represent a link between geometry and arithmetic. Figures 6, 7, and 8 account for the geometrical nomenclature of *triangular numbers, square numbers, pentagonal numbers,* and so on.

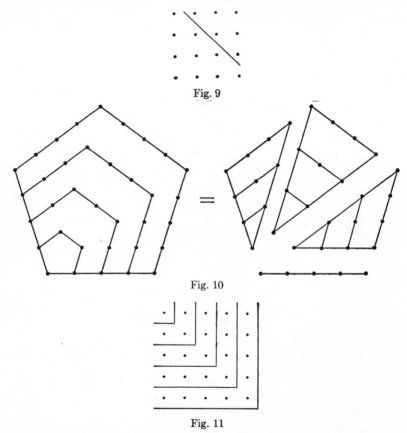

Fig. 9

Fig. 10

Fig. 11

Many interesting theorems concerning figurate numbers can be established in purely geometric fashion. For example, to show, Theorem I, that *any square number is the sum of two successive triangular numbers* we observe that a square number, in its geometric form, can be divided as in Figure 9. Again, Figure 10 illustrates Theorem II, that *the nth pentagonal number is equal to n plus three times the $(n-1)$th triangular number.* Theorem III, that *the sum of any number of consecutive odd integers, starting with 1, is a perfect square,* is exhibited geometrically by Figure 11.

Of course these theorems can also be established algebraically, once we obtain the algebraic representations of the general triangular, square, and pentagonal numbers. It is clear that the nth triangular number, T_n, is given by the sum of an arithmetic series,*

$$T_n = 1 + 2 + 3 + \cdots + n = n(n+1)/2,$$

and, of course, the nth square number, S_n, is n^2. Our first theorem may now be re-established algebraically by an identity as follows:

$$S_n = n^2 = n(n+1)/2 + (n-1)n/2 = T_n + T_{n-1}.$$

The nth pentagonal number, P_n, is also given by the sum of an arithmetic series,

$$\begin{aligned} P_n &= 1 + 4 + 7 + \cdots + (3n-2) \\ &= n(3n-1)/2 = n + 3n(n-1)/2 \\ &= n + 3T_{n-1}. \end{aligned}$$

This proves the second theorem. The third theorem is obtained algebraically by summing the arithmetic series

$$1 + 3 + 5 + \cdots + (2n-1) = n(2n)/2 = n^2.$$

As a last and very remarkable discovery about numbers, made by the Pythagoreans, we might mention the dependence of musical intervals upon numerical ratios. The Pythagoreans found that for strings under the same tension, the lengths should be 2 to 1 for the octave, 3 to 2 for the fifth, and 4 to 3 for the fourth. These results, the first recorded facts in mathematical physics, led the Pythagoreans to initiate the scientific study of musical scales.

3-4   Pythagorean Theorem and Pythagorean Triples

Tradition is unanimous in ascribing to Pythagoras the independent discovery of the theorem on the right triangle which now universally bears his name, that the square on the hypotenuse of a right triangle is equal to

---

*The sum of an arithmetic series is equal to the product of the number of terms and half the sum of the two extreme terms.

the sum of the squares on the two legs. We have seen that this theorem was known to the Babylonians of Hammurabi's time, more than a thousand years earlier, but the first general proof of the theorem may well have been given by Pythagoras. There has been much conjecture as to the proof Pythagoras might have offered, and it is generally felt that it probably was a dissection type of proof* like the following, illustrated in Figure 12. Let a, b, c denote the legs and hypotenuse of the given right triangle, and consider the

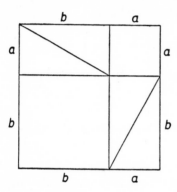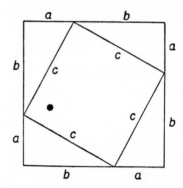

Fig. 12

two squares in the accompanying figure, each having $a + b$ as side. The first square is dissected into six pieces, namely the two squares on the legs and four right triangles congruent to the given triangle. The second square is dissected into five pieces, namely the square on the hypotenuse and again four right triangles congruent to the given triangle. By subtracting equals from equals, it now follows that the square on the hypotenuse is equal to the sum of the squares on the legs.

To prove that the central piece of the second dissection is actually a square of side c we need to employ the fact that the sum of the angles of a right triangle is equal to two right angles. But the *Eudemian Summary* attributes this theorem for the general triangle to the Pythagoreans. Since a proof of this theorem requires, in turn, a knowledge of some properties of parallels, the early Pythagoreans are also credited with the development of that theory.

Since Pythagoras' time many different proofs of the Pythagorean theorem have been supplied. In the second edition of his book *The Pythagorean Proposition*, E. S. Loomis has collected and classified 370 demonstrations of this famous theorem.

Closely allied to the Pythagorean theorem is the problem of finding

*See, however, Daniel Shanks, *Solved and Unsolved Problems in Number Theory*, Vol. 1 (Washington, D.C.: Spartan Books, 1962), pp. 124, 125.

integers a, b, c which can represent the legs and hypotenuse of a right triangle. A triple of numbers of this sort is known as a *Pythagorean triple* and, as we have seen in Section 2-6, the analysis of Plimpton 322 offers fairly convincing evidence that the ancient Babylonians knew how to calculate such triples. The Pythagoreans have been credited with the formula

$$m^2 + [(m^2 - 1)/2]^2 = [(m^2 + 1)/2]^2,$$

the three terms of which, for any odd m, yield a Pythagorean triple. The similar formula

$$(2m)^2 + (m^2 - 1)^2 = (m^2 + 1)^2,$$

where m may be even or odd, was devised for the same purpose and is attributed to Plato (*ca.* 380 B.C.). Neither of these formulas yields all Pythagorean triples and it is not until Euclid wrote his *Elements* that we find a complete solution of the problem.

3-5   Discovery of Irrational Magnitudes

The integers are abstractions arising from the process of counting finite collections of objects. The needs of daily life require us, in addition to counting individual objects, to measure various quantities, such as length, weight, and time. To satisfy these simple measuring needs fractions are required, for seldom will a length, to take an example, appear to contain an exact intégral number of linear units. Thus, if we define a *rational number* as the quotient of two integers, p/q, $q \neq 0$, this system of rational numbers, since it contains all the integers and fractions, is sufficient for practical measuring purposes.

The rational numbers have a simple geometrical interpretation. Mark two distinct points O and I on a horizontal straight line, I to the right of O, and choose the segment OI as a unit of length. If we let O and I represent the numbers 0 and 1 respectively, then the positive and negative integers can be represented by a set of points on the line spaced at unit intervals apart, the positive integers being represented to the right of O and the negative integers to the left of O. The fractions with denominator q may then be represented by the points which divide each of the unit intervals into q equal parts. Then for each rational number there is a point on the line. To the early mathematicians it seemed evident that all the points on the line would in this way be used up. It must have been something of a shock to learn that there are points on the line not corresponding to any rational number. This discovery was one of the greatest achievements of the Pythagoreans. In particular the Pythagoreans showed that there is no rational number corresponding to the point P on the line where the distance OP is

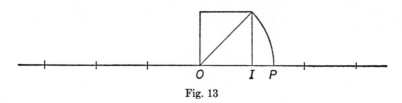

Fig. 13

equal to the diagonal of a square having unit side (see Figure 13). New numbers had to be invented to correspond to such points, and since these numbers cannot be rational numbers they came to be called *irrational numbers*. Their discovery marks one of the great milestones in the history of mathematics.

To prove that the length of the diagonal of a square of unit side cannot be represented by a rational number it suffices to show that $\sqrt{2}$ is irrational. To this end we first observe that, for a positive integer s, s^2 is even if and only if s is even. Now suppose for the purpose of argument, that $\sqrt{2}$ is rational, that is, $\sqrt{2} = a/b$, where a and b are relatively prime integers.* Then

$$a = \sqrt{2}b,$$

or
$$a^2 = 2b^2.$$

Since a^2 is twice an integer, we see that a^2, and hence a, must be even. Put $a = 2c$. Then the last equation becomes

$$4c^2 = 2b^2,$$

or
$$2c^2 = b^2,$$

from which we conclude that b^2, and hence b, must be even. But this is impossible since a and b were assumed to be relatively prime. Thus the assump-

*Two integers are *relatively prime* if they have no common positive integral factor other than unity. Thus 5 and 18 are relatively prime, whereas 12 and 18 are not relatively prime.

tion that $\sqrt{2}$ is rational has led us to this impossible situation, and must be abandoned.

Let us sketch an alternative, geometrical, demonstration by showing that a side and diagonal of a square are incommensurable (that is, have no common unit of measure). Suppose the contrary. Then, according to this

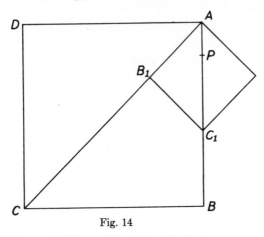

Fig. 14

supposition, there exists a segment AP (see Figure 14) such that both the diagonal AC and side AB of a square $ABCD$ are integral multiples of AP; that is, AC and AB are commensurable with respect to AP. On AC lay off $CB_1 = AB$ and draw B_1C_1 perpendicular to CA. One may easily prove that $C_1B = C_1B_1 = AB_1$. Then $AC_1 = AB - AB_1$ and AB_1 are commensurable with respect to AP. But AC_1 and AB_1 are a diagonal and a side of a square of dimensions less than half those of the original square. It follows that by repeating the process we may finally obtain a square whose diagonal AC_n and side AB_n are commensurable with respect to AP, and $AC_n < AP$. This absurdity proves the theorem.

The first proof is essentially the traditional one known to Aristotle (384–322 B.C.). This discovery of the irrationality of $\sqrt{2}$ caused some consternation in the Pythagorean ranks. Not only did it appear to upset the basic assumption that everything depends on the whole numbers, but, because the Pythagorean definition of proportion assumed any two like magnitudes to be commensurable, all the propositions in the Pythagorean theory of proportion had to be limited to commensurable magnitudes. So great was the "logical scandal" that efforts were made for a while to keep the matter secret, and one legend has it that the Pythagorean, Hippasus, drowned at sea for disclosing the secret to outsiders.

For some time $\sqrt{2}$ was the only known irrational.* Later, according to Plato, Theodorus of Cyrene (*ca.* 425 B.C.) showed that $\sqrt{3}$, $\sqrt{5}$, $\sqrt{6}$, $\sqrt{7}$, $\sqrt{8}$, $\sqrt{10}$, $\sqrt{11}$, $\sqrt{12}$, $\sqrt{13}$, $\sqrt{14}$, $\sqrt{15}$, $\sqrt{17}$ are also irrational. Then, about 370 B.C., the "scandal" was resolved by the brilliant Eudoxus, a pupil of Plato and of the Pythagorean, Archytas, by putting forth a new definition of proportion. Eudoxus' masterful treatment of incommensurables appears in the fifth book of Euclid's *Elements,* and coincides essentially with the modern exposition of irrational numbers that was given by Richard Dedekind in 1872.

The treatment of similar triangles in present-day high school geometry texts still reflects some of the difficulties and subtleties introduced by incommensurable magnitudes.

3-6   Algebraic Identities

Imbued with the representation of a number by a length, and completely lacking any adequate algebraic notation, the early Greeks devised ingenious geometrical processes for carrying out algebraic operations. Much of this geometrical algebra has been attributed to the Pythagoreans and can be found scattered through several of the earlier books of Euclid's *Elements.* Thus Book II of the *Elements* contains a number of propositions which in reality are algebraic identities couched in geometrical terminology. It seems quite certain that these propositions were developed, through means of a dissection method, by the early Pythagoreans. We may illustrate the method by considering a few of the propositions of Book II.

Proposition 4 of Book II establishes geometrically the identity

$$(a + b)^2 = a^2 + 2ab + b^2$$

by dissecting the square of side $a + b$ into two squares and two rectangles having areas a^2, b^2, ab, and ab, as indicated in Figure 15. Euclid's statement of the proposition is: *If a straight line is divided into any two parts, the square on the whole line is equal to the sum of the squares on the two parts together with twice the rectangle contained by the two parts.*

The statement of Proposition 5 of Book II is: *If a straight line is divided equally and also unequally, the rectangle contained by the unequal parts, together with the square on the line between the points of section, is equal to the square on*

*There is some possibility that $\sqrt{5}$ was the first known irrational.

half the line. Let AB be the given straight-line segment, and let it be divided equally at P, and unequally at Q. Then the proposition says that

$$(AQ)(QB) + (PQ)^2 = (PB)^2.$$

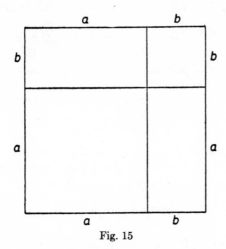

Fig. 15

If we set $AQ = 2a$ and $QB = 2b$ this leads to the algebraic identity

$$4ab + (a - b)^2 = (a + b)^2,$$

or, if we set $AB = 2a$ and $PQ = b$, to the identity

$$(a + b)(a - b) = a^2 - b^2.$$

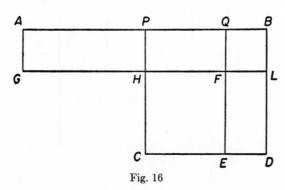

Fig. 16

The dissection given in the *Elements* for establishing this theorem is more complicated than that for Proposition 4 and appears in Figure 16. In the figure $PCDB$ and $QFLB$ are squares described on PB and QB as sides. Then

$$(AQ)(QB) + (PQ)^2 = AGFQ + HCEF = AGHP + PHFQ + HCEF$$
$$= PHLB + PHFQ + HCEF$$
$$= PHLB + FEDL + HCEF = (PB)^2.$$

The statement of Proposition 6 of Book II is: *If a straight line is bisected and produced to any point, the rectangle contained by the whole line thus produced and the part of it produced, together with the square on half the line*

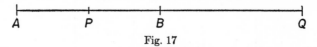

| | | | |
| A | P | B | Q |

Fig. 17

bisected, is equal to the square on the straight line made up of the half and the part produced. Here (see Figure 17), if the given straight-line segment AB with mid-point P is produced to Q, we are to show that

$$(AQ)(QB) + (PB)^2 = (PQ)^2.$$

If we set $AQ = 2a$ and $QB = 2b$ we are led again to the identity

$$4ab + (a - b)^2 = (a + b)^2,$$

and a similar dissection to that used for Proposition 5 may be used here.

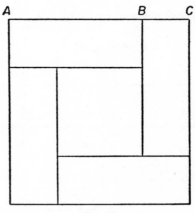

Fig. 18

Figure 18, with $AB = a$ and $BC = b$, suggests a less cumbrous proof of the identity

$$4ab + (a - b)^2 = (a + b)^2.$$

3-7   Geometric Solution of Quadratic Equations

Propositions 5 and 6 of Euclid's Book II, as given in the preceding section, were probably devised by the Pythagoreans to serve in the geometrical solution of quadratic equations. This clever solution of quadratic equations is to be found in Propositions 28 and 29 of Euclid's Book VI. These propositions are construction problems, the statement* of the first being: *To divide a given straight line so that the rectangle contained by its segments may be equal to a given space; but that space must not be greater than the square of half the given line.* To clarify the problem, let AB and q be two line

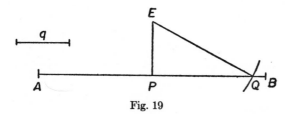

Fig. 19

segments, q not greater than half of AB. We are to divide AB by a point Q such that $(AQ)(QB) = q^2$. To accomplish this we mark off $PE = q$ on the perpendicular to AB at its mid-point P, and with E as center and PB as radius draw an arc cutting AB in the sought point Q, as in Figure 19. The proof is furnished by Proposition 5 of Book II, for by that proposition

$$(AQ)(QB) = (PB)^2 - (PQ)^2 = (EQ)^2 - (PQ)^2 = (EP)^2 = q^2.$$

Now, denoting the length of AB by p, we note that this construction is nothing but a disguised solution of the quadratic equation

$$x^2 - px + q^2 = 0.$$

For, if r and s are the roots of this quadratic equation we know from elementary algebra that

$$r + s = p \quad \text{and} \quad rs = q^2.$$

But it is AQ and QB whose sum is AB, or p, and whose product is q^2. There-

*We here state Propositions VI 28 and VI 29, not as given in the Theon revision of Euclid's *Elements*, but as simplified in later English editions of the *Elements*.

fore AQ and QB represent the roots r and s. The roots of the quadratic equation

$$x^2 + px + q^2 = 0$$

are represented by the negatives of the lengths of AQ and QB.

Proposition 29 of Book VI is: *To produce a given straight line so that the rectangle contained by the segments between the extremities of the given line and the point to which it is produced may be equal to a given space.* Again, let AB and q be two line segments. We are to produce AB to a point Q such that

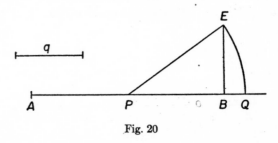

Fig. 20

$(AQ)(QB) = q^2$. To this end we mark off $BE = q$ on the perpendicular to AB at B, and with P, the mid-point of AB, as center and PE as radius draw an arc cutting AB produced in the sought point Q, as in Figure 20. This time the proof is furnished by Proposition 6 of Book II, for by that proposition

$$(AQ)(QB) = (PQ)^2 - (PB)^2 = (PE)^2 - (PB)^2 = (BE)^2 = q^2.$$

As before, we see that AQ and QB, where we take the first one as positive and the second one as negative, are the roots of the quadratic equation

$$x^2 - px - q^2 = 0,$$

p being the length of AB. The roots of

$$x^2 + px - q^2 = 0$$

are the same as those of $x^2 - px - q^2 = 0$, only with their signs changed.

The geometric algebra of the Pythagoreans, ingenious though it is, intensifies one's appreciation of the simplicity and convenience inherent in present-day algebraic notation.

3-8   Transformation of Areas

The Pythagoreans were interested in transforming an area from one rectilinear shape into another rectilinear shape. Their solution of the basic problem of constructing a square equal in area to that of a given polygon may be found in Propositions 42, 44, 45 of Book I and Proposition 14 of Book II of Euclid's *Elements*. A simpler solution, probably also known to the Pythagoreans, is the following. Consider any polygon $ABCD \cdots$ (see Figure 21). Draw BR parallel to AC to cut DC in R. Then, since triangles

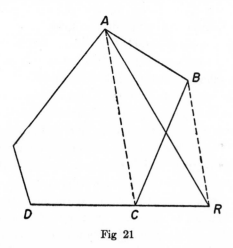

Fig 21

ABC and ARC have a common base AC and equal altitudes on this common base, these triangles have equal areas. It follows that polygons $ABCD \cdots$ and $ARD \cdots$ have equal areas. But the derived polygon has one fewer sides than the given polygon. By a repetition of this process we finally obtain a triangle having the same area as the given polygon. Now if b is any side of this triangle and h the altitude on b, the side of an equivalent square is given by $\sqrt{(bh)/2}$, that is, by the mean proportional between b and $h/2$. Since this mean proportional is easily constructed with straightedge and compasses, the entire problem can be carried out with these tools.

Many interesting area problems can be solved by this simple process of drawing parallel lines (see Problem Study 3.11).

3-9   The Regular Solids

A polyhedron is said to be *regular* if its faces are congruent regular polygons and if its polyhedral angles are all congruent. While there are regular polygons of all orders, it turns out that there are only five different regular polyhedra (see Problem Study 3.12). The regular polyhedra are named according to the number of faces each possesses. Thus there is the tetrahedron with 4 triangular faces, the hexahedron, or cube, with 6 square faces, the octahedron with 8 triangular faces, the dodecahedron with 12 pentagonal faces, and the icosahedron with 20 triangular faces.

The early history of these regular polyhedra is lost in the dimness of the past. A mathematical treatment of them is initiated in Book XIII of Euclid's *Elements*. The first scholium of this book remarks that the book "will treat of the so-called Platonic solids, incorrectly named, because three of them, the tetrahedron, cube, and dodecahedron are due to the Pythagoreans, while the octahedron and icosahedron are due to Theaetetus." This could well be the case. A description of all five was certainly given by Plato.

The tetrahedron, cube, and octahedron can be found in nature as crystals, for example as crystals of sodium sulphantimoniate, common salt, and chrome alum, respectively. The other two cannot occur in crystal form, but have been observed as skeletons of microscopic sea animals called *Radiolaria*. In 1885, a toy regular dodecahedron of Etruscan origin was unearthed on Monte Loffa, near Padua, and is held to date back to about 500 B.C.

3-10   Postulational Thinking

Sometime between Thales in 600 B.C. and Euclid in 300 B.C. was perfected the notion of a logical discourse as a sequence of rigorous deductions from some initial and explicitly stated assumptions. This process, the so-called postulational method, has become the very core of modern mathematics, and undoubtedly much of the development of geometry along this pattern is due to the Pythagoreans. Certainly one of the greatest contributions of the early Greeks was the development of this postulational method of thinking. We shall return, in Chapter V, to a fuller discussion of the subject.

Problem Studies

3.1 The Practical Problems of Thales

(a) There are two versions of how Thales calculated the height of an Egyptian pyramid by shadows. The earlier account, given by Hieronymus, a pupil of Aristotle, says that Thales noted the length of the shadow of the pyramid at the moment when his shadow was the same length as himself. The later version, given by Plutarch, says that he set up a stick and then made use of similar triangles. Both versions fail to mention the difficulty, in either case, of obtaining the length of the shadow of the pyramid, that is, the distance from the apex of the shadow to the center of the base of the pyramid.

Devise a method, based on similar triangles and independent of latitude and time of year, for determining the height of a pyramid *from two shadow observations.*

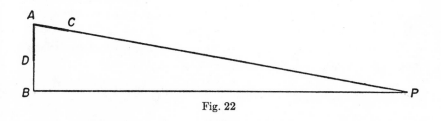

Fig. 22

(b) We are told that Thales measured the distance of a ship from shore using the fact that two triangles are congruent if two angles and the included side of one are equal to two angles and the included side of the other. Heath has conjectured that this was probably done by an instrument consisting of two rods AC and AD, hinged together at A, as shown in Figure 22. The rod AD was held vertically over point B on shore, while rod AC was pointed toward the ship P. Then, without changing the angle DAC, the instrument was revolved about AD, and point Q noted on the ground at which arm AC was directed. What distance must be measured in order to find the distance from B to the inaccessible point P?

3.2 Perfect and Amicable Numbers

(a) Show that in Euclid's formula for perfect numbers, n must be prime.

(b) What is the fourth perfect number furnished by Euclid's formula?

(c) Prove that the sum of the reciprocals of *all* the divisors of a perfect number is equal to 2.

(d) Show that if p is a prime then p^n is deficient.

(e) Show that Nicolo Paganini's numbers, 1184 and 1210, are amicable.

(f) Show that any multiple of an abundant or perfect number is abundant.

(g) Find the 21 abundant numbers less than 100. It will be noticed that they are all even numbers. To show that all abundant numbers are not even show that $945 = 3^3 \cdot 5 \cdot 7$ is abundant. This is the first odd abundant number.

3.3 Figurate Numbers

(a) List the first four hexagonal numbers.

(b) An *oblong number* is the number of dots in a rectangular array having one more column than rows. Show, geometrically and algebraically, that the sum of the first n positive even integers is an oblong number.

(c) Show, both geometrically and algebraically, that any oblong number is the sum of two equal triangular numbers.

(d) Show, geometrically and algebraically, that 8 times any triangular number, plus 1, is a square number.

3.4 Means

The *Eudemian Summary* says that in Pythagoras' time there were three means, the *arithmetic*, the *geometric*, and the *subcontrary*, the last name being later changed to *harmonic* by Archytas and Hippasus. We may define these three means of two positive numbers a and b as

$$A = (a + b)/2, \qquad G = \sqrt{ab}, \qquad H = 2ab/(a + b),$$

respectively.

(a) Show that $A \geqq G \geqq H$, equality holding if and only if $a = b$.

(b) Show that $a : A = H : b$. This was known as the "musical" proportion.

(c) Show that H is the harmonic mean between a and b if there exists a number n such that $a = H + a/n$ and $H = b + b/n$. This was the Pythagorean definition of the harmonic mean of a and b.

(d) Show that $1/(H - a) + 1/(H - b) = 1/a + 1/b$.

(e) Since 8 is the harmonic mean of 12 and 6, Philolaus, a Pythagorean of about 425 B.C., called the cube a "geometrical harmony." Explain this.

(f) Show that if a, b, c are in harmonic progression, so also are $a/(b + c)$, $b/(c + a)$, $c/(a + b)$.

3.5 Dissection Proofs of the Pythagorean Theorem

Two areas, or two volumes, P and Q, are said to be *congruent by addition* if they can be dissected into corresponding pairs of congruent pieces. They are said to be *congruent by subtraction* if corresponding pairs of congruent pieces can be added to P and Q to give two new figures which are congruent by addition. There are many proofs of the Pythagorean theorem which achieve their end by showing that the square on the hypotenuse of the right triangle is congruent either by addition or subtraction to the combined squares on the legs of the right triangle. The proof given in Section 3-4 is a congruency-by-subtraction proof. Give two congruency-by-addition proofs of the Pythagorean theorem suggested by Figures 23 and 24, the first given by H. Perigal in 1873* and the second by H. E. Dudeney in 1917.

(a)   **(b)**

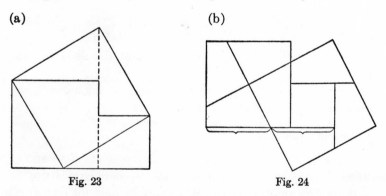

Fig. 23   Fig. 24

*This was a rediscovery, for the dissection was known to T̲âbit ibn Qorra (826–901).

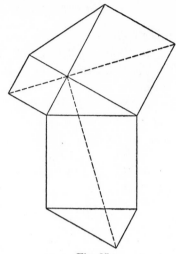

(c) Give a congruency-by-subtraction proof of the Pythagorean theorem suggested by Figure 25, said to have been devised by Leonardo da Vinci (1452–1519).

Fig. 25

It is interesting that *any two equal polygonal areas are congruent by addition, and that the dissection can always be carried out with straightedge and compasses.* On the other hand, in 1901, Max Dehn showed that two equal polyhedral volumes are not necessarily congruent by either addition or subtraction. In particular, it is impossible to dissect a regular tetrahedron into polyhedral pieces which can be reassembled to form a cube.

3.6   Pythagorean Triples

(a) What is the relation between the hypotenuse and the longer leg of the integral-sided right triangles given by the Pythagorean formula of Section 3-4?

(b) Find the Pythagorean triples given by the Pythagorean formula of Section 3-4 for which the hypotenuse does not exceed 100.

(c) Prove that there exists no isosceles right triangle whose sides are integers.

(d) Prove that no Pythagorean triple exists in which one integer is a mean proportional between the other two.

3.7   Irrational Numbers

(a) Prove that the straight line through the points (0,0) and (1,$\sqrt{2}$) passes through no point, other than (0,0), of the coordinate lattice.

(b) Show how the coordinate lattice may be used for finding rational approximations of $\sqrt{2}$.

(c) If p is a prime number, show that $\sqrt{p}$ is irrational.

(d) Show that $\log_{10} 2$ is irrational.

(e) Generalize part (d) by showing that $\log_a b$ is irrational if a and b are positive integers and one of them contains a prime factor not contained in the other.

3.8 Algebraic Identities

Indicate how each of the following algebraic identities might be established geometrically.

(a) $(a - b)^2 = a^2 - 2ab + b^2$

(b) $a(b + c) = ab + ac$

(c) $(a + b)(c + d) = ac + bc + ad + bd$

(d) The statement of Proposition 9 of Book II of Euclid's *Elements* is: *If a straight line is divided equally and also unequally, the sum of the squares on the two unequal parts is twice the sum of the squares on half the line and on the line between the points of section.* From this theorem obtain the algebraic identity

$$(a + b)^2 + (a - b)^2 = 2(a^2 + b^2).$$

3.9 Geometric Algebra

Draw three. unequal line segments. Label the longest one a, the medium one b, and take the smallest one as 1 unit. With straightedge and compasses construct line segments of lengths

(a) $a + b$ and $a - b$,

(b) ab,

(c) a/b,

(d) $\sqrt{a}$,

(e) a/n, n a positive integer,

(f) $x = (a^2 + b^2 - ab)^{1/2}$. If we form a triangle with sides a, b, x, what is the size of the angle between sides a and b?

3.10 Geometric Solution of Quadratic Equations

(a) Given a unit segment, solve the quadratic equation $x^2 - 7x + 12 = 0$ by the Pythagorean method.

(b) Given a unit segment, solve the quadratic equation $x^2 + 4x - 21 = 0$ by the Pythagorean method.

(c) With straightedge and compasses divide a segment a into two parts such that the difference of their squares shall be equal to their product.

(d) Show that, in part (c), the longer segment is a mean proportional between the shorter segment and the whole line. The line segment is said to be divided in *extreme and mean ratio*, or in *golden section*.

(e) Let us be given a quadratic equation $x^2 - gx + h = 0$. On a rectangular Cartesian frame of reference plot the points $B:(0,1)$ and $Q:(g,h)$. Draw the circle on BQ as diameter and let it cut the x-axis in M and N. Show that the signed lengths of OM and ON represent the roots of the given quadratic equation. This geometrical solution of quadratic equations appeared in Leslie's *Elements of Geometry* with the remark: "The solution of this important problem now inserted in the text, was suggested to me by Mr. Thomas Carlyle, an ingenious young mathematician, and formerly my pupil."

(f) Solve the quadratic equations $x^2 - 7x + 12 = 0$ and $x^2 + 4x - 21 = 0$ by Carlyle's method.

3.11   Transformation of Areas

(a) Draw an irregular hexagon and then construct, with straightedge and compasses, a square having the same area.

(b) With straightedge and compasses divide a quadrilateral $ABCD$ in three equivalent parts by straight lines drawn through vertex A.

(c) Bisect a trapezoid by a line drawn from a point P in the smaller base.

(d) Transform triangle ABC so that the angle A is not altered, but the side opposite the angle A becomes parallel to a given line MN.

(e) Transform a given triangle into an isosceles triangle having a given vertex angle.

3.12   Regular Solids

(a) Show that there can be no more than five regular polyhedra.

(b) Find the volume and surface of a regular octahedron of edge e.

(c) For each of the five regular polyhedra enumerate the number of vertices v, edges e, and faces f, and then evaluate the quantity $v - e + f$. One of the most interesting theorems relating to any convex (or more generally any *simply-connected*) polyhedron, is that $v - e + f = 2$. This may have been known to Archimedes (*ca.* 225 B.C.), but was first explicitly stated by Descartes about 1635. Since Euler later independently announced it in 1752, the result is often referred to as the Euler-Descartes formula.

(d) A *cuboctahedron* is a solid whose edges are obtained by joining together the mid-points of adjacent edges of a cube. Enumerate v, e, and f for a cuboctahedron.

(e) Consider a solid cube with regular pyramids built on a pair of opposite faces as bases. Now let a hole with square cross section, and with its axis on the line joining the vertices of the pyramids, be cut from the solid. Evaluate $v - e + f$ for this ring-shaped solid.

Construction patterns for 100 different solids can be found in Miles C. Hartley, *Patterns of Polyhedrons*, revised edition, Ann Arbor, Mich.: priv. ptd., Edwards Brothers, 1957.

Bibliography

ALLMAN, G. J., *Greek Geometry from Thales to Euclid*. Dublin: University Press, 1889.

BELL, E. T., *The Magic of Numbers*. New York: McGraw-Hill Book Company, Inc., 1946.

COOLIDGE, J. L., *A History of Geometrical Methods*. New York: Oxford University Press, 1940.

COURANT, RICHARD, AND H. E. ROBBINS, *What Is Mathematics?* New York: Oxford University Press, 1941.

DANTZIG, T., *Number, The Language of Science*, 3d ed. New York: The Macmillan Company, 1939.

————, *The Bequest of the Greeks*. New York: Charles Scribner's Sons, 1955.

GOW, JAMES, *A Short History of Greek Mathematics*. New York: Hafner Publishing Company, 1923.

HEATH, T. L., *History of Greek Mathematics*, Vol. 1. New York: Oxford University Press, 1921.

————, *A Manual of Greek Mathematics*. New York: Oxford University Press, 1931.

————, *The Thirteen Books of Euclid's Elements*, 2d ed., 3 vols. New York: Cambridge University Press, 1926.

LOOMIS, E. S., *The Pythagorean Proposition*, 2d ed. Ann Arbor, Mich.: priv. ptd., Edwards Brothers, Inc., 1940.

MUIR, JANE, *Of Men and Numbers*. New York: Dodd, Mead & Company, Inc., 1961.

TURNBULL, H. W., *The Great Mathematicians*. New York: New York University Press, 1961.

VAN DER WAERDEN, B. L., *Science Awakening*, tr. by Arnold Dresden. Groningen, Netherlands: P. Noordhoff, N.V., 1954.

4

Duplication, Trisection, and Quadrature

4-1 The Period from Thales to Euclid

The first three centuries of Greek mathematics, commencing with the initial efforts at demonstrative geometry by Thales in about 600 B.C. and culminating with the remarkable *Elements* of Euclid in about 300 B.C., constitute a period of extraordinary achievement. In the last chapter we considered some of the Pythagorean contributions to this achievement. Besides the Ionian school founded by Thales at Miletus and the early Pythagorean school at Crotona, a number of mathematical centers arose and flourished at places and for periods which were largely governed by the background of Greek political history.

It was about 1200 B.C. that the primitive Dorian tribes moved southward into the Greek peninsula, leaving their northern mountain fastnesses for more favorable territory, and subsequently their chief tribe, the Spartans, developed the city of Sparta. A large section of the former inhabitants of the invaded region fled, for self-preservation, to Asia Minor and the Ionian islands of the Aegean Sea, where in time they established Greek trading colonies. It was in these colonies, in the sixth century B.C., that the Ionian school was founded and Greek philosophy blossomed and demonstrative geometry was born.

Meanwhile Persia had become a great military empire, and, following the inevitable expansionist program induced by a slave-based economy, conquered in 546 B.C. the Ionian cities and the Greek colonies of Asia Minor. As a result a number of Greek philosophers, like Pythagoras and Xenophanes, abandoned their native land and moved to the prospering Greek colonies in southern Italy. Schools of philosophy and mathematics developed at Crotona, under Pythagoras, and at Elea, under Xenophanes, Zeno, and Parmenides.

The yoke of oppression rested uneasily on the conquered Ionian cities, and in 499 B.C. a revolt was fomented. Athens, which was becoming a center of western civilization with political progress toward democracy, aided the revolution by sending armies. Although the revolt was crushed, the incensed King Darius of Persia decided to punish Athens. In 492 B.C. he organized a huge army and navy to attack the mainland of Greece, but his fleet was destroyed in a storm and his land forces suffered expeditionary difficulties. Two years later the Persian armies penetrated Attica where they were decisively defeated by the Athenians at Marathon. Athens assumed the mantle of Greek leadership.

In 480 B.C. Xerxes, son of Darius, attempted another land and sea invasion of Greece. The Athenians met the Persian fleet in the great naval battle of Salamis, and won, and although the Greek land forces under Spartan leadership were defeated and wiped out at Thermopylae, the Greeks overcame the Persians the following year at Plataea and forced the invaders out of Greece. The hegemony of Athens was consolidated and the following half century of peace was a brilliant period in Athenian history. This city of Pericles and Socrates became the center of democratic and intellectual development. Mathematicians were attracted from all parts of the Greek world. Anaxagoras, the last eminent member of the Ionian school, settled there. Many of the dispersed Pythagoreans found their way to Athens, and Zeno and Parmenides, of the Eleatic school, went to Athens to teach. Hippocrates,* from the Ionian island of Chios, visited Athens and is reputed by ancient writers to have published the first connected geometry there.

Peace came to an end in 431 B.C. with the start of the Peloponnesian War between Athens and Sparta. This proved to be a long-drawn-out conflict. Athens, at first successful, later suffered a devastating plague which killed off a fourth of its population, and finally, in 404 B.C., had to accept humiliating defeat. Sparta assumed political leadership only to lose it, in 371 B.C., by defeat at the hands of a league of rebellious city-states. During these struggles little progress was made in geometry at Athens, and once again development came from the more peaceful regions of Magna Graecia. The Pythagoreans of southern Italy had been allowed to return, purified of political association, and a new Pythagorean school at Tarentum arose, under the influence of the gifted and much admired Archytas.

*Not to be confused with Hippocrates of Cos, the famous Greek physician of antiquity.

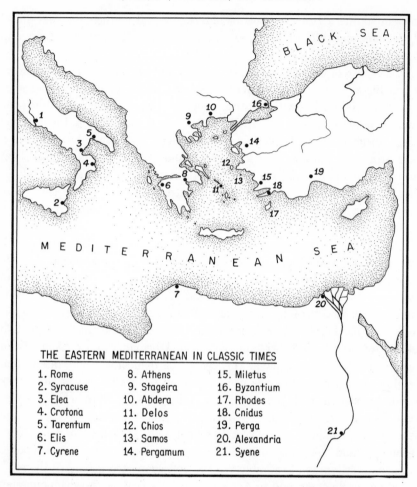

THE EASTERN MEDITERRANEAN IN CLASSIC TIMES

1. Rome	8. Athens	15. Miletus
2. Syracuse	9. Stageira	16. Byzantium
3. Elea	10. Abdera	17. Rhodes
4. Crotona	11. Delos	18. Cnidus
5. Tarentum	12. Chios	19. Perga
6. Elis	13. Samos	20. Alexandria
7. Cyrene	14. Pergamum	21. Syene

Abdera: Democritus, 400 B.C.

Alexandria: Euclid, 300 B.C.; Eratosthenes, 230 B.C.; Apollonius, 225 B.C.; Aristarchus, 280 B.C.; Hypsicles, 180 B.C.; Heron, 50; Menelaus, 100; Ptolemy, 150; Diophantus, 50(?); Pappus, 300; Theon, 390; Hypatia, 410.

Athens: Socrates, 425 B.C.; Theaetetus, 375 B.C.; Plato, 380 B.C.; Ptolemy, 150.

Byzantium: Proclus, 460.

Chios: Hippocrates, 460 B.C.

Cnidus: Eudoxus, 370 B.C.

Crotona: Pythagoras, 540 B.C.; Philolaus, 425 B.C.

Cyrene: Theodorus, 425 B.C.; Eratosthenes, 230 B.C.

Delos: The Delian Problem.

Elea: Zeno, 450 B.C.; Parmenides, 460 B.C.

Elis: Hippias, 425 B.C.

Miletus: Thales, 600 B.C.

Perga: Apollonius, 225 B.C.

Pergamum: Library.

Rhodes: Eudemus, 335 B.C.; Hipparchus, 140 B.C.

Rome: Menelaus, 100.

Samos: Pythagoras, 540 B.C.; Conon, 260 B.C.; Aristarchus, 280 B.C.

Stageira: Aristotle, 340 B.C.

Syene: Eratosthenes, 230 B.C.

Syracuse: Archimedes, 225 B.C.

Tarentum: Pythagoras, 540 B.C.; Archytas, 400 B.C.

With the end of the Peloponnesian War, Athens, although reduced to a minor political power, regained her cultural leadership, and the Athenian school again flourished. Plato was born in Athens during the year of the great plague, studied philosophy under Socrates there and mathematics under Theodorus at Cyrene on the African coast, was an intimate friend of Archytas, and, upon his return to Athens in 380 B.C., founded his famous Academy there. Eudoxus, who studied under both Archytas and Plato, founded a school at Cyzicus in northern Asia Minor. Menaechmus, an associate of Plato and a pupil of Eudoxus, invented the conic sections. Dinostratus, brother of Menaechmus, was an able geometer and a pupil of Plato. Theaetetus, a man of unusual natural gifts and to whom we are probably indebted for much of the material of Euclid's tenth and thirteenth books, was another Athenian pupil of Theodorus. Mention should also be made of Aristotle who, though not a professed mathematician, was the systematizer of deductive logic and a writer on physical subjects; some parts of his *Analytica posteriora* show an unusual grasp of the mathematical method.

4-2   Lines of Mathematical Development

One can notice three important and distinct lines of development during the first 300 years of Greek mathematics. In the first place, we have the development of the material that ultimately was organized into the *Elements*, ably begun by the Pythagoreans and then added to by Hippocrates, Eudoxus, Theodorus, Theaetetus, and others. We have already considered portions of this development, and shall return to it in the next chapter. Secondly, there is the development of notions connected with infinitesimals and with limit and summation processes which did not attain final clarification until after the invention of the calculus in modern times. The paradoxes of Zeno, the method of exhaustion of Antiphon and Eudoxus, and the atomistic theory associated with the name of Democritus belong to this second line of development, and will be discussed in a later chapter devoted to the origins of the calculus. The third line of development is that of higher geometry, or the geometry of curves other than the circle and straight line and of surfaces other than the sphere and plane. Curiously enough, most of this higher geometry originated in continued attempts to solve three now famous construction problems. The present chapter is devoted to these three famous problems.

4-3   The Three Famous Problems

The three famous problems are

(1) *The duplication of the cube,* or the problem of constructing the edge of a cube having twice the volume of a given cube.

(2) *The trisection of an angle,* or the problem of dividing a given arbitrary angle into three equal parts.

(3) *The quadrature of the circle,* or the problem of constructing a square having an area equal to that of a given circle.

The importance of these problems lies in the fact that they cannot be solved, except by approximation, with straightedge and compasses, although these tools successfully serve for the solution of so many other construction problems. The energetic search for solutions to these three problems profoundly influenced Greek geometry and led to many fruitful discoveries, such as that of the conic sections, of many cubic and quartic curves, and of several transcendental curves. A much later outgrowth was the development of portions of the theory of equations concerning domains of rationality, algebraic numbers, and group theory. The impossibility of the three constructions under the self-imposed limitation that only the straightedge and compasses could be used was not established until the nineteenth century, more than 2000 years after the problems were first conceived.

4-4   The Euclidean Tools

It is important to be clear as to just what we are permitted to do with the straightedge and compasses. *With the straightedge we are permitted to draw a straight line of indefinite length through any two given distinct points; with the compasses we are permitted to draw a circle with any given point as center and passing through any given second point.* The drawing of constructions with straightedge and compasses, viewed as a game played according to these two rules, has proved to be one of the most fascinating and absorbing games ever devised. One is surprised at the really intricate constructions that can be accomplished in this manner, and accordingly it is hard to believe that the seemingly simple construction problems announced in the last section cannot also be so accomplished.

Since the postulates of Euclid's *Elements* restrict the use of the straightedge and compasses in accordance with the above rules, these instruments, so used, have become known as *Euclidean tools*. It is to be noted that the straightedge is to be *unmarked*. We shall see that with a marked straightedge it is possible to trisect a given angle. Also, we notice that the Euclidean compasses differ from our modern compasses, for with the modern compasses we are permitted to draw a circle having any point C as center and any segment AB as radius. In other words, we are permitted to transfer the distance AB to the center C, using the compasses as dividers. The Euclidean compasses, on the other hand, may be supposed to collapse if either leg is lifted from the paper. It might seem that the modern compasses are somewhat more powerful than the Euclidean, or collapsing, compasses. Curiously enough, the two are equivalent tools (see Problem Study 4.1).

4-5   Duplication of the Cube

There is evidence that the problem of duplicating a cube may have originated in the words of some mathematically unschooled and obscure ancient Greek poet who represented the mythical King Minos as dissatisfied with the size of a tomb erected to his son Glaucus. Minos ordered that the tomb be doubled in size. The poet then had Minos add, incorrectly, that this can be accomplished by doubling each dimension of the tomb. This faulty mathematics on the part of the poet led the geometers to take up the problem of finding how one can double a given solid while keeping the same shape. No progress seems to have been made on the problem until some time later, when Hippocrates discovered his famous reduction which we give below. Again, still later, it is told that the Delians were instructed by their oracle that, to get rid of a certain pestilence, they must double the size of Apollo's cubical altar. The problem was reputedly taken to Plato who submitted it to the geometers. It is this latter story which has led the duplication problem frequently to be referred to as the *Delian problem*. Whether the story is true or not, the problem was studied in Plato's Academy, and there are higher geometry solutions attributed to Eudoxus, Menaechmus, and even (though probably erroneously) to Plato himself.

The first real progress in the duplication problem was, no doubt, the reduction of the problem by Hippocrates (*ca.* 440 B.C.) to the construction of two mean proportionals between two given line segments of lengths s

and 2s. If we denote the two mean proportionals by x and y, then

$$s : x = x : y = y : 2s.$$

From these proportions we have $x^2 = sy$ and $y^2 = 2sx$. Eliminating y we find that $x^3 = 2s^3$. Thus x is the edge of a cube having twice the volume of the cube on edge s.

After Hippocrates made his reduction, subsequent attempts at duplicating the cube took the form of constructing two mean proportionals between two given line segments. One of the earliest, and certainly one of the most remarkable, higher geometry solutions in this form was given by Archytas (*ca.* 400 B.C.). His solution rests on finding a point of intersection of a right circular cylinder, a torus of zero inner diameter, and a right circular cone! The solution sheds some light on the unusual extent to which geometry must have been developed at this early date. The solution by Eudoxus (*ca.* 370 B.C.) is lost. Menaechmus (*ca.* 350 B.C.) gave two solutions of the problem, and, as far as is known, invented the conic sections for the purpose. A later solution using a mechanical contrivance is credited to Eratosthenes (*ca.* 230 B.C.), and another of about the same time to Nicomedes. A still later solution was offered by Apollonius (*ca.* 225 B.C.). Diocles (*ca.* 180 B.C.) invented the cissoid curve to obtain the desired end. And of course many solutions using higher plane curves have been devised in more recent times.

A number of the solutions alluded to above may be found in the Problem Studies at the end of the chapter. To illustrate the spirit of the attempts let us reproduce the one credited by Eutocius to Plato. Since the solution is by mechanical means and since it is known that Plato objected to such methods, it is felt that the ascription to Plato is erroneous.

Consider two triangles (see first part of Figure 26), CBA and DAB, right angled at B and A respectively, and lying on the same side of the common leg AB. Let the hypotenuses AC and BD of the triangles intersect perpendicularly in P. From the similar triangles CPB, BPA, APD it follows that

$$PC : PB = PB : PA = PA : PD.$$

Thus PB and PA are the two mean proportionals between PC and PD. It follows that the problem is solved if a figure can be constructed having $PD = 2(PC)$. The second part of Figure 26 shows how such a figure can be drawn by mechanical means. Draw two perpendicular lines intersecting in P and mark off PC and PD on them, with $PD = 2(PC)$. Now place a carpenter's square, with inner edge *RST*, on the figure so that *SR* passes through *D* and the vertex *S* of the right angle lies on *CP* produced. On *ST* slide a right triangle *UVW*, with leg *VW* on *ST*, until leg *VU* passes through *C*. Now manipulate the apparatus* until *V* falls on *DP* produced.

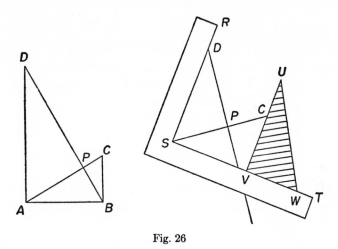

Fig. 26

4-6　Trisection of an Angle

Of the three famous problems of Greek antiquity the trisection of an angle is pre-eminently the most popular among the mathematically uninitiated in America today. Every year the mathematics journals and the members of the mathematics teaching profession of the country receive many communications from "angle trisectors," and it is not unusual to read in a newspaper that someone has finally "solved" the elusive problem. The problem is certainly the simplest one of the three famous problems to comprehend, and since the bisection of an angle is so very easy it is natural to wonder why trisection is not equally easy.

The multisection of a line segment with Euclidean tools is a simple matter, and it may be that the ancient Greeks were led to the trisection problem in an effort to solve the analogous problem of multisecting an angle.

*For an improved form of this apparatus see, for example, Richard Courant and H. E. Robbins, *What Is Mathematics?* p. 147.

Or perhaps, more likely, the problem arose in efforts to construct a regular nine-sided polygon, where the trisection of a 60° angle is required.

In dealing with the trisection problem the Greeks seem first to have reduced it to what they called a *verging* problem. Any acute angle ABC (see Figure 27) may be taken as the angle between a diagonal BA and a side BC

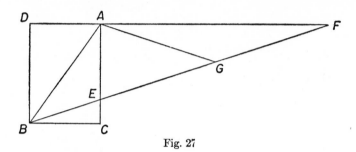

Fig. 27

of a rectangle $BCAD$. Consider a line through B cutting CA in E and DA produced in F, and such that $EF = 2(BA)$. Let G be the mid-point of EF. Then

$$EG = GF = GA = BA,$$

whence

$$\angle ABG = \angle AGB = \angle GAF + \angle GFA = 2\angle GFA = 2\angle GBC,$$

and BEF trisects angle ABC. Thus the problem is reduced to that of constructing a straight-line segment EF of given length $2(BA)$ between AC and AF so that FE *verges* toward B.

If, contrary to Euclidean assumptions, we permit ourselves to mark, on our straightedge, a segment $E'F' = 2(BA)$, and then to adjust the straightedge so that it passes through B and has the marked points E' and F' on AC and the prolongation of DA, the angle ABC will be trisected. This disallowed use of the straightedge may be referred to as an application of "the insertion principle." For other applications of the principle see Problem Study 4.6.

Various higher plane curves have been discovered which will solve the *verging* problem to which the trisection problem may be reduced. One of the oldest of these is the conchoid invented by Nicomedes (*ca.* 240 B.C.). Let c be a straight line and O any point not on c. On the prolongation of OP, where P is any point on c, mark off PQ equal to a given fixed length k. Then the locus of Q, as P moves along c, is (one branch of) the *conchoid* of c for the pole O and the constant k. It is not difficult to devise an apparatus which will draw conchoids, and with such an apparatus one may easily trisect angles. Thus, let AOB be any given acute angle. Draw line MN perpendicular to OA, cutting OA and OB in D and L, as shown in Figure 28. Now draw the conchoid of MN for pole O and constant $2(OL)$. At L draw the parallel to OA to cut the conchoid in C. Then OC trisects angle AOB.

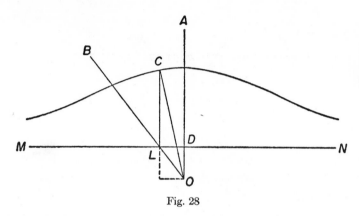

Fig. 28

A general angle may be trisected with the aid of a conic. The early Greeks were not familiar enough with the conics to accomplish this and the earliest proof of this type was given by Pappus (*ca.* 300 A.D.), using the focus and directrix property of conics. Two trisections using conics may be found in Problem Study 4.8.

There are transcendental (nonalgebraic) curves which will not only trisect a given angle but, more generally, multisect it into any number of equal parts. Among such curves are the *quadratrix*, invented by Hippias (*ca.* 425 B.C.), and the *spiral of Archimedes*. These two curves will also solve the problem of the quadrature of the circle. Applications of the quadratrix to both trisection and quadrature occur in Problem Study 4.10.

Over the years many mechanical contrivances, linkage machines, and compound compasses, have been devised to solve the trisection problem. An interesting and elementary implement of this kind is the so-called *tomahawk*. The inventor of the tomahawk is not known, but the instrument was described in a book in 1835. To construct a tomahawk, start with a line segment RU trisected at S and T (see Figure 29). Draw a semicircle on SU as diameter and draw SV perpendicular to RU. Complete the instrument as indicated in the accompanying figure. To trisect an angle ABC with the tomahawk, place the implement on the angle so that R falls on BA, SV

passes through B, and the semicircle touches BC, at D say. Then, since we may show that triangles RSB, TSB, DTB are all congruent, BS and BT trisect the given angle. The tomahawk may be constructed with straightedge and compasses on tracing paper and then adjusted on the given angle. By this subterfuge we may trisect an angle with straightedge and compasses.

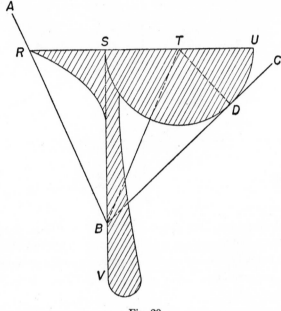

Fig. 29

Although an arbitrary angle cannot be trisected exactly with Euclidean tools, there are constructions with these tools which give remarkably good approximate trisections. An excellent example is the construction given in 1525 by the famous etcher and painter, Albrecht Dürer. Take the given angle AOB as a central angle of a circle (see Figure 30). Let C be that trisection point of the chord AB which is nearer to B. At C erect the perpendicular to AB to cut the circle in D. With B as center and BD as radius draw an arc to cut AB in E. Let F be the trisection point of EC which is nearer to E. Again, with B as center, and BF as radius, draw an arc to cut the circle in G. Then OG is an approximate trisecting line of angle AOB. It can be shown that the error in trisection increases with the size of the angle AOB, but is only about $1''$ for angle $AOB = 60°$ and about $18''$ for angle $AOB = 90°$.

Problem Study 4.9 describes an approximate trisection, using Euclidean tools, which may be made just as close to exact trisection as may be desired.

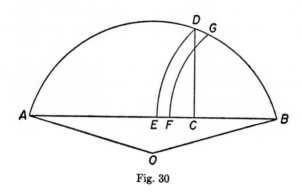

Fig. 30

4-7 Quadrature of the Circle

Probably no other problem has exercised a greater or a longer attraction than that of constructing a square equal in area to a given circle. As far back as 1800 B.C. the ancient Egyptians "solved" the problem by taking the side of the square equal to 8/9 the diameter of the given circle. Since then literally thousands of people have worked on the problem, and in spite of the present existence of a proof that the construction cannot be made with Euclidean tools, not a year passes without its crop of "circle squarers."

The first Greek known to be connected with the problem is Anaxagoras (*ca.* 499–*ca.* 427 B.C.), but what his contribution was is not known. Hippocrates of Chios, who was a contemporary of Anaxagoras, succeeded in squaring certain special lunes, or moon-shaped figures bounded by two circular arcs, probably in the hope that his investigations might lead toward a solution of the quadrature problem. Some years later Hippias of Elis (*ca.* 425 B.C.) invented the curve which became known as the *quadratrix.* This curve solves both the trisection and the quadrature problems, but traditions vary as to who first used it in the quadrature role. It may be that Hippias used it for trisecting angles, and that Dinostratus (*ca.* 350 B.C.), or some later geometer, realized its application to the quadrature problem. Some of the lunes of Hippocrates are considered in Problem Study 4.12; the quadratrix, in its dual role, is considered in Problem Study 4.10; and a few approximate quadratures are described in Problem Study 4.11.

A neat solution of the quadrature problem can be achieved with the *spiral of Archimedes*, and we are told that Archimedes (*ca.* 225 B.C.) actually used his spiral for this purpose. We may define the spiral, in dynamical terms, as the locus of a point P moving uniformly along a ray which, in turn, is uniformly rotating in a plane about its origin. If we take for the polar frame of reference the position OA of the rotating ray when P coincides with the origin O of the ray, we have that OP is proportional to angle AOP, and the polar equation of the spiral is $r = a\theta$, a being the constant of proportionality.

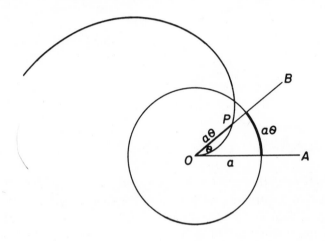

Let us draw the circle with center at O and radius equal to a. Then OP and the arc on this circle between the lines OA and OP are equal, since each is given by $a\theta$ (see Figure 31). It follows that if we take OP perpendicular to OA, then OP will have a length equal to one-fourth the circumference of the circle. Since the area K of the circle is half the product of its radius and its circumference we have

$$K = (a/2)(4OP) = (2a)(OP).$$

The side of the required square is thus the mean proportional between $2a$ and OP, or between the diameter of the circle and the length of that radius vector of the spiral which is perpendicular to OA.

We may trisect (or more generally multisect) an angle AOB with the spiral of Archimedes. Let OB cut the spiral in P and trisect the segment OP

by points P_1 and P_2. If the circles with O as center and OP_1 and OP_2 as radii cut the spiral in T_1 and T_2, then OT_1 and OT_2 trisect the angle AOB.

4-8 A Chronology of π*

Closely allied to the quadrature problem is the computation of π, the ratio of the circumference of a circle to its diameter. We have seen that in the ancient orient the value of π was frequently taken as 3,† and for the Egyptian quadrature of the circle given in the Rhind papyrus we have $\pi = (4/3)^4 = 3.1604 \cdots$. The first scientific attempt to compute π, however, seems to be that of Archimedes, and we shall commence our chronology with his achievement.

ca. 240 B.C. To simplify matters, suppose we choose a circle with unit diameter. Now the (length of the) circumference of a circle lies between the perimeter of any inscribed polygon and that of any circumscribed polygon. Since it is a simple matter to compute the perimeters of the regular inscribed and circumscribed six-sided polygons, we easily obtain bounds for π. Now there are formulas (see Problem Study 4.13) which tell us how, from the perimeters of given regular inscribed and circumscribed polygons, we may obtain the perimeters of the regular inscribed and circumscribed polygons having twice the number of sides. By successive applications of this process, starting with the regular inscribed and circumscribed six-sided polygons, we can compute the perimeters of the regular inscribed and circumscribed polygons of 12, 24, 48, and 96 sides, in this way obtaining ever closer bounds for π. This is essentially what Archimedes did, finally obtaining the fact that π is between $223/71$ and $22/7$, or that, to two decimal places, π is given by 3.14. The work is found in Archimedes' *Measurement of a Circle*, a treatise containing only three propositions. The treatise as it has come down to us is not in its original form and may be only a fragment of a larger discussion. One inescapable conclusion, in view of the poor number system in use at the time, is that Archimedes was a very able computer. In the work are found some remarkable rational approximations to irrational square roots.

*For a fuller chronology of π, containing over 120 entries, see H. C. Schepler, "The Chronology of Pi," *Mathematics Magazine*, January–February 1950, pp. 165–170; March–April 1950, pp. 216–228; May–June 1950, pp. 279–283.

†See the Biblical references: I Kings 7 : 23; II Chron. 4 : 2.

The above method of computing π by using regular inscribed and circumscribed polygons is known as the *classical method* of computing π.

ca. 150 A.D. The first notable value for π after that of Archimedes was given by Claudius Ptolemy of Alexandria in his famous *Syntaxis mathematica* (more popularly known by its Arabian title of the *Almagest*), the greatest ancient Greek work on astronomy. In this work π is given, in sexagesimal notation, as 3 8'30'', which is 377/120, or 3.1416. Undoubtedly this value was derived from the table of chords, which appears in the treatise. The table gives the lengths of the chords of a circle subtended by central angles of each degree and half degree. If the length of the chord of the 1° central angle is multiplied by 360, and the result divided by the length of the diameter of the circle, the above value for π is obtained.

ca. 480. The early Chinese worker in mechanics, Tsu Ch'ung-chih, gave the interesting rational approximation $355/113 = 3.1415929 \cdots$, which is correct to six decimal places. See Problem Study 4.11 (c) for an application of this ratio to the quadrature problem.

ca. 530. The early Hindu mathematician Āryabhata gave 62,832/20,000 = 3.1416 as an approximate value for π. It is not known how this result was obtained. It may have come from some earlier Greek source or, perhaps, from calculating the perimeter of a regular inscribed polygon of 384 sides.

ca. 1150. The later Hindu mathematician, Bhāskara, gave several approximations for π. He gave 3927/1250 as an accurate value, 22/7 as an inaccurate value, and $\sqrt{10}$ for ordinary work. The first value may have been taken from Āryabhata. Another value, $754/240 = 3.1416$, given by Bhāskara is of uncertain origin; it is the same as that given by Ptolemy.

1579. The eminent French mathematician François Viète found π correct to 9 decimal places by the classical method, using polygons having $6(2^{16}) = 393,216$ sides. He also discovered the equivalent of the interesting infinite product (see Problem Study 4.13)

$$\frac{2}{\pi} = \frac{\sqrt{2}}{2} \frac{\sqrt{(2+\sqrt{2})}}{2} \frac{\sqrt{\{2+\sqrt{(2+\sqrt{2})}\}}}{2} \cdots$$

1585. Adriaen Anthoniszoon rediscovered the ancient Chinese ratio 355/113. This was apparently a lucky accident since all he showed was that $377/120 > \pi > 333/106$. He then averaged the numerators and the denominators to obtain the "exact" value of π. There is evidence that Valentin Otho, a pupil of the early table maker Rhaeticus, may have introduced this ratio for π into the western world at the slightly earlier date of 1573.

1593. Adriaen van Roomen, more commonly referred to as Adrianus Romanus, of the Netherlands, found π correct to 15 decimal places by the classical method, using polygons having 2^{30} sides.

1610. Ludolph van Ceulen of Germany computed π to 35 decimal places by the classical method, using polygons having 2^{62} sides. He spent a large part of his life on this task and his achievement was considered so extraordinary that the number was engraved on his tombstone, and to this day is frequently referred to in Germany as "the Ludolphine number."

1621. The Dutch physicist Willebrord Snell, best known for his discovery of the law of refraction, devised a trigonometrical improvement of the classical method for computing π so that from each pair of bounds on π given by the classical method he was able to obtain considerably closer bounds. By his method he was able to get van Ceulen's 35 decimal places by using polygons having only 2^{30} sides. With such polygons the classical method yields only 15 places. For polygons of 96 sides the classical method yields 2 decimal places whereas Snell's improvement gives 7 places. A correct proof of Snell's refinement was furnished in 1654 by the Dutch mathematician and physicist Christiaan Huygens.

1630. Grienberger, using Snell's refinement, computed π to 39 decimal places. This was the last major attempt to compute π by the method of perimeters.

1650. The English mathematician John Wallis obtained the curious expression

$$\pi/2 = \frac{2 \cdot 2 \cdot 4 \cdot 4 \cdot 6 \cdot 6 \cdot 8 \cdots}{1 \cdot 3 \cdot 3 \cdot 5 \cdot 5 \cdot 7 \cdot 7 \cdots}.$$

Lord Brouncker, the first president of the Royal Society, converted Wallis' result into the continued fraction

$$4/\pi = 1 + \cfrac{1^2}{2 + \cfrac{3^2}{2 + \cfrac{5^2}{2 + \cdots}}}$$

Neither of these expressions, however, has served for an extensive calculation of π.

1671. The Scotch mathematician James Gregory obtained the infinite series

$$\arctan x = x - x^3/3 + x^5/5 - x^7/7 + \cdots, \qquad (-1 \leqq x \leqq 1).$$

Not noted by Gregory is the fact that for $x = 1$ the series becomes

$$\pi/4 = 1 - 1/3 + 1/5 - 1/7 + \cdots.$$

This very slowly converging series was known to Leibniz in 1674. Gregory attempted to prove that a Euclidean solution of the quadrature problem is impossible.

1699. Abraham Sharp found 71 correct decimal places by using Gregory's series with $x = \sqrt{1/3}$.

1706. John Machin obtained 100 decimal places by using Gregory's series in connection with the relation (see Problem Study 4.13)

$$\pi/4 = 4 \arctan (1/5) - \arctan (1/239).$$

1719. The French mathematician De Lagny obtained 112 correct places by using Gregory's series with $x = \sqrt{1/3}$.

1737. The symbol π was used by the early English mathematicians William Oughtred, Isaac Barrow, and David Gregory to designate the circumference, or periphery, of a circle. The first to use the symbol for the ratio of the circumference to the diameter was the English writer, William Jones, in a publication in 1706. The symbol was not generally used in this sense, however, until Euler adopted it in 1737.

1754. Jean Étienne Montucla, an early French historian of mathematics, wrote a history of the quadrature problem.

1755. The French Academy of Sciences declined to examine any more solutions of the quadrature problem.

1760. Comte de Buffon devised his famous *needle problem* by which π may be determined by probability methods. Suppose a number of parallel lines, distance a apart, are ruled on a horizontal plane, and suppose a homogeneous uniform rod of length $l < a$ is dropped at random onto the plane. Buffon showed that the probability* that the rod will fall across one of the lines in the plane is given by

$$p = 2l/\pi a.$$

By actually performing this experiment a given large number of times and noting the number of successful cases, thus obtaining an empirical value

---

*If a given event can happen in h ways and fail to happen in f ways, and if each of the $h + f$ ways is equally likely to occur, the *mathematical probability* p of the event happening is $p = h/(h + f)$.

for p, we may use the above formula to compute π. The best result obtained in this way was given by the Italian, Lazzerini, in 1901. From only 3408 tosses of the rod he found π correct to 6 decimal places! His result is so much better than those obtained by other experimenters that it is sometimes regarded with suspicion. There are other probability methods for computing π. Thus, in 1904, R. Chartres reported an application of the known fact that if two positive integers are written down at random, the probability that they will be relatively prime is $6/\pi^2$.

1767. Johann Heinrich Lambert showed that π is irrational.

1794. Adrien-Marie Legendre showed that π^2 is irrational.

1841. William Rutherford of England calculated π to 208 places, of which 152 were later found to be correct, by using Gregory's series in connection with the relation

$$\pi/4 = 4 \arctan (1/5) - \arctan (1/70) + \arctan (1/90).$$

1844. Zacharias Dase, the lightning calculator, found π correct to 200 places using Gregory's series in connection with the relation

$$\pi/4 = \arctan (1/2) + \arctan (1/5) + \arctan (1/8).$$

Dase, who was born in Hamburg in 1824, died at the early age of 37. He was perhaps the most extraordinary mental calculator who ever lived. Among his performances were the mental calculation of the product of two 8-digit numbers in 54 seconds, of two 20-digit numbers in 6 minutes, of two 40-digit numbers in 40 minutes, and of two 100-digit numbers in 8 hours and 45 minutes. He mentally computed the square root of a 100-digit number in 52 minutes. Dase used his powers more worthily when he constructed a seven-place table of natural logarithms and a factor table of all numbers between 7,000,000 and 10,000,000.

1853. Rutherford returned to the problem and obtained 400 correct decimal places.

1873. William Shanks of England, using Machin's formula, computed π to 707 places. For a long time this remained the most fabulous piece of calculation ever performed. It occupied Shanks for more than 15 years.

1882. A number is said to be *algebraic* if it is a root of some polynomial having rational coefficients; otherwise it is said to be *transcendental*. F. Lindemann showed that π is transcendental. This fact proves (see the following section) that the quadrature problem cannot be solved by Euclidean tools.

1906. Among the curiosities connected with π are various mnemonics that have been devised for the purpose of remembering π to a large number of decimal places. The following, by A. C. Orr, appeared in the *Literary Digest.* One has merely to replace each word by the number of letters it contains to obtain π correct to 30 decimal places.

> Now I, even I, would celebrate
> In rhymes unapt, the great
> Immortal Syracusan, rivaled nevermore,
> Who in his wondrous lore,
> Passed on before,
> Left men his guidance
> How to circles mensurate.

A few years later, in 1914, the following similar mnemonic appeared in the *Scientific American*: "See, I have a rhyme assisting my feeble brain, its tasks ofttimes resisting."

1948. In 1946, D. F. Ferguson of England discovered errors, starting with the 528th place, in Shanks' value for π, and in January 1947 gave a corrected value to 710 places. In the same month J. W. Wrench, Jr., of America, published an 808-place value of π, but Ferguson soon found an error in the 723rd place. In January 1948, Ferguson and Wrench jointly• published the corrected and checked value of π to 808 places. Wrench used Machin's formula whereas Ferguson used the formula

$$\pi/4 = 3 \arctan (1/4) + \arctan (1/20) + \arctan (1/1985).$$

1949. The electronic computer, the ENIAC, at the Army Ballistic Research Laboratories in Aberdeen, Maryland, calculated π to 2037 decimal places.

1959. François Genuys, in Paris, computed π to 16,167 decimal places, using an IBM 704.

1961. Wrench and Daniel Shanks, of Washington, D.C., computed π to 100,265 decimal places, using an IBM 7090.

We have not placed in the above chronology of π any items from the vast literature supplied by sufferers of *morbus cyclometricus*, the circle-squaring disease. These contributions, often amusing and at times almost unbelievable, would require a publication all to themselves. To illustrate their tenor consider the instance, in 1892, when a writer announced in the *New York Tribune* the rediscovery of a long lost secret which leads to 3.2 as the exact value of π. The lively discussion following this announcement won many advocates for the new value. Again, since its publication in 1934, a great many college and public libraries throughout the United States have received, from the obliging author, complimentary copies of a thick book devoted to the demonstration that $\pi = 3\ 13/81$. And then there is House Bill No. 246 of the Indiana State Legislature which attempted, in 1897, to determine the value of π by legislation. In Section I of the bill we read: "Be it enacted by the General Assembly of the State of Indiana: It has been found that a circular area is to the square on a line equal to the quadrant of the circumference, as the area of an equilateral rectangle is to the square on one side $\cdots$." The bill passed the House but, because of some newspaper ridicule, was shelved by the Senate, in spite of the energetic backing of the State Superintendent of Public Instruction.*

4-9   Impossibility of Solving the Three Famous Problems with Euclidean Tools

It was not until the nineteenth century that the three famous problems of antiquity were finally shown to be impossible with Euclidean tools. Proofs of this fact can now be found in many of the present-day textbooks dealing with the theory of equations, where it is shown that needed criteria for constructibility are essentially algebraic in nature. In particular, the following two theorems are established:

(1) *The magnitude of any length constructible with Euclidean tools from a given unit length is an algebraic number.*

(2) *From a given unit length it is impossible to construct with Euclidean tools a segment the magnitude of whose length is a root of a cubic equation with rational coefficients but with no rational root.*

The quadrature problem is disposed of by the first theorem. For if we take the radius of the given circle as our unit of length, the side of the sought equivalent square is $\sqrt{\pi}$. Thus, if the problem were possible with Euclidean tools, we could construct from the unit segment another segment of length $\sqrt{\pi}$. But this is impossible, since π, and hence $\sqrt{\pi}$, was shown by Lindemann in 1882 to be nonalgebraic.

The second theorem disposes of the other two problems. Thus, in the duplication problem, take for our unit of length the edge of the given cube

---

*See W. E. Edington, "House Bill No. 246, Indiana State Legislature, 1897," *Proceedings of the Indiana Academy of Science*, 45 (1935), 206–210.

and let x denote the edge of the sought cube. Then we must have $x^3 = 2$. If the problem is solvable with Euclidean tools we could construct from the unit segment another segment of length x. But this is impossible since $x^3 = 2$ is a cubic equation with rational coefficients but without any rational root.*

We may prove that the *general* angle cannot be trisected with Euclidean tools by showing that some *particular* angle cannot be so trisected. Now, from trigonometry, we have the identity

$$\cos \theta = 4 \cos^3 (\theta/3) - 3 \cos (\theta/3).$$

Taking $\theta = 60°$ and setting $x = \cos (\theta/3)$ this becomes

$$8x^3 - 6x - 1 = 0.$$

Let OA be a given unit segment. Describe the circle with center O and radius OA, and with A as center and AO as radius draw an arc to cut the circle in B (see Figure 32). Then angle $BOA = 60°$. Let trisector OC, which makes

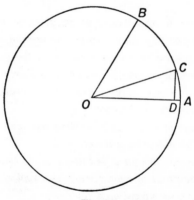

Fig. 32

angle $COA = 20°$, cut the circle in C, and let D be the foot of the perpendicular from C on OA. Then $OD = \cos 20° = x$. It follows that if a 60° angle can be trisected with Euclidean tools, in other words if OC can be drawn with

---

*It will be recalled that if a polynomial equation

$$a_0 x^n + a_1 x^{n-1} + \cdots + a_n = 0,$$

with integral coefficients $a_0, a_1, \cdots, a_n$, has a reduced rational root a/b, then a is a factor of a_n and b is a factor of a_0. Thus any rational roots of $x^3 - 2 = 0$ are among $1, -1, 2, -2$. Since by direct testing none of these numbers satisfies the equation, the equation has no rational roots.

these tools, we can construct from a unit segment OA another segment of length x. But this is impossible, by the second theorem, since the above cubic equation has rational coefficients but no rational root.

It should be noted that we have not proved that *no* angle can be trisected with Euclidean tools, but only that *not all* angles can be so trisected. The truth of the matter is that 90° and an infinite number of other angles can be trisected by the use of Euclidean tools.

4-10 Compasses or Straightedge Alone

.The eighteenth-century Italian geometer Lorenzo Mascheroni made the surprising discovery that all Euclidean constructions, insofar as the given and required elements are points, can be made with the compasses alone, and that the straightedge is thus a redundant tool. Of course, straight lines cannot be drawn with the compasses, but any straight line arrived at in a Euclidean construction can be determined by the compasses alone by finding two points of the line. This discovery appeared in 1797 in Mascheroni's *Geometria del compasso*.

Since in a Euclidean construction new points are found from old points by (1) finding an intersection of two circles, (2) finding an intersection of a straight line and a circle, or (3) finding the intersection of two straight lines, all Mascheroni had to do was to show how, with compasses alone, problems (2) and (3) might be solved, where for a straight line we have given some two points of the line.

Shortly before 1928, a student of the Danish mathematician J. Hjelmslev, while browsing in a bookstore in Copenhagen, came across a copy of an old book, *Euclides Danicus*, published in 1672 by an obscure writer named Georg Mohr. Upon examining the book Hjelmslev was surprised to find that it contained Mascheroni's discovery, with a proof, arrived at a hundred and twenty-five years before Mascheroni's publication.

Inspired by Mascheroni's discovery, the French mathematician Jean-Victor Poncelet considered constructions with straightedge alone. Now not all Euclidean constructions can be achieved with only the straightedge, but, curiously enough, in the presence of one circle and its center drawn on the plane of construction, all Euclidean constructions can be carried out with straightedge alone. This remarkable theorem was conceived by Poncelet in 1822 and then later, in 1833, fully developed by the Swiss-German geometry genius Jacob Steiner. Here it is necessary to show that, in the presence of a circle and its center, constructions (1) and (2) can be solved with straightedge alone, where now a circle is considered as given by its center and a point on its circumference.

It was about 980 that the Arabian mathematician Abû'l-Wefâ proposed using the straightedge along with compasses with a fixed opening. In view of the Poncelet-Steiner theorem we need, in fact, use the compasses only once, after which they may be discarded. In 1904, the Italian Francesco Severi went still further, and showed that all we need is an arc, no matter how small, of one circle, and its center, in order thenceforth to accomplish all Euclidean constructions with straightedge alone. It has also been shown that any Euclidean construction can be carried out with a two-edged straightedge, whether the edges are parallel or not. There are many intriguing construction theorems of this sort, the proofs of which require considerable ingenuity.

The problem of finding the "best" Euclidean solution to a required construction has also been considered, and a science of *geometrography* was developed in 1907 by Émile Lemoine for quantitatively comparing one construction with another. To this end, Lemoine considered the following five operations:

S_1: to make the straightedge pass through one given point.
S_2: to rule a straight line,
C_1: to make one compass leg coincide with a given point,
C_2: to make one compass leg coincide with any point of a given locus,
C_3: to describe a circle.

If the above operations are performed m_1, m_2, n_1, n_2, n_3 times in a construction, then $m_1 S_1 + m_2 S_2 + n_1 C_1 + n_2 C_2 + n_3 C_3$ is regarded as the *symbol* of the construction. The total number of operations, $m_1 + m_2 + n_1 + n_2 + n_3$, is called the *simplicity* of the construction, and the total number of coincidences, $m_1 + n_1 + n_2$, is called the *exactitude* of the construction. The total number of loci drawn is $m_2 + n_3$, the difference between the simplicity and the exactitude of the construction. The symbol for drawing the straight line through points A and B is $2S_1 + S_2$, and that for drawing the circle with center C and radius AB is $3C_1 + C_3$.

Problem Studies

4.1 Euclidean and Modern Compasses

A student reading Euclid's *Elements* for the first time might experience some surprise at the opening propositions of Book I. The first three propositions are the construction problems

1. To describe an equilateral triangle upon a given finite straight line.

2. From a given point to draw a straight line equal to a given straight line.

3. From the greater of two given straight lines to cut off a part equal to the less.

These three constructions are trivial with straightedge and *modern* compasses, but require some ingenuity with straightedge and *Euclidean* compasses.

(a) Solve Proposition 1 of Book I with Euclidean tools.

(b) Solve Proposition 2 of Book I with Euclidean tools.

(c) Solve Proposition 3 of Book I with Euclidean tools.

(d) Show that Proposition 2 of Book I proves that the straightedge and *Euclidean* compasses are equivalent to the straightedge and *modern* compasses.

4.2 Duplication by Conics

Menaechmus (*ca.* 350 B.C.) gave the following two solutions to the duplication problem. They utilize certain conic sections which, apparently, were first invented by Menaechmus for the problem at hand.

(a) Draw two parabolas having a common vertex, perpendicular axes, and such that the latus rectum of one is double that of the other. Denote by x the length of the perpendicular dropped from the other intersection of the two parabolas upon the axis of the smaller parabola. Then x is the edge of a cube having twice the volume of the cube having the smaller latus rectum for edge. Prove this construction correct, using modern analytic geometry.

(b) Draw a parabola of latus rectum s, then a rectangular hyperbola with transverse axis equal to $4s$ and having for asymptotes the axis of the parabola and the tangent to the parabola at its vertex. Let x be the length of the perpendicular dropped from the intersection of the two curves upon the axis of the parabola. Then $x^3 = 2s^3$. Prove this construction correct, using modern analytic geometry.

4.3   Duplication by Apollonius and Eratosthenes

Apollonius (*ca.* 225 B.C.) solved the duplication problem as follows. Draw a rectangle $OADB$, and then a circle concentric with the rectangle cutting OA and OB produced in A' and B' such that A', D, B' are collinear. Actually, it is impossible to construct this circle with Euclidean tools, but Apollonius gave a mechanical way of describing it.

(a) Show that BB' and AA' are two mean proportionals between OA and OB.

(b) If $OB = 2(OA)$, show that $(BB')^3 = 2(OA)^3$.

(c) Eratosthenes (*ca.* 230 B.C.) devised a mechanical "mean-finder" consisting of three equal rectangular frames, with a set of corresponding diagonals, capable of sliding in grooves so that the second frame can be slid under the first one, and the third frame under the second. Suppose the

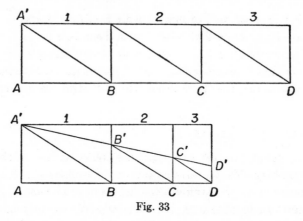

Fig. 33

frames are slid, as indicated in Figure 33, so that points A', B', C' are collinear. Show that BB' and CC' are the two mean proportionals between AA' and DD'. A "mean-finder" of this sort is easily made from a set of equal paper rectangles and can be generalized so as to insert n means between two given segments.

4.4   The Cissoid of Diocles

Diocles (*ca.* 180 B.C.) invented the *cissoid* in order to solve the duplication problem. A general cissoid may be defined as follows: Let C_1 and C_2

be two given curves and let O be a fixed point. Let P_1 and P_2 be the intersections of a variable line through O with the given curves. The locus of P on this line such that $OP = OP_2 - OP_1 = P_1P_2$ is called the *cissoid of* C_1 *and* C_2 *for the pole O*. If C_1 is a circle, C_2 a tangent to C_1 at point A, and O is the point on C_1 diametrically opposite A, then the cissoid of C_1 and C_2 for the pole O is the *cissoid of Diocles*.

(a) Taking O as origin and OA as the positive x-axis, show that the Cartesian equation of the cissoid of Diocles is $y^2 = x^3/(2a - x)$, where a is the radius of C_1. Show that the corresponding polar equation is $r = 2a \sin \theta \tan \theta$.

(b) On the positive y-axis lay off $OD = n(OA)$. Draw DA to cut the cissoid in P. Let OP cut line C_2 in Q. Show that $(AQ)^3 = n(OA)^3$. When $n = 2$ we have a solution of the duplication problem.

(c) Newton has shown how the cissoid of Diocles may be generated by a carpenter's square. Let the outside edge of the square be ACB, AC being the shorter arm. Draw a line MN and mark a point R at distance AC from MN. Move the square so that A always lies on MN and BC always passes through R. Show that the mid-point P of AC describes a cissoid of Diocles.

(d) What is the cissoid of two concentric circles with respect to their common center? Of a pair of parallel lines with respect to any point not on either line?

(e) If C_1 and C_2 intersect in P, show that OP is a tangent at O to the cissoid of C_1 and C_2 for the pole O.

4.5 Some Seventeenth-Century Duplications

Many eminent seventeenth-century mathematicians, like Huygens, Descartes, Grégoire de Saint-Vincent, and Newton, devised constructions for duplicating a cube. Following are two of these constructions.

(a) Grégoire de Saint-Vincent (1647) gave a construction for finding the two mean proportionals between two given line segments based on the following theorem: *The hyperbola drawn through a vertex of a rectangle and having the two sides opposite this vertex for asymptotes meets the circumcircle of the rectangle in a point whose distances from the asymptotes are the mean proportionals between the adjacent sides of the rectangle.* Prove this theorem.

(b) Descartes (1659) pointed out that the curves

$$x^2 = ay, \qquad x^2 + y^2 = ay + bx$$

intersect in a point (x,y) such that x and y are the two mean proportionals between a and b. Show this.

4.6   Applications of the Insertion Principle

Let us be given two curves m and n, and a point O. Suppose we permit ourselves to mark, on a given straightedge, a segment MN, and then to adjust the straightedge so that it passes through O and cuts the curves m and n with M on m and N on n. The line drawn along the straightedge is then said to have been drawn by "the insertion principle." Problems beyond the Euclidean tools can often be solved with these tools if we also permit ourselves to use the insertion principle. Establish the correctness of the following constructions, each of which uses the insertion principle.

(a) Let AB be a given segment. Draw angle $ABM = 90°$ and the angle $ABN = 120°$. Now draw ACD cutting BM in C and BN in D and such that $CD = AB$. Then $(AC)^3 = 2(AB)^3$. Essentially this construction was given in publications by Viète (1646) and Newton (1728).

(b) Let AOB be any central angle in a given circle. Through B draw a line BCD cutting the circle again in C, AO produced in D, and such that $CD = OA$, the radius of the circle. Then angle $ADB = 1/3$ angle AOB. This solution of the trisection problem is implied by a theorem given by Archimedes (ca. 240 B.C.).

4.7   The Conchoid of Nicomedes

Little is known about Nicomedes (ca. 240 B.C.) beyond his invention of the *conchoid*, a curve with which one may solve both the trisection and the duplication problems. A general conchoid may be defined as follows: Let c be a given curve and O a fixed point. On the radius vector OP from O to a point P on c mark off $PQ = \pm k$, where k is a constant. Then the locus of Q is called *the conchoid of c for pole O and constant k*. The complete curve consists of two branches, one corresponding to $PQ = +k$ and the other to $PQ = -k$. If c is a straight line and O is any point not on c, we get a *conchoid of Nicomedes*.

(a) Taking O as origin and the line through O parallel to the given line c as x-axis, show that the Cartesian equation of the conchoid of Nicomedes for constant k is $(y - a)^2(x^2 + y^2) = k^2y^2$, where a is the distance of O from c.

(b) Show how the conchoid of Nicomedes may be used to solve the duplication problem.

(c) A conchoid of a circle for a fixed point on the circle is called a *limaçon of Pascal* (named after the discoverer Etienne Pascal (1588–1640), father of the famous Blaise Pascal). If $k = a$, the radius of the given circle, we obtain a special limacon known as the *trisectrix*. Establish the following construction for trisecting an angle with the trisectrix. Let AOB be any central angle in a circle with center O and radius OA. Draw the trisectrix for the circle with pole at A, and let BO produced cut the trisectrix in C. Then angle $ACB = 1/3$ angle AOB.

(d) Show that the two branches of the conchoid of curve c for pole O and constant k constitute the cissoid of s and c for the pole O, where s is the circle with center O and radius k (see Problem Study 4.4).

4.8 Trisection by Conics

A general angle is easily trisected by the aid of conics. Establish the following constructions of this sort.

(a) Let the given angle be AOB. Draw the branch of an equilateral hyperbola having O as center and OA as an asymptote, cutting OB in P. With P as center and $2(PO)$ as radius draw a circle cutting the hyperbola in R. Draw PM parallel to OA and RM perpendicular to OA, to intersect in M. Then angle $AOM = 1/3$ angle AOB.

(b) Let angle AOB be taken as a central angle of a circle, and let OC be the bisector of angle AOB. Draw the branch of the hyperbola of eccentricity 2 having A for focus and OC for corresponding directrix, and let this branch cut arc AB in P. Then angle $AOP = 1/3$ angle AOB. This construction was quoted by Pappus (*ca.* 300 A.D.).

4.9 Asymptotic Euclidean Constructions

A construction using Euclidean tools but requiring an infinite number of operations is called an *asymptotic Euclidean construction*. Establish the following two constructions of this type for solving the trisection and the quadrature problems. (For an asymptotic Euclidean solution of the duplication problem, see T. L. Heath, *History of Greek Mathematics*, vol. 1, pp. 268–270.)

(a) Let OT_1 be the bisector of angle AOB, OT_2 that of angle AOT_1, OT_3 that of angle T_2OT_1, OT_4 that of angle T_3OT_2, OT_5 that of angle T_4OT_3, and so forth. Then $\lim_{i \to \infty} OT_i = OT$, one of the trisectors of angle AOB. (This construction was given by Fialkowski, 1860.)

(b) On the segment AB_1 produced mark off $B_1B_2 = AB_1$, $B_2B_3 = 2(B_1B_2)$, $B_3B_4 = 2(B_2B_3)$, and so forth. With B_1, B_2, B_3, $\cdots$ as centers draw the circles $B_1(A)$, $B_2(A)$, $B_3(A)$, $\cdots$. Let M_1 be the mid-point of the semicircle on AB_2. Draw B_2M_1 to cut circle $B_2(A)$ in M_2, B_3M_2 to cut circle $B_3(A)$ in M_3, $\cdots$. Let N_i be the projection of M_i on the common tangent of the circles at A. Then $\lim_{i \to \infty} AN_i =$ quadrant of circle $B_1(A)$.

4.10 The Quadratrix

Hippias (*ca.* 425 B.C.) invented a transcendental curve, called the *quadratrix*, by means of which one can multisect angles and square the circle. The quadratrix may be defined as follows: Let the radius OX of a circle rotate uniformly about the center O from OC to OA, at right angles to OC. In the same time let a line MN parallel to OA move uniformly parallel to itself from CB to OA. The locus of the intersection P of OX and MN is the quadratrix.

(a) Taking $OA = 1$ and the positive x-axis along OA, show that the Cartesian equation of the quadratrix is $y = x \tan(\pi y/2)$.

(b) Show how an angle may be multisected with the quadratrix.

(c) Find the x-intercept of the quadratrix, and show how the curve may be used for squaring the circle.

4.11 Approximate Rectification

Many approximate constructions have been given for finding a line segment equal in length to the circumference of a given circle. An approximate quadrature of the circle is then easily obtained by constructing the square on the mean proportional between the radius of the circle and a segment equal in length to half the circumference of the circle.

(a) Show that the circumference of a circle is given approximately by three times the diameter of the circle increased by 1/5 the side of the inscribed square. This leads to what approximation for π?

(b) Let AOB be a diameter of the given circle. Find C on the tangent at B such that angle $COB = 30°$. Mark off CBD on the tangent equal to three times the radius of the circle. Then $2(AD)$ is approximately the circumference of the circle. This leads to what approximation for π? This construction was given in 1685 by the Polish Jesuit Kochanski.

(c) Let $AB = 1$ be a diameter of the given circle. Draw $BC = 7/8$, perpendicular to AB at B. Mark off $AD = AC$ on AB produced. Draw $DE = 1/2$, perpendicular to AD at D, and let F be the foot of the perpendicular from D on AE. Draw EG parallel to FB to cut BD in G. Then GB is approximately the decimal part of π. Find the length of GB to seven decimal places.

4.12 Lunes of Hippocrates

Hippocrates of Chios (*ca.* 440 B.C.) squared certain lunes, perhaps hoping that his investigations might throw some light on the quadrature problem. Following are two of Hippocrates' lune quadratures.

(a) Let AOB be a quadrant of a circle. On AB as diameter draw a semicircle lying outside the quadrant. Show that the lune bounded by the quadrant and the semicircle has the same area as triangle AOB.

(b) Let $ABCD$ be half of a regular hexagon inscribed in a circle of diameter AD. Construct a lune by describing, exterior to the circle, a semicircle on AB as diameter. Show that the area of the trapezoid $ABCD$ is equal to three times the area of the lune plus the area of the semicircle on AB as diameter.

4.13   Computation of π

(a) If a is the side of a regular polygon inscribed in a circle of radius r, show that $b = \{2r^2 - r(4r^2 - a^2)^{1/2}\}^{1/2}$ is the side of a regular inscribed polygon having twice the number of sides. This is a basic formula for the classical method of computing π.

(b) Prove that $\pi/4 = 4 \tan^{-1} (1/5) - \tan^{-1} (1/239)$. This is the formula utilized by Machin in 1706 to compute π to 100 decimal places.

(c) Establish Viète's formula given under the date *1579* in Section 4-8.

(d) Show that $\pi/6 = \sqrt{1/3}\{1 - 1/(3)(3) + 1/(3^2)(5) - 1/(3^3)(7) + \cdots\}$.

(e) A common approximation, in the Middle Ages, for a square root was $\sqrt{n} = \sqrt{a^2 + b} = a + b/(2a + 1)$. By taking $n = 10 = 3^2 + 1$, show why it may be that $\sqrt{10}$ was so frequently used for π.

(f) Show that the theorem in House Bill No. 246, Indiana State Legislature, 1897 (see Section 4-8) makes the incorrect assumption that a circle and a square have equal areas if they have equal perimeters. This assumption leads to what value for π?

4.14   Approximation of Cusa and Snell

Let angle AOP be an acute central angle in a circle of unit radius. Produce diameter AOB to point S so that $BS = AO$. Draw SP to cut, in the point T, the tangent to the circle at A. Cusa and Snell noticed that if angle AOP is sufficiently small, *the tangential segment AT is approximately equal in length to the arc AP.*

(a) Find the error in the Cusa-Snell approximation when angle $AOP = 90°$.

(b) Designating angle AOP by θ and angle AST by ϕ show that

$$AT = 3 \sin \theta/(2 + \cos \theta) = 3 \tan \phi.$$

(c) Show that $\phi < \theta/3$, whence

$$\sin \theta/(2 + \cos \theta) < \tan (\theta/3).$$

(d) Show how the Cusa-Snell approximation may be used for multi-secting angles.

(e) Show how the Cusa-Snell approximation may be used for dividing a circumference into n equal parts.

(f) Show how the Cusa-Snell approximation may be used for squaring the circle.

4.15 Impossible Constructions

(a) Establish the identity: $\cos \theta = 4 \cos^3 (\theta/3) - 3 \cos (\theta/3)$.

(b) Show that it is impossible with Euclidean tools to construct a regular polygon of nine sides.

(c) Show that it is impossible with Euclidean tools to construct an angle of $1°$.

(d) Show that it is impossible with Euclidean tools to construct a regular polygon of seven sides.

(e) Let us be given an angle AOB and a point P within the angle. The line through P cutting OA and OB in C and D so that $CE = PD$, where E is the foot of the perpendicular from O on CD, is known as *Philon's line for angle AOB and the point P*. It can be shown that Philon's line is the minimum chord CD that can be drawn through P. Show that in general it is impossible to construct with Euclidean tools Philon's line for a given angle and a given point.

4.16 Mascheroni Construction Theorem

Let us designate the circle with center at point C and passing through the point A by the symbol $C(A)$, and the circle with center at point C and with radius equal to the segment AB by the symbol $C(AB)$. Prove the following chain of constructions and show that they establish the Mascheroni construction theorem: *Any Euclidean construction, insofar as the given and required elements are points, may be accomplished with the Euclidean compasses alone.* The constructions are recorded in a tabular form in which the upper line indicates what is to be drawn, while the lower line indicates the new points that are thus constructed.

(a) To construct with Euclidean compasses the circle $C(AB)$.

$C(A), A(C)$	$M(B), N(B)$	$C(X)$
M, N	X	

(*Note.* This construction shows that the Euclidean and modern compasses are equivalent tools.)

(b) To construct with modern compasses the intersection of $C(D)$ with the line determined by the points A and B.

Case 1. C not on AB.

$A(C)$, $B(C)$	$C(D)$, $C_1(CD)$
C_1	X, Y

Case 2. C on AB.

$A(D)$, $C(D)$	$C(DD_1)$, $D(C)$	$C(DD_1)$, $D_1(C)$	$F(D_1)$, $F_1(D)$	$F(CM)$, $C(D)$
D_1	F, fourth vertex of parallelogram CD_1DF	F_1, fourth vertex of parallelogram CDD_1F_1	M	X, Y

(c) To construct with modern compasses the point of intersection of the lines determined by the pairs of points A, B and C, D.

$A(C)$, $B(C)$	$A(D)$, $B(D)$	$C(DD_1)$, $D_1(CD)$	$C_1(G)$, $G(D_1)$	$C_1(C)$, $G(CE)$	$C(F)$, $C_1(CF)$
C_1	D_1	G, collinear with C, C_1	E, either intersection	F, collinear with C_1, E	X

(d) On page 268 of Cajori's *A History of Mathematics* we read: "Napoleon proposed to the French mathematicians the problem, to divide the circumference of a circle into four equal parts by the compasses only. Mascheroni does this by applying the radius three times to the circumference; he obtains the arcs AB, BC, CD; then AD is a diameter; the rest is obvious." Complete the "obvious" part of the construction.

4.17 Lemoine's Geometrography

Find the *symbol*, *simplicity*, and *exactitude* for the following familiar constructions of a line through a given point A and parallel to a given line MN.

(a) Through A draw any line to cut MN in B. With any radius r draw the circle $B(r)$ to cut MB in C and AB in D. Draw circle $A(r)$ to cut AB in E. Draw circle $E(CD)$ to cut circle $A(r)$ in X. Draw AX, obtaining the required parallel.

(b) With any suitable point D as center draw circle $D(A)$ to cut MN in B and C. Draw circle $C(AB)$ to cut circle $D(A)$ in X. Draw AX.

(c) With any suitable radius r draw the circle $A(r)$ to cut MN in B. Draw circle $B(r)$ to cut MN in C. Draw circle $C(r)$ to cut circle $A(r)$ in X. Draw AX.

Bibliography

ALLMAN, G. J., *Greek Geometry from Thales to Euclid.* Dublin: University Press, 1889.

BALL, W. W. R., AND H. S. M. COXETER, *Mathematical Recreations and Essays,* 11th ed. New York: The Macmillan Company, 1939.

BOROFSKY, SAMUEL, *Elementary Theory of Equations.* New York: The Macmillan Company, 1950.

BRAUMBAGH, R. S., *Plato's Mathematical Imagination; the Mathematical Passages in the Dialogues and Their Interpretation.* Bloomington, Ind.: Indiana University Press, 1954.

COOLIDGE, J. L., *The Mathematics of Great Amateurs.* New York: Oxford University Press, 1949.

DE MORGAN, AUGUSTUS, *A Budget of Paradoxes,* 2 vols., 2d ed., ed. by D. E. Smith. Chicago: The Open Court Publishing Company, 1915.

DICKSON, L. E., *New First Course in the Theory of Equations.* New York: John Wiley & Sons, Inc., 1939.

GOW, JAMES, *A Short History of Greek Mathematics.* New York: Hafner Publishing Company, 1923.

HEATH, T. L., *History of Greek Mathematics,* Vol. 1. New York: Oxford University Press, 1921.

———, *A Manual of Greek Mathematics.* New York: Oxford University Press, 1931.

———, *Mathematics in Aristotle.* New York: Oxford University Press, 1949.

HOBSON, E. W., *"Squaring the Circle," a History of the Problem.* New York: Chelsea Publishing Company, 1953.

HUDSON, H. P., *Ruler and Compasses.* New York: David McKay Company, Inc., 1916. Reprinted in *Squaring the Circle, and Other Monographs,* Chelsea Publishing Company, 1953.

KEMPE, A. B., *How to Draw a Straight Line; a Lecture on Linkages.* New York: The Macmillan Company, 1887. Reprinted in *Squaring the Circle, and Other Monographs,* Chelsea Publishing Company, 1953.

KLEIN, FELIX, *Famous Problems of Elementary Geometry*, tr. by W. W. Beman and D. E. Smith. New York: Stechert-Hafner, Inc., 1930. Reprinted in *Famous Problems, and Other Monographs*, Chelsea Publishing Company, 1955.

KOSTOVSKII, A. N., *Geometrical Constructions Using Compasses Only*, tr. by Halina Moss. New York: Blaisdell Publishing Company, 1961.

LOVITT, W. V., *Elementary Theory of Equations*. Englewood Cliffs, N. J.: Prentice-Hall, Inc., 1939.

PHIN, JOHN, *The Seven Follies of Science*, 3d ed. Princeton, N. J.: D. Van Nostrand Company, Inc., 1911.

SMOGORZHEVSKII, A. S., *The Ruler in Geometrical Constructions*, tr. by Halina Moss. New York: Blaisdell Publishing Company, 1961.

STEINER, JACOB, *Geometrical Constructions with a Ruler*, tr. by M. E. Stark, ed. by R. C. Archibald. New York: *Scripta Mathematica*, Yeshiva University, 1950.

VAN DER WAERDEN, B. L., *Science Awakening*, tr. by Arnold Dresden. Groningen, Netherlands: P. Noordhoff, N.V., 1954.

WEDBERG, ANDERS, *Plato's Philosophy of Mathematics*. Stockholm: Almqvist & Wiksell, 1955.

WEISNER, LOUIS, *Introduction to the Theory of Equations*. New York: The Macmillan Company, 1938.

YATES, R. C., *Geometrical Tools*. St. Louis: Educational Publishers, Inc., 1949.
———, *The Trisection Problem*. Ann Arbor, Mich.: Edwards Bros., Inc.,1947.

YOUNG, J. W. A., ed., *Monographs on Topics of Modern Mathematics Relevant to the Elementary Field*. New York: David McKay Company, Inc., 1911. Reprinted by Dover Publications, Inc., 1955.

5

Euclid's Elements

5-1 Alexandria*

The period following the Peloponnesian War was one of political disunity among the Greek states, rendering them an easy prey for the now strong kingdom of Macedonia which lay to the north. King Philip of Macedonia was gradually extending his power southward and Demosthenes thundered his unheeded warnings. The Greeks rallied too late for a successful defense and, with the Athenian defeat at Chaeronea in 338 B.C., Greece became a part of the Macedonian empire.

Two years after the fall of the Greek states, ambitious Alexander the Great succeeded his father Philip and set out upon his unparalleled career of conquest which added vast portions of the civilized world to the growing Macedonian domains. Behind him, wherever he led his victorious army, he created, at well-chosen places, a string of new cities. It was in this way, when Alexander had entered Egypt, that the city of Alexandria was founded in 332 B.C.

It is said that the choice of the site, the drawing of the ground plan, and the process of colonization for Alexandria were directed by Alexander himself, and that the actual building of the city was assigned to the eminent architect Dinocrates. From its inception, Alexandria showed every sign of fulfilling a remarkable future. In an incredibly short time, largely due to its very fortunate location at a natural intersection of many important trade routes, it grew in wealth and became the most magnificent and cosmopolitan center of the world.

After Alexander the Great died in 323 B.C. his empire was partitioned among some of his military leaders, resulting in the eventual emergence of

*See R. E. Langer, "Alexandria—Shrine of Mathematics," *American Mathematical Monthly*, 48 (February, 1941), pp. 109–125.

three empires, under separate rule, but nevertheless united by the bonds of the Hellenistic civilization that had followed Alexander's conquests. Egypt fell to the lot of Ptolemy. It was not until about 306 B.C. that Ptolemy actually began his reign. He selected Alexandria as his capital and, to attract learned men to his city, immediately began the erection of the famed University of Alexandria. This was the first institution of its kind and in its scope and setup was much like the universities of today. Report has it that it was highly endowed and that its attractive and elaborate plan contained lecture rooms, laboratories, gardens, museums, library facilities, and living quarters. The core of the institution was the great library, which for a long time was the largest repository of learned works to be found anywhere in the world, boasting, within forty years of its founding, of over 600,000 papyrus rolls. It was about 300 B.C. that the university opened its doors and Alexandria became, and remained for a thousand years, the intellectual metropolis of the Greek race.

For recognized scholars to staff the university Ptolemy turned to Athens and invited the distinguished Demetrius Phalereus to take charge of the great library. Able and talented men were selected to develop the various fields of study. Euclid, who also may have come from Athens, was probably chosen to head the department of mathematics.

5-2   Euclid

Disappointingly little is known about the life and personality of Euclid except that he was a professor of mathematics at the University of Alexandria and apparently the founder of the illustrious and long-lived Alexandrian School of Mathematics. Even his dates and his birthplace are not known, but it seems probable that he received his mathematical training in the Platonic school at Athens. Many years later, when comparing Euclid with Apollonius for the latter's discredit, Pappus praised Euclid for his modesty and consideration of others. Proclus augmented his *Eudemian Summary* with the frequently told story of Euclid's reply to Ptolemy's request for a short cut to geometric knowledge that "there is no royal road in geometry." But the same story has been told of Menaechmus when serving as instructor to Alexander the Great. Another pretty story has been told by Stobaeus. Some student studying geometry under Euclid questioned what he would get from learning the subject, whereupon Euclid ordered a slave to give the fellow a penny, "since he must make gain from what he learns."

5-3   Euclid's *Elements*

Although Euclid was the author of at least ten works, and fairly complete texts of five of these have come down to us, his reputation rests mainly on his *Elements*. It appears that this remarkable work immediately and completely superseded all previous Elements; in fact, no trace remains of the earlier efforts. As soon as the work appeared it was accorded the highest respect, and from Euclid's successors on up to modern times the mere citation of Euclid's book and proposition numbers was regarded as sufficient to identify a particular theorem or construction. No work, except the Bible, has been more widely used, edited, or studied, and probably no work has exercised a greater influence on scientific thinking. Over a thousand editions of Euclid's *Elements* have appeared since the first one printed in 1482, and for more than two millennia this work has dominated all teaching of geometry.

No copy of Euclid's *Elements* actually dating from the author's time has been found. The modern editions of the *Elements* are based upon a revision prepared by Theon of Alexandria almost 700 years after the original work had been written. It was not until the beginning of the nineteenth century that an older copy, showing only minor differences from Theon's recension, was discovered in the Vatican library. A careful study of citations and commentary by early writers indicates that the definitions, axioms, and postulates of the original treatise differed some from the subsequent revisions but that the propositions and their proofs have remained essentially as Euclid wrote them.

The first complete Latin translations of the *Elements* were not made from the Greek but from the Arabic. In the eighth century, a number of Byzantine manuscripts of Greek works were translated by the Arabians, and in 1120 the English scholar, Adelard of Bath, made a Latin translation of the *Elements* from one of these older Arabian translations. Other Latin translations were made from the Arabic by Gherardo of Cremona (1114–1187) and, 150 years after Adelard, by Johannes Campanus. The first printed edition of the *Elements* was made at Venice in 1482 and contained Campanus' translation. This very rare book was beautifully executed and was the first mathematical book of any consequence to be printed. An important Latin translation from the Greek was made by Commandino in 1572. This translation served as a basis for many subsequent translations, including the very influential work by Robert Simson, from which, in turn, so many of the English editions were derived. The first complete English translation of the *Elements* was the monumental Billingsley translation issued in 1570.*

5-4   Content of the *Elements*

Contrary to popular impressions, Euclid's *Elements* is not devoted to geometry alone but contains a good deal of number theory and elementary (geometric) algebra. It is, for the most part, a compilation and systematic arrangement of works of earlier writers. Although some of the proofs and propositions were no doubt invented by Euclid, the chief merit of the work lies in the skillful selection of propositions and their arrangement into a logical sequence. The work is composed of 13 books with a total of 465 propositions. American high school plane and solid geometry texts contain much of the material found in Books I, III, IV, VI, XI, and XII.

Book I commences, of course, with the necessary preliminary definitions, postulates, and axioms; we shall return to these in the next section. The 48 propositions of Book I fall into three groups. The first 26 deal mainly with properties of triangles and include the three congruence theorems. Propositions I 27 through I 32 establish the theory of parallels and prove that the sum of the angles of a triangle is equal to two right angles. The remaining propositions of the book deal with parallelograms, triangles, and squares, with special reference to area relations. Proposition I 47 is the Pythagorean theorem, with a proof universally credited to Euclid himself, and the final proposition, I 48, is the converse of the Pythagorean theorem. The material of this book was developed by the early Pythagoreans.

Book II deals with the transformation of areas and the geometric algebra

*See R. C. Archibald, "The First Translation of Euclid's *Elements* into English and Its Source," *American Mathematical Monthly*, 57 (August–September, 1950), pp. 443–452, and W. F. Shenton, "The First English Euclid," *American Mathematical Monthly*, 35 (December, 1928), pp. 505–512.

of the Pythagorean school. We have considered some of the propositions of this book in Chapter 3. It is in this book that we find the geometrical equivalents of a number of algebraic identities. At the end of the book are two propositions which establish the generalization of the Pythagorean theorem that we today refer to as the "law of cosines."

Book III contains those familiar theorems about circles, chords, tangents, and the measurement of associated angles which we find in high school geometry texts. Since little of this geometry of the circle is found in Pythagorean work, the material of this book was probably furnished by the early sophists and the researchers on the three famous problems discussed in Chapter IV.

In Book IV are found discussions of the Pythagorean constructions, with straightedge and compasses, of regular polygons of three, four, five, six, and fifteen sides (see Problem Study 5.3). By successive angle, or arc, bisections, we may then with Euclidean tools construct regular polygons having 2^n, $3(2^n)$, $5(2^n)$, or $15(2^n)$ sides. Not until almost the nineteenth century was it known that any other regular polygons could be constructed with these limited tools. In 1796, the eminent German mathematician Carl Friedrich Gauss developed the theory that showed that a regular polygon having a *prime* number of sides can be constructed with Euclidean tools if and only if that number is of the form $f(n) = 2^{2^n} + 1$. For $n = 0, 1, 2, 3, 4$ we find $f(n) = 3, 5, 17, 257, 65,537$, all prime numbers. Thus, unknown to the Greeks, regular polygons of 17, 257, and 65,537 sides can be constructed with straightedge and compasses. For no other value of n, than those listed above, is it known that $f(n)$ is a prime number.

Many Euclidean constructions of the regular polygon of 17 sides have been given. In 1832, Richelot published an investigation of the regular polygon of 257 sides, and a Professor Hermes of Lingen gave up ten years of his life to the problem of constructing a regular polygon of 65,537 sides. It has been said that it was Gauss's discovery, at the age of 19, that a regular polygon of 17 sides can be constructed with straightedge and compasses that decided him to devote his life to mathematics. His pride in this discovery is evidenced by his request that a regular polygon of 17 sides be engraved on his tombstone. Although this request was never fulfilled, such a polygon is found on a monument to Gauss erected at his birthplace in Brunswick.

Book V is a masterly exposition of Eudoxus' theory of proportion. It was this theory, applicable to incommensurable as well as to commensurable magnitudes, that resolved the "logical scandal" created by the Pythagorean discovery of irrational numbers. The Eudoxian definition of proportion, or equality of two ratios, is remarkable, and worth repeating here. *Magnitudes are said to be in the same ratio, the first to the second and the third to the fourth, when, if any equimultiples whatever be taken of the first and third, and any equimultiples whatever of the second and fourth, the former equimultiples alike exceed, are alike equal to, or are alike less than the latter equimultiples taken in corresponding order.* In other words, if A, B, C, D are any four unsigned magnitudes, A and B being of the same kind (both line segments, or angles, or areas, or volumes) and C and D being of the same kind, then the ratio of A to B is equal to that of C to D when, for arbitrary positive integers m and n, $mC \gtreqless nD$ according as $mA \gtreqless nB$. The Eudoxian theory of proportion provided a foundation, later developed by Dedekind and Weierstrass, for the real number system of mathematical analysis.

Book VI applies the Eudoxian theory of proportion to plane geometry. Here we find the fundamental theorems on similar triangles, constructions giving third, fourth, and mean proportionals; the geometric solution of quadratic equations which we considered in Chapter III; the proposition that the internal bisector of an angle of a triangle divides the opposite side into segments proportional to the other two sides; a generalization of the Pythagorean theorem where, instead of squares, three similar and similarly described figures are drawn on the three sides of a right triangle; and many other theorems. There probably is no theorem in this book that was not known to the early Pythagoreans, but the pre-Eudoxian proofs of many of them were at fault since they were based upon an incomplete theory of proportion.

It might be interesting to indicate the differences between the Pythagorean, the Eudoxian, and the modern textbook proofs of a simple proposition involving proportions. Let us select Proposition VI 1: *The areas of triangles having the same altitude are to one another as their bases.* We shall permit ourselves to use Proposition I 38, which says that *triangles having equal bases and equal altitudes have equal areas*, and a consequence of I 38 to the effect that *of any two triangles having the same altitude, that one has the greater area which has the greater base.*

Let the triangles be ABC and ADE, the bases BC and DE lying on the same straight line MN, as in Figure 34. Now the Pythagoreans, before the discovery of irrational numbers, tacitly assumed that any two line segments are commensurable. Thus BC and DE were assumed to have some common unit of measure, going, say, p times into BC and q times into DE. Mark off these points of division on BC and DE and connect them with vertex A. Then triangles ABC and ADE are divided, respectively, into p and q smaller triangles, all having, by I 38, the same area. It follows that $\triangle ABC : \triangle ADE$ $= p : q = BC : DE$, and the proposition is established. With the later discovery that two line segments need not be commensurable this proof, along with others, became inadequate, and the very disturbing "logical scandal" came into existence.

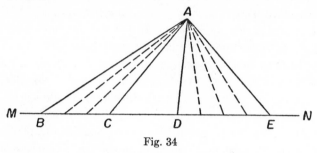

Fig. 34

Eudoxus' theory of proportion cleverly resolved the "scandal," as we shall now illustrate by reproving VI 1 in the manner found in the *Elements*. On CB produced mark off, successively from B, $m - 1$ segments equal to CB, and connect the points of division, B_2, B_3, $\cdots$, B_m, with vertex A, as shown in Figure 35. Similarly, on DE produced mark off, successively from E,

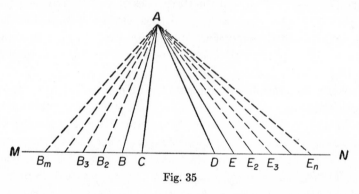

Fig. 35

$n - 1$ segments equal to DE, and connect the points of division, E_2, E_3, $\cdots$, E_n, with vertex A. Then $B_mC = m(BC)$, $\triangle AB_mC = m(\triangle ABC)$, $DE_n = n(DE)$, $\triangle ADE_n = n(\triangle ADE)$. Also, by I 38 and its corollary, $\triangle AB_mC \gtreqless \triangle ADE_n$ according as $B_mC \gtreqless DE_n$. That is, $m(\triangle ABC) \gtreqless n(\triangle ADE)$ according as $m(BC) \gtreqless n(DE)$, whence, by the Eudoxian definition of proportion, $\triangle ABC : \triangle ADE = BC : DE$, and the proposition is established. No mention was made of commensurable and incommensurable quantities since the Eudoxian definition applies equally to both situations.

Many present-day high school textbooks advocate a proof of this theorem involving two cases, according as BC and DE are or are not commensurable. The commensurable case is handled as in the Pythagorean solution above, and simple limit notions are used to deal with the incommensurable case. Thus, suppose BC and DE are incommensurable. Divide BC into n equal parts, BR being one of the parts (see Figure 36). On DE

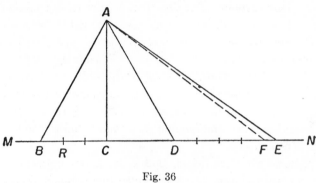

Fig. 36

mark off a succession of segments equal to BR, finally arriving at a point F on DE such that $FE < BR$. By the commensurable case, already established, $\triangle ABC : \triangle ADF = BC : DF$. Now let $n \to \infty$. Then $DF \to DE$ and $\triangle ADF \to \triangle ADE$. Hence, in the limit, $\triangle ABC : \triangle ADE = BC : DE$. This approach uses the fact that any irrational number may be regarded as the limit of a sequence of rational numbers, an approach that was rigorously developed in modern times by Georg Cantor (1845–1918).

Books VII, VIII, and IX, containing a total of 102 propositions, deal with elementary number theory. Book VII commences with the process, referred to today as the *Euclidean algorithm*, for finding the greatest common integral divisor of two or more integers and uses it as a test for two integers to be relatively prime (see Problem Study 5.1). We also find an exposition of the numerical, or Pythagorean, theory of proportion. Many basic number properties are established in this book.

Book VIII concerns itself largely with continued proportions and related geometric progressions. If we have the continued proportion $a : b = b : c = c : d$, then a, b, c, d form a geometric progression.

A number of significant theorems are found in Book IX. Proposition IX 14 is equivalent to the important *fundamental theorem of arithmetic*, namely that *any integer greater than 1 can be expressed as a product of primes in one, and essentially only one, way.* Proposition IX 35 gives a geometric derivation of the formula for the sum of the first n terms of a geometric progression, and the last proposition, IX 36, establishes the remarkable formula for perfect numbers that was stated in Section 3-3.

Euclid's proof of IX 20, that *the number of prime numbers is infinite*, has been universally regarded by mathematicians as a model of mathematical elegance. The proof employs the indirect method,* or *reductio ad absurdum*, and runs essentially as follows. Suppose there is only a finite number of prime numbers, which we shall denote by a, b, $\cdots$, k. Set $P = ab \cdots k$. Then $P + 1$ is either prime or composite. But, since a, b, $\cdots$, k are all the primes, $P + 1$, which is greater than each of a, b, $\cdots$, k, cannot be a prime. On the other hand, if $P + 1$ is composite it must be divisible by some prime p. But p must be a member of the set a, b, $\cdots$, k of all primes, which means that p is a divisor of P. Consequently, p cannot divide $P + 1$, since $p > 1$. Thus our initial hypothesis that the number of primes is finite is untenable, and the theorem is established.

Book X deals with irrationals, that is, with line segments which are incommensurable with respect to some given line segment. Many scholars regard this book as perhaps the most remarkable book in the *Elements*. Much of the subject matter of this book is considered as due to Theaetetus, but the extraordinary completeness, elaborate classification, and finish are usually credited to Euclid. It taxes one's credulity to realize that the results of this book were arrived at by abstract reasoning unassisted by any convenient algebraic notation. The opening proposition (X 1) is the basis of the method of exhaustion later employed in Book XII, namely that, *if from any magnitude there be subtracted a part not less than its half, from the remainder another part not less than its half, and so on, there will at length remain a magnitude less than any assigned magnitude of the same kind.* In this book we also find the formulas yielding all primitive Pythagorean triples, formulas that the ancient Babylonians may have known over a thousand years earlier (see Section 2-6).

The remaining three books, XI, XII, and XIII, concern themselves with solid geometry, covering much of the material, with the exception of that on spheres, commonly found in high school texts. The definitions, the theorems about lines and planes in space, and theorems concerning parallelepipeds are found in Book XI. The method of exhaustion plays an important role in the treatment of volumes in Book XII, and will be reconsidered

*It is easy to formulate the proof so that the indirect method is avoided.

in some detail in Chapter XI. In Book XIII constructions are developed for inscribing the five regular polyhedra in a sphere.

The frequently stated remark that Euclid's *Elements* was really intended to serve merely as a drawn-out account of the five regular polyhedra appears to be a lopsided evaluation. A better appraisal would seem to be that it was intended, for its time, to serve as a beginning text in general mathematics. Euclid also wrote texts on higher mathematics.

5-5   Formal Aspect of the *Elements*

Important as are the contents of the *Elements*, perhaps still more important is the formal manner in which those contents are presented. In fact, Euclid's *Elements* has become the prototype of modern mathematical form.

Certainly one of the greatest achievements of the early Greek mathematicians was the creation of the postulational form of thinking. In order to establish a statement in a deductive system one must show that the statement is a necessary logical consequence of some previously established statements. These, in their turn, must be established from some still more previously established statements, and so on. Since the chain cannot be continued backward indefinitely one must, at the start, accept some finite body of statements without proof, or else commit the unpardonable sin of circularity by deducing statement *A* from statement *B* and then later *B* from *A*. These initially assumed statements are called the *postulates*, or *axioms*, of the discourse, and all other statements of the discourse must be logically implied by them. When the statements of a discourse are so arranged, the discourse is said to be presented in postulational form.

So great was the impression made by the formal aspect of Euclid's *Elements* on following generations that the work became a model for rigorous mathematical demonstration. In spite of a considerable abandonment of the Euclidean form during the seventeenth and eighteenth centuries, the postulational method has today penetrated into almost every field of mathematics, and many mathematicians adhere to the thesis that not only is mathematical thinking postulational thinking but, conversely, postulational thinking is mathematical thinking. A relatively modern outcome has been the creation of a field of study called *axiomatics*, devoted to an examination of the general properties of sets of postulates and of postulational thinking. We shall return to this in Section 5-8.

It is not certain precisely what statements Euclid assumed for his postulates, nor, for that matter, exactly how many he had, for changes and additions were made by subsequent editors. The early Greeks made a distinction between "postulates" and "axioms." The latter were probably considered as general truths common to all studies, whereas the former were thought of as pertaining to the special study at hand. This distinction was altered in later times, and a "postulate" was regarded as a permissible construction, while all other initial assumptions were considered as "axioms." In modern mathematics no distinction is made, and the words *axiom* and *postulate* are considered as synonymous. There is fair evidence that Euclid probably assumed the equivalents of the following ten statements, five "axioms," or common notions, and five geometric "postulates":

A 1. *Things which are equal to the same thing are also equal to one another.*

A 2. *If equals be added to equals, the wholes are equal.*

A 3. *If equals be subtracted from equals, the remainders are equal.*

A 4. *Things which coincide with one another are equal to one another.*

A 5. *The whole is greater than the part.*

P 1. *It is possible to draw a straight line from any point to any other point.*

P 2. *It is possible to produce a finite straight line indefinitely in that straight line.*

P 3. *It is possible to describe a circle with any point as center and with a radius equal to any finite straight line drawn from the center.*

P 4. *All right angles are equal to one another.*

P 5. *If a straight line intersects two straight lines so as to make the interior angles on one side of it together less than two right angles, these straight lines will intersect, if indefinitely produced, on the side on which are the angles which are together less than two right angles.*

The *Elements* purports to derive all its 465 propositions from the above ten statements! The development is the *synthetic* one of proceeding from the known and simpler to the unknown and more complex. Without a doubt the reverse process, called *analysis*,* of reducing the unknown and more complex to the known, played a part in the discovery of the proofs of many of the theorems, but it plays no part in the exposition of the subject.

---

*The words *analysis* and *analytic* are used in several senses in mathematics. Thus we have *analytic* geometry, the large branch of mathematics called *analysis*, *analytic* functions, and so forth.

5-6   Logical Shortcomings of the *Elements*

It would be very remarkable indeed if Euclid's *Elements*, being such an early and colossal attempt at the postulational method of presentation, should be free of logical blemishes. The searchlight of subsequent criticism has revealed many defects in the logical structure of the work. Perhaps the gravest of these defects are numerous tacit assumptions made by Euclid, assumptions not granted by his postulates. Thus, although Postulate P 2 asserts that a straight line may be produced indefinitely, it does not necessarily imply that a straight line is infinite, but merely that it is endless, or boundless. The arc of a great circle joining two points on a sphere may be produced indefinitely along the great circle, making the prolonged arc endless, but certainly not infinite. The great German mathematician Riemann, in his famous probationary lecture, *Über die Hypothesen welche der Geometrie zu Grunde liegen*, of 1854, distinguished between the boundlessness and the infinitude of straight lines. There are numerous occasions, for instance in the proof of Proposition I 16, where Euclid unconsciously assumed the infinitude of straight lines. Again, Euclid tacitly assumed, in his proof of Proposition I 21 for example, that if a straight line enters a triangle at a vertex it must, if sufficiently produced, intersect the opposite side. Moritz Pasch (1843–1930) recognized the necessity of a postulate to take care of this situation. Another oversight of Euclid's geometry is the assumption of the existence of points of intersection of certain lines and circles. Thus in Proposition I 1 it is assumed that circles with centers at the ends of a line segment and having the line segment as a common radius intersect, and do not, somehow or other, slip through each other with no common point. Some sort of continuity postulate, such as one later furnished by R. Dedekind, is needed to assure us of the existence of such a point of intersection. Also, Postulate P 1 guarantees the existence of at least one straight line joining two points A and B, but does not assure us that there cannot be more than one such joining line. Euclid frequently assumed there is a unique line joining two distinct points. Objections can also be raised to the principle of superposition, used by Euclid with apparent reluctance to establish some of his early congruence theorems, although these objections can partially be met by Axiom A 4.

Many of Euclid's definitions are also open to criticism. Euclid made some sort of attempt to define, or at least explain, all of his important terms. Now some of the terms of a logical discourse must deliberately be chosen as primitive or undefined terms, for it is as impossible to define all of the elements and relations between elements of the discourse as it is to prove all the statements. The postulates of the discourse are then assumed statements about the primitive terms.* All other important terms of the discourse must ultimately be defined by means of the primitive ones.

In Euclid's development of geometry the elements *point* and *straight line*, for example, could well have been included in a set of primitive terms for the discourse. At any rate, Euclid's definition of a point as "that which has no part" and of a straight line as "a line which lies evenly with the points on itself" are, from a logical viewpoint, woefully inadequate. To make matters worse, the term *line*, in turn, is somewhat circularly defined by Euclid as "breadthless length."

It was not until the end of the nineteenth century and the early part of the twentieth century, after the foundations of geometry had been subjected to an intensive study, that satisfactory postulate sets were supplied for Euclidean plane and solid geometry. Prominent among such sets are those of O. Veblen, D. Hilbert, M. Pieri, and E. V. Huntington. Veblen's set contains 16 postulates and has *point* and *order* as primitive terms; Hilbert's set contains 21 postulates and has *point, straight line, plane, on, congruent,* and *between* as primitive terms; Pieri's set contains 20 postulates and has *point* and *motion* as primitive terms; Huntington's set contains 23 postulates and has *sphere* and *inclusion* as primitive terms.

5-7   Non-Euclidean Geometries

There is evidence that a logical development of the theory of parallels gave the early Greeks considerable trouble. Euclid met the difficulties by defining parallel lines as coplanar straight lines which do not meet one another however far they may be produced in either direction, and by adopting as an assumption his now famous parallel postulate. This postulate lacks the terseness of the others and in no sense possesses the characteristic of being "self-evident." Actually it is the converse of I 17, and it seemed more like a proposition than a postulate. Moreover, Euclid made no use of the parallel postulate until he reached Proposition I 29. It was natural to wonder

---

*The primitive terms may be considered as defined, at least implicitly, in the sense that they are any things which satisfy the postulates.

if the postulate was really needed at all, and to think that perhaps it could be derived as a theorem from the remaining nine "axioms" and "postulates," or, at least, that it could be replaced by a more acceptable equivalent.

Of the many substitutes that have been devised to replace Euclid's parallel postulate, the one most commonly used is that made well known in modern times by the Scottish physicist and mathematician, John Playfair (1748–1819), although this particular alternative had been used by others and had been stated as early as the fifth century by Proclus. It is the substitute usually encountered in high school texts, namely: *Through a given point can be drawn only one line parallel to a given line.* Some other proposed alternatives for the parallel postulate are: (1) *There exists at least one triangle having the sum of its three angles equal to two right angles*, (2) *There exists a pair of similar noncongruent triangles*, (3) *There exists a pair of straight lines everywhere equally distant from one another*, (4) *A circle can be passed through any three noncollinear points*, and (5) *Through any point within an angle less than 60° there can always be drawn a straight line intersecting both sides of the angle.*

The attempts to derive the parallel postulate as a theorem from the remaining nine "axioms" and "postulates" occupied geometers for over two thousand years and culminated in some of the most far-reaching developments of modern mathematics. Many "proofs" of the postulate were offered, but each was sooner or later shown to rest upon a tacit assumption equivalent to the postulate itself.

Not until 1733 was the first really scientific investigation of the parallel postulate printed. That year the Italian Jesuit priest Girolamo Saccheri (1667–1733), while Professor of Mathematics at the University of Pavia, received permission to print a little book entitled *Euclides ab omni naevo vindicatus* (Euclid Freed of Every Flaw). In an earlier work on logic, Saccheri had become charmed with the powerful method of *reductio ad absurdum* and conceived the idea of applying this method to an investigation of the parallel postulate. Without using the parallel postulate Saccheri easily showed, as can any high school geometry student, that if, in a quadrilateral $ABCD$ (see Figure 37), angles A and B are right angles and sides AD and BC are equal, then angles D and C are equal. There are, then, three possibilities: angles D and C are equal acute angles, equal right angles, or equal obtuse angles. These three possibilities were referred to by Saccheri as the *hypothesis of the acute angle*, the *hypothesis of the right angle*, and the *hypothesis of the obtuse angle*. The plan of the work was to show that the assumption of

*Proposition I 27 guarantees the existence of at least *one* parallel.

either the hypothesis of the acute angle or the hypothesis of the obtuse angle would lead to a contradiction. Then, by *reductio ad absurdum*, the hypothesis of the right angle must hold, and this hypothesis, Saccheri showed, carried with it a proof of the parallel postulate. Tacitly assuming the infinitude of the straight line, Saccheri readily eliminated the hypothesis of the obtuse angle, but the case of the hypothesis of the acute angle proved to be much more difficult. After obtaining many of the now classical theorems of so-

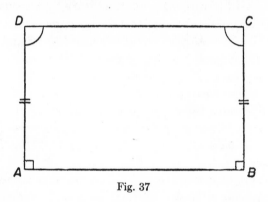

Fig. 37

called non-Euclidean geometry, Saccheri lamely forced into his development an unconvincing contradiction involving hazy notions about infinite elements. Had he not been so eager to exhibit a contradiction here, but rather had admitted his inability to find one, Saccheri would today unquestionably be credited with the discovery of non-Euclidean geometry. His work was little regarded by his contemporaries and soon became forgotten,* and it was not until 1889 that it was resurrected by his countryman, Eugenio Beltrami (1835–1900).

Thirty-three years after Saccheri's publication, Johann Heinrich Lambert (1728–1777) of Germany wrote a similar investigation entitled *Die Theorie der Parallellinien*, which, however, was not published until after his death. Lambert chose a quadrilateral containing three right angles (half of a Saccheri quadrilateral) as his fundamental figure, and considered three hypotheses according as the fourth angle is acute, right, or obtuse. He went considerably beyond Saccheri in deducing propositions under the hypotheses of the acute and obtuse angles. Thus, with Saccheri, he showed that in the three hypotheses the sum of the angles of a triangle is less than, equal to, or

*There is a narrative-rich alternative explanation, involving an unpleasant insinuation of suppression, that has been offered to account for the long period of disregard of Saccheri's masterpiece. See, for example, E. T. Bell, *The Magic of Numbers* (New York: McGraw-Hill Book Company, Inc., 1946), Chap. 25.

greater than two right angles respectively, and then, in addition, that the deficiency below two right angles in the hypothesis of the acute angle, or the excess above two right angles in the hypothesis of the obtuse angle, is proportional to the area of the triangle. He observed the resemblance of the geometry following from the hypothesis of the obtuse angle to spherical geometry, where the area of a triangle is proportional to its spherical excess, and conjectured that the geometry following from the hypothesis of the acute angle could perhaps be verified on a sphere of imaginary radius. The hypothesis of the obtuse angle was eliminated by making the same tacit assumption as had Saccheri, but his conclusions with regard to the hypothesis of the acute angle were indefinite and unsatisfactory.

Adrien-Marie Legendre (1752–1833), one of the eminent eighteenth-century French analysts, began anew and considered three hypotheses according as the sum of the angles of a triangle is less than, equal to, or greater than two right angles. Tacitly assuming the infinitude of a straight line he was able to eliminate the third hypothesis, but, although he made several attempts, he could not dispose of the first hypothesis. In 1794, Legendre published his very popular *Eléments de géométrie*, which was generally adopted in continental Europe and in the United States as a substitute for Euclid's *Elements*.

It is no wonder that no contradiction was found under the hypothesis of the acute angle, for it is now known that the geometry developed from a collection of axioms comprising a basic set plus the acute angle hypothesis is as consistent as the Euclidean geometry developed from the same basic set plus the hypothesis of the right angle; that is, the parallel postulate is independent of the remaining postulates and therefore cannot be deduced from them. The first to suspect this fact were Gauss of Germany, Johannes Bolyai (1802–1860) of Hungary, and Nicolai Ivanovitch Lobachevsky (1793–1856) of Russia. These men approached the subject through the Playfair form of the parallel postulate by considering the three possibilities: Through a given point can be drawn *more than one*, or *just one*, or *no* line parallel to a given line. These situations are equivalent, respectively, to the hypotheses of the acute, the right, and the obtuse angles. Again, assuming the infinitude of a straight line, the third case was easily eliminated. Suspecting, in time, a consistent geometry under the first possibility, each of these three mathematicians independently carried out extensive geometric and trigonometric developments of the hypothesis of the acute angle.

It is likely that Gauss was the first to reach penetrating conclusions concerning the hypothesis of the acute angle, but since throughout his life he failed to publish anything on the matter, the honor of discovering this particular non-Euclidean geometry must be shared with Bolyai and Lobachevsky. Bolyai published his findings in 1832 in an appendix to a mathematical work of his father. Later it was learned that Lobachevsky had published similar findings as early as 1829–1830, but, because of language barriers and the slowness with which information of new discoveries traveled in those days, Lobachevsky's work did not become known in western Europe for some years. There seems little point in discussing here the intricate, and probably unfounded, theories explaining how various of these men might have obtained and appropriated information of the findings of some other. There was considerable suspicion and incrimination of plagiarism at the time.

The actual independence of the parallel postulate from the other postulates of Euclidean geometry was not unquestionably established until consistency proofs of the hypothesis of the acute angle were furnished. These were now not long in coming and were supplied by Beltrami, Arthur Cayley, Felix Klein, Henri Poincaré, and others. The method was to set up a model in Euclidean geometry so that the abstract development of the hypothesis of the acute angle could be given a concrete interpretation in a part of Euclidean space. Then any inconsistency in the non-Euclidean geometry would imply a corresponding inconsistency in Euclidean geometry (see Problem Study 5.8).

In 1854, Riemann showed that if the infinitude of a straight line be discarded and merely its boundlessness be assumed, then, with some other slight adjustments of the remaining postulates, another consistent non-Euclidean geometry can be developed from the hypothesis of the obtuse angle. The three geometries, that of Bolyai and Lobachevsky, that of Euclid, and that of Riemann were given, by Klein in 1871, the names *hyperbolic geometry, parabolic geometry,* and *elliptic geometry.*

5-8 Axiomatics

It was largely the modern search for a logically acceptable postulate set for Euclidean geometry, and the revelation furnished by the discovery of equally consistent non-Euclidean geometries, that led to the development of *axiomatics,* or the study of postulate sets and their properties.

One of the pitfalls of working with a deductive system is too great a familiarity with the subject matter of the system. It is this pitfall that accounts for most of the blemishes in Euclid's *Elements.* In order to escape this pitfall it is advisable to replace the primitive or undefined terms of the discourse by symbols, like x, y, z, and so forth. Then the postulates of the discourse become statements about these symbols and are thus devoid of concrete meaning; conclusions, therefore, are obtained upon a strictly logical basis without the intrusion of intuitive factors. The study of axiomatics considers properties of such sets of postulates.

Clearly we cannot take as a postulate set *any* set of statements about the primitive terms. There are certain required and certain desired properties that a postulate set should possess. It is required, for example, that the postulates be *consistent*, that is, that no contradictions can be deduced from the set. Also, the postulates should be fertile, that is, be such that at least one additional statement, or theorem, can be deduced from them. Tests for consistency and fertility of a postulate set constitute important subject matter of axiomatics. The only known test for consistency is the establishment of a concrete interpretation of the postulate set. That is, one must try to assign concrete meanings to the symbols representing the primitive terms so that the postulates now become true statements about these concrete replacements. Whether a postulate set might be consistent without our being able to establish the fact is one of the interesting open questions of axiomatics. Studies upon consistency have led to several disturbing and controversial results for those who are concerned with the foundations of mathematical knowledge. The only known test for fertility is the actual deduction of a theorem from the postulates. Whether there can exist a non-fertile postulate set is another interesting query of the subject.

A set of postulates is said to be *independent* if no postulate can be deduced from the others. To show that any particular postulate of the set cannot be deduced from the others one must devise a concrete interpretation of the primitive terms which verifies all the postulates of the set except the one under consideration; this postulate must be falsified by the interpretation. Thus, if a postulate set contains n postulates, n separate tests of this sort will be needed to show the independence of the set. It was this matter of independence that was so important in connection with the non-Euclidean geometries.

A given body of material may be deducible from more than one postulate set. All that is required of two postulate sets $P^{(1)}$ and $P^{(2)}$, in order that they lead to the same development, is that set $P^{(1)}$ implies set $P^{(2)}$ and set $P^{(2)}$ implies set $P^{(1)}$. This is what is involved when trying to find equivalent substitutes for the parallel postulate.

Axiomatics is closely connected with symbolic logic and with the philosophy of mathematics. There have been, and are at present, many contributors to this field. Prominent among such contributors are Hilbert, Peano, Pieri, Veblen, Huntington, Russell, Whitehead, Gödel, and many others.

5-9　Sequel to Euclid

Until modern times it had been thought that the Greeks had pretty well exhausted the elementary synthetic geometry of the triangle and the circle. Such proved to be far from the case, for the nineteenth century witnessed an astonishing reopening of this study. It now seems that this field of investigation must be unlimited, for an enormous number of papers have appeared, and are continuing to appear, concerned with the synthetic examination of the triangle and associated points, lines, and circles. Much of the material has been extended to the tetrahedron and its associated points, planes, lines, and spheres. It would be too great a task here to enter into any sort of a detailed history of this rich and extensive subject. Many of the special points, lines, circles, planes, and spheres have been named after original or subsequent investigators. Among these names are Gergonne, Nagel, Feuerbach, Hart, Casey, Brocard, Lemoine, Tucker, Neuberg, Simson, McCay, Euler, Gauss, Bodenmiller, Furhmann, Schoute, Spieker, Taylor, Droz-Farny, Morley, Miquel, Hagge, Peaucellier, Steiner, Tarry, and many others. Central Europe seems to be the most active center of investigation in this field today, although contributions to journals are received from almost all parts of the world. Large portions of the material have been summarized and organized in numerous recent texts bearing the title modern, or college, geometry. It is not too much to say that a course in this material is very desirable for every prospective teacher of high school geometry. The material is definitely elementary, but not easy, and is extremely fascinating.

5-10　Euclid's Other Works

Euclid wrote several treatises besides the *Elements*, some of which have survived to the present day. One of the latter, called the *Data*, is concerned with the material of the first six books of the *Elements*. A *datum* may be defined as a set of parts or relations of a figure such that if all but one are given, then that remaining one is determined. Thus the parts A, a, R of a triangle, where A is one angle, a the opposite side, and R the circumradius, constitute a datum, for, given any two of these parts, the third is thereby determined. This is clear either geometrically or from the relation $a = 2R \sin A$. It is apparent that a collection of data of this sort could be useful in the analysis which precedes the discovery of a construction or a proof, and this is undoubtedly the purpose of the work.

Another work in geometry by Euclid, which has come down to us through an Arabian translation, is the book *On Divisions*. Here we find construction problems requiring the division of a figure by a restricted straight line so that the parts will have areas in a prescribed ratio. An example is the problem of dividing a given triangle into two equal areas by a line drawn through a given point within the triangle. Other examples occur in Problem Study 3.11 (b) and (c).

Other geometrical works of Euclid that are now lost to us and are known only from subsequent commentaries are the *Pseudaria*, or book of geometrical fallacies, *Porisms*, about which there has been considerable speculation,* *Conics*, a treatise in four books which was later completed and then added to by Apollonius, and *Surface Loci*, about which nothing certain is known.

Euclid's other works concern applied mathematics, and two of these are extant: the *Phaenomena*, dealing with the spherical geometry required for observational astronomy, and the *Optics*, an elementary treatise on perspective. Euclid is supposed also to have written a work on the *Elements of Music*.

Problem Studies

5.1 The Euclidean Algorism

The *Euclidean algorism*, or process, for finding the greatest common integral divisor (g.c.d.) of two positive integers is so named because it is found at the start of Book VII of Euclid's *Elements*, although the process no doubt was known considerably earlier. This algorism is at the foundation of several developments in modern mathematics. Stated in the form of a rule the process is this: *Divide the larger of the two positive integers by the smaller*

---

*A *porism* is thought to be a proposition stating a condition that renders a certain problem solvable, and then the problem has infinitely many solutions. For example, if r and R are the radii of two circles and d is the distance between their centers, the problem of inscribing a triangle in the circle of radius R which will be circumscribed about the circle of radius r is solvable if and only if $R^2 - d^2 = 2Rr$, and then there are infinitely many triangles of the desired sort.

one. Then divide the divisor by the remainder. Continue this process, of dividing the last divisor by the last remainder, until the division is exact. The final divisor is the sought g.c.d. of the two original positive integers.

(a) Find, by the Euclidean algorism, the g.c.d. of 5913 and 7592.

(b) Find, by the Euclidean algorism, the g.c.d. of 1827, 2523, and 3248.

(c) Prove that the Euclidean algorism does lead to the g.c.d.

(d) Let h be the g.c.d. of the positive integers a and b. Show that there exist integers p and q (not necessarily positive) such that $pa + qb = h$.

(e) Find p and q for the integers of part (a).

(f) Prove that a and b are relatively prime if and only if there exist integers p and q such that $pa + qb = 1$.

5.2   Applications of the Euclidean Algorism

(a) Prove, using Problem Study 5.1 (f), that if p is a prime and divides the product uv, then either p divides u or p divides v.

(b) Prove, from part (a), the fundamental theorem of arithmetic: *Every integer greater than 1 can be uniquely factored into a product of primes.*

(c) Find integers a, b, c such that $65/273 = a/3 + b/7 + c/13$.

5.3   Regular Polygons

(a) Suppose $n = rs$, where n, r, s are positive integers. Show that if a regular n-gon is constructible with Euclidean tools, then so also are a regular r-gon and a regular s-gon.

(b) Show that it is impossible to construct with Euclidean tools a regular 27-gon.

(c) Suppose r and s are relatively prime positive integers and that a regular r-gon and a regular s-gon are constructible with Euclidean tools. Show that a regular rs-gon is also so constructible.

(d) Of the regular polygons having less than 20 sides, one can with Euclidean tools construct those having 3, 4, 5, 6, 8, 10, 12, 15, 16, and 17 sides. Actually construct these polygons, with the exception of the regular 17-gon.

(e) Construct a regular 17-gon by the following method (H. W. Richmond, "To construct a regular polygon of seventeen sides," *Mathematische Annalen*, vol. 67 (1909), p. 459).

Let OA and OB be two perpendicular radii of a given circle with center O. Find C on OB such that $OC = OB/4$. Now find D on OA such that angle $OCD = (\text{angle } OCA)/4$. Next find E on AO produced such that angle $DCE = 45°$. Draw the circle on AE as diameter, cutting OB in F, and then draw the circle $D(F)$, cutting OA and AO produced in G_4 and G_6. Erect perpendiculars to OA at G_4 and G_6, cutting the given circle in P_4 and P_6. These last points are the fourth and sixth vertices of the regular 17-gon whose first vertex is A.

5.4 The Eudoxian Theory of Proportion

(a) Prove, by the Eudoxian method and by the modern textbook method, Proposition VI 33: *Central angles in the same or equal circles are to each other as their intercepted arcs.*

(b) Prove, by the Pythagorean method and then complete by the modern textbook method, Proposition VI 2: *A line parallel to one side of a triangle divides the other two sides proportionally.*

(c) Prove Proposition VI 2 by using Proposition VI 1 (see Section 5-4).

5.5 Applications of the Fundamental Theorem of Arithmetic

The fundamental theorem of arithmetic says that, for any given positive integer a, there are unique nonnegative integers $a_1, a_2, a_3, \cdots$, only a finite number of which are different from zero, such that

$$a = 2^{a_1} 3^{a_2} 5^{a_3} \cdots,$$

where $2, 3, 5, \cdots$ are the consecutive primes. This suggests a useful notation. We shall write

$$a = (a_1, a_2, \cdots, a_n),$$

where a_n is the last nonzero exponent. Thus we have $12 = (2,1)$, $14 = (1,0,0,1)$, $27 = (0,3)$, and $360 = (3,2,1)$.

Prove the following theorems:

(a) $ab = (a_1 + b_1, a_2 + b_2, \cdots)$.

(b) b is a divisor of a if and only if $b_i \leqq a_i$ for each i.

(c) The number of divisors of a is $(a_1 + 1)(a_2 + 1) \cdots (a_n + 1)$.

(d) A necessary and sufficient condition for a number n to be a perfect square is that the number of divisors of n be odd.

(e) Set g_i equal to the smaller of a_i and b_i if $a_i \neq b_i$ and equal to either a_i or b_i if $a_i = b_i$. Then $g = (g_1, g_2, \cdots)$ is the g.c.d. of a and b.

(f) If a and b are relatively prime and b divides ac, then b divides c.

(g) If a and b are relatively prime and if a divides c and b divides c, then ab divides c.

(h) Show that $\sqrt{2}$ and $\sqrt{3}$ are irrational.

5.6 Saccheri Quadrilaterals

A *Saccheri quadrilateral* is a quadrilateral $ABCD$ in which the sides AD and BC are equal and the angles at A and B are right angles. Side AB is known as the *base*, the opposite side, DC, as the *summit*, and the angles at D and C as the *summit angles*. Prove, by simple congruence theorems (which do not require the parallel postulate), the following relations:

(a) The summit angles of a Saccheri quadrilateral are equal.

(b) The line joining the mid-points of the base and summit of a Saccheri quadrilateral is perpendicular to both of them.

(c) If perpendiculars are drawn from the extremities of the base of a triangle upon the line passing through the mid-points of the two sides, a Saccheri quadrilateral is formed.

(d) The line joining the mid-points of the equal sides of a Saccheri quadrilateral is perpendicular to the line joining the mid-points of the base and summit.

5.7 The Hypothesis of the Acute Angle

The *hypothesis of the acute angle* assumes that the equal summit angles of a Saccheri quadrilateral are acute, or that the fourth angle of a Lambert quadrilateral is acute. In the following we shall assume the hypothesis of the acute angle.

(a) Let ABC be any right triangle and let M be the mid-point of the hypotenuse AB. At A construct angle BAD = angle ABC. From M draw MP perpendicular to CB. On AD mark off $AQ = PB$ and draw MQ. Prove triangles AQM and BPM congruent, thus showing that angle AQM is a right angle and points Q, M, P are collinear. Then $ACPQ$ is a Lambert quadrilateral with acute angle at A. Now show that, *under the hypothesis of the acute angle, the sum of the angles of any right triangle is less than two right angles*.

(b) Let angle A of triangle ABC be not smaller than either angle B or angle C. Draw the altitude through A and show, by part (a), that, *under the hypothesis of the acute angle, the sum of the angles of any triangle is less than two right angles*. The difference between two right angles and the sum of the angles of a triangle is known as the *defect* of the triangle.

(c) Consider two triangles, ABC and $A'B'C'$, in which corresponding angles are equal. If $A'B' = AB$ then these triangles are congruent. Suppose $A'B' < AB$. On AB mark off $AD = A'B'$, and on AC mark off $AE = A'C'$. Then triangles ADE and $A'B'C'$ are congruent. Show that E cannot fall on C, since then angle BCA would be greater than angle DEA. Show also that E

cannot fall on AC produced, since then DE would cut BC in a point F and the sum of the angles of triangle FCE would exceed two right angles. Therefore E lies between A and C and $BCED$ is a convex quadrilateral. Show that the sum of the angles of this quadrilateral is equal to four right angles. But this is impossible under the hypothesis of the acute angle. It thus follows that we cannot have $A'B' < AB$ and that, *under the hypothesis of the acute angle, two triangles are congruent if the three angles of one are equal to the three angles of the other.* In other words, in hyperbolic geometry similar figures of different sizes do not exist.

(d) A line segment joining a vertex of a triangle to a point on the opposite side is called a *transversal*. A transversal divides a triangle into two subtriangles, each of which may be similarly subdivided, and so on. Show that if a triangle is partitioned by transversals into a finite number of subtriangles, the defect of the original triangle is equal to the sum of the defects of the triangles in the partition.

5.8   A Euclidean Model for Hyperbolic Geometry

Take a fixed circle, Σ, in the Euclidean plane and interpret the hyperbolic plane as the interior of Σ, a "point" of the hyperbolic plane as a Euclidean point within Σ, and a "line" of the hyperbolic plane as that part of a Euclidean line which is contained within Σ. Verify, in this model, the following statements:

(a) Two "points" determine one and only one "line."

(b) Two distinct "lines" intersect in at most one "point."

(c) Given a "line" l and a "point" P not on l. Through P can be passed infinitely many "lines" not meeting "line" l.

(d) Let the Euclidean line determined by the two "points" P and Q intersect Σ in S and T, in the order S, P, Q, T. Then we interpret the hyperbolic "distance" from P to Q as $\log[(QS)(PT)/(PS)(QT)]$. If P, Q, R are three "points" on a "line," show that

$$\text{"distance" } PQ + \text{"distance" } QR = \text{"distance" } PR.$$

(e) Let "point" P be fixed and let "point" Q move along a fixed "line" through P toward T. Show that "distance" $PQ \to \infty$.

This model was devised by Felix Klein. With the above interpretations, along with a suitable interpretation of "angle" between two "lines," it can be shown that all of Hilbert's postulates for Euclidean plane geometry, except the parallel postulate, are true theorems in the geometry of the model. We have seen, in part (c), that the Euclidean parallel postulate is not such a theorem, but that the Lobachevskian parallel postulate holds instead. The model thus proves that the Euclidean parallel postulate cannot be deduced from the other postulates of Euclidean geometry, for if it were implied by the other postulates it would have to be a true theorem in the geometry of the model.

5.9   An Abstract Mathematical System

Consider a set K of undefined elements, which we shall denote by lower case letters, and let R denote an undefined dyadic relation which may or may not hold between a given pair of elements of K. If element a of K is related to element b of K by the R relation we shall write $R(a,b)$. We now assume the following four postulates concerning the elements of K and the dyadic relation R.

P 1. If a and b are any two distinct elements of K, then we have either $R(a,b)$ or $R(b,a)$.

P 2. If a and b are any two elements of K such that we have $R(a,b)$, then a and b are distinct elements.

P 3. If a, b, c are any three elements of K such that we have $R(a,b)$ and $R(b,c)$, then we have $R(a,c)$. (In other words, the R relation is *transitive*.)

P 4. K consists of exactly four distinct elements.

Deduce the following seven theorems from the above four postulates:

T 1. If we have $R(a,b)$ and if c is distinct from a, then we have either $R(a,c)$ or $R(c,b)$.

T 2. If we have $R(a,b)$ and if c is distinct from both a and b, then we have either $R(a,c)$ or $R(c,b)$.

T 3. There is at least one element of K not R related to any element of K. (This is an existence theorem.)

T 4. There is at most one element of K not R related to any element of K. (This is a uniqueness theorem.)

Definition 1. If we have $R(b,a)$, then we shall say we have $D(a,b)$.

T 5. If we have $D(a,b)$ and $D(b,c)$, then we have $D(a,c)$.

Definition 2. If we have $R(a,b)$ and there is no element c such that we also have $R(a,c)$ and $R(c,b)$, then we shall say we have $F(a,b)$.

T 6. If we have $F(a,c)$ and $F(b,c)$, then a is identical with b.

T 7. If we have $F(a,b)$ and $F(b,c)$, then we do not have $F(a,c)$.

Definition 3. If we have $F(a,b)$ and $F(b,c)$, then we shall say we have $G(a,c)$.

5.10   Axiomatics

(a) Establish the consistency of the postulate set of Problem Study 5.9 by means of each of the following interpretations:

1. Let K consist of a man, his father, his father's father, and his father's father's father, and let $R(a,b)$ mean "a is an ancestor of b."

2. Let K consist of four distinct points on a horizonal line, and let $R(a,b)$ mean "a is to the left of b."

3. Let K consist of the four integers 1, 2, 3, 4, and let $R(a,b)$ mean "$a < b$."

The postulates of this set are those for *sequential relation among four elements*. Any R which interprets the postulates is called a *sequential relation*, and the elements of K are said to form a *sequence*. The interpretations suggested above furnish three applications of the abstract branch of mathematics developed in Problem Study 5.9.

(b) Write out the statements of the theorems and definitions in Problem Study 5.9 for each of the interpretations of part (a).

(c) Establish the independence of the postulate set of Problem Study 5.9 by means of the following four partial interpretations:

1. Let K consist of two brothers, their father, and their father's father, and let $R(a,b)$ mean "a is an ancestor of b." This establishes the independence of Postulate P 1.

2. Let K consist of the four integers 1, 2, 3, 4, and let $R(a,b)$ mean "$a \leqq b$." This establishes the independence of Postulate P 2.

3. Let K consist of the four integers 1, 2, 3, 4, and let $R(a,b)$ mean "$a \neq b$." This establishes the independence of Postulate P 3.

4. Let K consist of the five integers 1, 2, 3, 4, 5, and let $R(a,b)$ mean "$a < b$." This establishes the independence of Postulate P 4.

(d) Show that P 1, T 1, P 3, P 4 constitute a postulate set equivalent to P 1, P 2, P 3, P 4.

5.11   Associated Hypothetical Propositions

(a) Prove the proposition: *If a triangle is isosceles, then the bisectors of its base angles are equal.*

(b) State the *converse* of the proposition of part (a). (This converse, which is somewhat troublesome to establish, has become known as the *Steiner-Lehmus problem.*)

(c) State the *opposite* of the proposition of part (a).

(d) State the *contradictory* of the proposition of part (a).

(e) If a proposition of the form *If A then B* is true, does it necessarily follow that its converse is true? Its opposite? Its contradictory?

(f) Show that if a proposition of the form *If A then B*, and its opposite, are both true, then the converse is also true.

(g) State the propositions that must be true if A is a necessary condition for B; a sufficient condition for B; a necessary and sufficient condition for B. (If A is both necessary and sufficient for B, then A is called a *criterion* for B.)

5.12   The Feuerbach Configuration

Prominent in the modern elementary geometry of the triangle is the *nine-point circle*. In a given triangle $A_1A_2A_3$, of circumcenter O and orthocenter (intersection of the three altitudes) H, let O_1, O_2, O_3 be the mid-points of the sides, H_1, H_2, H_3 the feet of the three altitudes, and C_1, C_2, C_3 the mid-points of the segments HA_1, HA_2, HA_3. Then the nine points, O_1, O_2, O_3, H_1, H_2, H_3, C_1, C_2, C_3, lie on a circle, known as the *nine-point circle* of the given triangle. This circle, due to later misplaced credit for earliest discovery, is sometimes referred to as *Euler's circle*. In Germany it is called *Feuerbach's circle*, because Karl Wilhelm Feuerbach (1800–1834) published a pamphlet in which he not only arrived at the nine-point circle but also proved that it is tangent to the inscribed and the three escribed circles of the given triangle. This last fact is known as *Feuerbach's theorem* and is justly regarded as one of the more elegant theorems in the modern geometry of the triangle. The four points of contact of the nine-point circle with the inscribed and escribed circles are known as the *Feuerbach points* of the triangle and have received considerable study. The center, F, of the nine-point circle is at the mid-point of OH. The centroid (intersection of the three medians of the triangle), G, also lies on OH such that $HG = 2(GO)$. The line of collinearity of O, F, G, H is known as the *Euler line* of the given triangle. Let H_2H_3, H_3H_1, H_1H_2 intersect the opposite sides A_2A_3, A_3A_1, A_1A_2 in P_1, P_2, P_3. Then P_1, P_2, P_3 lie on a line known as the *polar axis* of triangle $A_1A_2A_3$, and the polar axis is perpendicular to the Euler line. If the nine-point circle and the circumcircle intersect, then the polar axis is the line of the common chord of these two circles, and the circle on HG as diameter, the so-called *orthocentroidal circle* of the given triangle, also passes through the same points of intersection.

Draw a large and carefully constructed figure of an obtuse triangle with its centroid, orthocenter, circumcenter, incenter, three excenters, Euler line, polar axis, nine-point circle, Feuerbach points, circumcircle, and ortho-centroidal circle.

5.13 Data

A set of three parts of a triangle such that any two of the parts determine the third is known as a *datum* for the triangle. Thus the three angles A, B, C constitute a datum, since if any two are given, the third is thereby determined (because $A + B + C = 180°$). A datum may be useful in solving a construction problem if any one part of the datum can be constructed from the other two parts.

(a) Show that the side b, the angle A, and the altitude h_c on side c constitute a datum.

(b) Show that $b + c$, angle A, and $h_b + h_c$ constitute a datum.

(c) Construct a triangle given a, angle A, and $h_b + h_c$.

Bibliography

ALTSHILLER-COURT, NATHAN, *College Geometry*, 2d ed. New York: Barnes & Noble, Inc., 1952.

————, *Modern Pure Solid Geometry*. New York: The Macmillan Company, 1935.

ARCHIBALD, R. C., *Euclid's Book on Division of Figures*. New York: Cambridge University Press, 1915.

BOLYAI, JOHN, "The Science of Absolute Space," tr. by G. B. Halsted (1895). See BONOLA.

BONOLA, ROBERTO, *Non-Euclidean Geometry*, tr. by H. S. Carslaw. New York: Dover Publications, Inc., 1955. Containing BOLYAI and LOBACHEVSKY.

CARSLAW, H. S., *Non-Euclidean Plane Geometry and Trigonometry*. New York: David McKay Company, Inc., 1916. Reprinted in *String Figures, and Other Monographs*, Chelsea Publishing Company, 1960.

COOLIDGE, J. L., *A History of Geometrical Methods*. New York: Oxford University Press, 1940.

COXETER, H. S. M., *Non-Euclidean Geometry*, 2d ed. Toronto: University of Toronto Press, 1947.

DAUS, P. H., *College Geometry*. New York: Prentice-Hall, Inc., 1941.

DAVIS, H. T., *Alexandria, the Golden City*, 2 vols. Evanston, Ill.: Principia Press of Illinois, 1957.

DUNNINGTON, G. W., *Carl Friedrich Gauss: Titan of Science*. New York: Exposition Press, 1955.

FORDER, H. G., *The Foundations of Euclidean Geometry*. New York: Cambridge University Press, 1927.

GOW, JAMES, *A Short History of Greek Mathematics*. New York: Hafner Publishing Company, 1923.

HEATH, T. L., *History of Greek Mathematics*, Vol. 1. New York: Oxford University Press, 1921.

————, *A Manual of Greek Mathematics*. New York: Oxford University Press, 1931.

————, *The Thirteen Books of Euclid's Elements*, 2d ed., 3 vols. New York: Cambridge University Press, 1926.

HILBERT, DAVID, *The Foundations of Geometry*, tr. by E. J. Townsend. Chicago: The Open Court Publishing Company, 1902.

JAMES, GLENN, ed., *The Tree of Mathematics*. Pacoima, Calif.: The Digest Press, 1957.

JOHNSON, R. A., *Modern Geometry*. Boston: Houghton Mifflin Company, 1929.

KULCZYCKI, STEFAN, *Non-Euclidean Geometry*, tr. by Stanislaw Knapowski. New York: Pergamon Press, Inc., 1961.

LIEBER, H. G. AND L. R., *Non-Euclidean Geometry, or Three Moons in Mathesis*, 2d ed. Lancaster, Pa.: The Science Press Printing Company, 1940.

LOBACHEVSKY, N. I., *Geometrical Researches on the Theory of Parallels*, tr. by G. B. Halsted. See BONOLA.

MANNING, H. P., *Non-Euclidean Geometry*. Boston: Ginn & Company, 1901. Reprinted by Dover Publications, Inc., 1963.

POINCARÉ, HENRI, *The Foundations of Science*, tr. by G. B. Halsted. Lancaster, Pa.: The Science Press Printing Company, 1913.

ROBINSON, G. DE B., *The Foundations of Geometry*, 2d ed. Toronto: University of Toronto Press, 1946.

SARTON, GEORGE, *Ancient Science and Modern Civilization*. Lincoln, Neb.: The University of Nebraska Press, 1954.

SHIVELY, L. S., *An Introduction to Modern Geometry*. New York: John Wiley & Sons, Inc., 1939.

SMITH, D. E., *A Source Book in Mathematics*. New York: McGraw-Hill Book Company, Inc., 1929.

SOMMERVILLE, D. M. Y., *The Elements of Non-Euclidean Geometry*. Chicago: The Open Court Publishing Company, 1916.

VAN DER WAERDEN, B. L., *Science Awakening*, tr. by Arnold Dresden. Groningen, Netherlands: P. Noordhoff, N.V., 1954.

WOLFE, H. E., *Introduction to Non-Euclidean Geometry*. New York: Holt, Rinehart and Winston, Inc., 1945.

YOUNG, J. W. A., ed., *Monographs on Topics of Modern Mathematics Relevant to the Elementary Field*. New York: David McKay Company, Inc., 1911. Reprinted by Dover Publications, Inc., 1955.

6

Greek Mathematics after Euclid

6-1 Historical Setting

The city of Alexandria enjoyed many advantages, and not the least enviable of these was long-lasting peace. During the reign of the Ptolemies, which lasted for almost 300 years, the city remained free from internal and external strife. This was ended by a short period of conflict when Egypt became a part of the Roman Empire, after which the Pax Romana settled over the land. It is no wonder that Alexandria became a haven for scholars and that for close to a thousand years so much of ancient scholastic attainment emanated from that city. Almost every mathematician of antiquity to be discussed in this chapter was either a professor or a student at the University of Alexandria.

The closing period of ancient times was dominated by Rome. In 212 B.C., Syracuse yielded to a Roman siege; in 146 B.C., Carthage fell before the power of imperial Rome; and in the same year, the last of the Greek cities, Corinth, also fell, and Greece became a province of the Roman Empire. Mesopotamia was not conquered until 65 B.C., and Egypt remained under the Ptolemies until 30 B.C. Greek civilization spread through Roman life, and Christianity began to develop, especially among the slaves and the poor. The Roman administrators collected heavy taxes but otherwise did not interfere with the underlying economic organization of the eastern colonies.

Constantine the Great was the first Roman emperor to embrace Christianity and he pronounced it the official religion. In 330 A.D., Constantine moved his capital from Rome to Byzantium, which he renamed Constantinople. In 395 A.D., the Roman Empire was divided into the Eastern and the Western Empires, with Greece as a part of the eastern division.

141

The economic structure of both empires was based essentially on agriculture, with a spreading use of slave labor. Such conditions were stifling to original scientific work and a gradual decline in productive thinking set in, more pronounced in the western part where slavery was employed on a larger scale. The eventual decline of the slave market, with its disastrous effect on Roman economy, found science reduced to a mediocre level. The Alexandrian school gradually faded along with the breakup of ancient society. Creative thinking gave way to compilation and commentarization. Hectic days followed the fight of Christianity against paganism, and finally, in 641 A.D., Alexandria was taken by the Arabs.

6-2   Archimedes

One of the greatest mathematicians of all time, and certainly the greatest of antiquity, was Archimedes, a native of the Greek city of Syracuse on the island of Sicily. He was born about 287 B.C. and died during the Roman pillage of Syracuse in 212 B.C. He was the son of an astronomer and was in high favor with (perhaps even related to) King Hieron of Syracuse. There is a report that he spent time in Egypt, in all likelihood at the University of Alexandria, for he numbered among his friends Conon, Dositheus, and Eratosthenes; the first two were successors to Euclid, the last was a librarian, at the University. Many of Archimedes' mathematical discoveries were communicated to these men.

Roman historians have related many picturesque stories about Archimedes. Very familiar among these are the descriptions of the ingenious contrivances devised by Archimedes to aid in the defense of Syracuse against the siege directed by the Roman general Marcellus. There were catapults with adjustable ranges, movable projecting poles for dropping heavy weights on enemy ships which approached too near the city walls, and great grappling cranes that hoisted enemy ships from the water. The story that he used large burning-glasses to set the enemy's vessels afire is of later origin, but could be true. There also is the story of how he lent credence to his statement, "Give me a place to stand on and I will move the earth," by effortlessly and singlehandedly moving with a compound pulley a heavily weighted ship which had with difficulty been drawn up by a large contingent of laborers.

Apparently Archimedes was capable of strong mental concentration, and tales are told of his obliviousness to surroundings when engrossed by a problem. Typical is the frequently told story of King Hieron's crown and the suspected goldsmith. It seems King Hieron had a gold crown which he feared contained hidden silver, and he referred the matter to Archimedes, who, one day during a bath, hit upon the solution by discovering the first law of hydrostatics. Forgetting to clothe himself, he rose from his bath and ran home through the streets shouting, "Eureka, eureka."

Archimedes worked much of his geometry from figures drawn in the ashes or in the after-bathing oil smeared on his body. In fact it is related that he met his end during the sack of Syracuse while preoccupied with a diagram drawn on a sand tray. According to one version, he ordered a pillaging Roman soldier to stand clear of his diagram, whereupon the incensed looter ran a spear through the old man.

The works of Archimedes are masterpieces of mathematical exposition and resemble to a remarkable extent modern journal articles. They are written with a high finish and an economy of presentation and exhibit great originality, computational skill, and rigor in demonstration. Some ten treatises have come down to us and there are various traces of lost works. Probably the most important contribution made to mathematics in these works is the early development of some of the methods of the integral calculus. We shall return to this in a later chapter.

Three of Archimedes' extant works are devoted to plane geometry. They are *Measurement of a Circle, Quadrature of the Parabola,* and *On Spirals.* It was in the first of these that Archimedes inaugurated the classical method of computing π, which we have already described in Section 4-8. In the second work, which contains 24 propositions, it is shown that the area of a parabolic segment is 4/3 that of the inscribed triangle having the same base and having its opposite vertex at the point where the tangent is parallel to the base. The summation of a convergent geometric series is involved. The third work contains 28 propositions devoted to properties of the curve which is now known as the spiral of Archimedes and which has $r = k\theta$ for a polar equation. In particular, the area enclosed by the curve and two radii vectors is found essentially as would be done today as a calculus exercise. There are allusions to many lost works on plane geometry by Archimedes, and there is reason to believe that some of the theorems of these works have been preserved in the *Liber assumptorum,* a collection which has reached us through the Arabians (see Problem Study 6.4). One Arabian writer claims that Archimedes was the discoverer of the celebrated formula,

$$K = \sqrt{\{s(s - a)(s - b)(s - c)\}},$$

for the area of a triangle in terms of its three sides. This formula had hitherto been attributed to Heron of Alexandria.

Two of Archimedes' extant works are devoted to geometry of three dimensions, namely *On the Sphere and Cylinder* and *On Conoids and Spheroids*. In the first of these, written in two books and containing 60 propositions, appear theorems giving the areas of a sphere and of a zone of one base and the volumes of a sphere and of a segment of one base (see Problem Study 6.2). In Book II appears the problem of dividing a sphere by a plane into two segments whose volumes shall be in a given ratio. This problem leads to a cubic equation whose solution is not given in the text as it has come down to us, but was found by Eutocius in an Archimedean fragment. There is a discussion concerning the conditions under which the cubic may have a real and positive root. Similar considerations do not appear again in European mathematics for over a thousand years. The treatise ends with the two interesting theorems: (1) *If V, V' and S, S' are the volumes of the segments and the areas of the zones into which a sphere is cut by a nondiametral plane, V and S pertaining to the greater piece, then*

$$S^{3/2} : S'^{3/2} < V : V' < S^2 : S'^2,$$

and, (2) *Of all spherical segments of one base having equal zonal areas, the hemisphere has the greatest volume.* The treatise *On Conoids and Spheroids* contains 40 propositions, which are concerned chiefly with an investigation of the volumes of quadrics of revolution. Pappus has ascribed to Archimedes 13 semiregular polyhedra, but unfortunately Archimedes' own account of them is lost.*

Archimedes wrote two related essays on arithmetic, one of which is lost. The extant paper, entitled *The Sand-Reckoner*, is addressed to Gelon, son of King Hieron, and applies an arithmetical system for the representation of large numbers to the finding of an upper limit to the number of grains of sand which would fill a sphere with center at the earth and radius reaching to the sun. It is here, among related remarks pertaining to astronomy, that we learn that Aristarchus (*ca.* 310–230 B.C.) had put forward the Copernican theory of the solar system. In addition to the two arithmetical essays, there is the so-called Cattle Problem which, from a salutation, appears to have been communicated by Archimedes to Eratosthenes. It is a difficult indeterminate

---

*Construction patterns for the Archimedean solids can be found in Miles C. Hartley, *Patterns of Polyhedra*, rev. ed. (Ann Arbor, Mich.: Edwards Bros., Inc., 1957).

problem involving eight integral unknowns connected by seven linear equations and subjected to the two additional conditions that the sum of a certain pair of the unknowns be a perfect square while the sum of another certain pair be a triangular number. Without the two additional conditions the smallest values of the unknowns are numbers in the millions, and with the two additional conditions one of the unknowns must be a number of more than 206,500 digits!

There are two extant treatises by Archimedes on applied mathematics, *On Plane Equilibriums* and *On Floating Bodies*. The first of these is in two books containing 25 propositions. Here, following a postulational treatment, are found the elementary properties of centroids and the determination of the centroids of a variety of plane areas, culminating with that of a parabolic segment and of an area bounded by a parabola and two parallel chords. The work *On Floating Bodies* is also in two books, containing 19 propositions, and is the first application made of mathematics to hydrostatics. The treatise, resting on two postulates, first develops those familiar laws of hydrostatics which nowadays are encountered in an elementary physics course. It then considers some rather difficult problems, concluding with a remarkable investigation of the positions of rest and stability of a right segment of a paraboloid of revolution floating in a fluid. Archimedes wrote other, but now lost, treatises on mathematical physics. Thus Pappus mentions a work *On Levers*, and Theon of Alexandria quotes a theorem from another purported work on the properties of mirrors. It may be that there was originally a larger work by Archimedes of which the two books *On Plane Equilibriums* formed only a part. It was not until the sixteenth-century work of Simon Stevin that the science of statics and the theory of hydrostatics were appreciably advanced beyond the points reached by Archimedes.

One of the most thrilling discoveries of modern times in the history of mathematics was the discovery by Heiberg, in Constantinople, as late as 1906, of Archimedes' long lost treatise entitled *Method*. This work is in the form of a letter addressed to Eratosthenes and is important because of the information it furnishes concerning a "method" which Archimedes used in discovering many of his theorems. Although the "method" can today be made rigorous by the modern integration process, Archimedes used the "method" only to discover results which he then established rigorously by the method of exhaustion. Since the "method" is so closely connected with the ideas of the integral calculus we shall reserve its consideration for a later chapter, devoted specially to the origin and development of the calculus.

6-3   Eratosthenes

Eratosthenes was a native of Cyrene on the south coast of the Mediterranean Sea and was only a few years younger than Archimedes. He spent many years of his early life in Athens and, when about forty, was invited by Ptolemy III of Egypt to come to Alexandria as tutor to his son and to serve as chief librarian at the University there. It is told that in old age, about 194 B.C., he became almost blind from ophthalmia and committed suicide by voluntary starvation.

Eratosthenes was gifted in all the branches of knowledge of his time but failed to top his contemporaries in any one branch. This is the reason, some suppose, that he was nicknamed "Beta." He was an outstanding athlete, poet, astronomer, geographer, historian, and mathematician, and various of his works are mentioned by later writers. We have already seen, in Problem Study 4.3 (c), his mechanical solution of the duplication problem. His most scientific achievement, the measurement of the earth, is considered in Problem Study 6.1 (c).

In arithmetic Eratosthenes is noted for the following device, known as the *sieve*, for finding all the prime numbers less than a given number n. One writes down, in order and starting with 3, all the odd numbers less than n. The composite numbers in the sequence are then sifted out by crossing off, from 3, every third number, then from the next remaining number, 5, every fifth number, then from the next remaining number, 7, every seventh number, from the next remaining number, 11, every eleventh number, and so on. In the process some numbers will be crossed off more than once. All the remaining numbers, along with the number 2, constitute the list of primes less than n.

6-4   The Prime Numbers

The fundamental theorem of arithmetic says that the prime numbers are building bricks from which all other integers may be made. Accordingly, the prime numbers have received much study, and considerable efforts have been spent trying to determine the nature of their distribution in the sequence of positive integers. The chief results obtained in antiquity are

Euclid's proof of the infinitude of the primes and Eratosthenes' sieve for finding all primes below a given integer n.

From the sieve of Eratosthenes can be obtained a cumbersome formula which will determine the number of primes below n when the primes below $\sqrt{n}$ are known. This formula was considerably improved in 1870 by Ernst Meissel, who succeeded in showing that the number of primes below 10^8 is 5,761,455. The Danish mathematician Bertelsen continued these computations and announced, in 1893, that the number of primes below 10^9 is 50,847,478. In 1959, the American mathematician D. H. Lehmer showed that this last result is incorrect and that it should read 50,847,534; he also showed that the number of primes below 10^{10} is 455,052,511.

No practicable procedure is yet known for testing large numbers for primality, and the effort spent on testing certain special numbers has been enormous. For more than 75 years the largest number actually verified as a prime was the 39-digit number

$$2^{127} - 1 = 170{,}141{,}183{,}460{,}469{,}231{,}731{,}687{,}303{,}715{,}884{,}105{,}727,$$

given by the French mathematician Anatole Lucas in 1876. In 1952, the EDSAC machine, in Cambridge, England, established the primality of the much larger (79-digit) number

$$180(2^{127} - 1)^2 + 1,$$

and since then other digital computers have shown the primality of the enormous numbers $2^n - 1$ for $n = 521$, 607, 1279, 2203, 2281, 3217, 4253, and 4423.

A dream of number theorists is the finding of a function $f(n)$ which, for positive integral n, will yield only prime numbers, the sequence of primes so obtained containing infinitely many different primes. Thus $f(n) = n^2 - n + 41$ yields primes for all such $n < 41$, but $f(41) = (41)^2$, a composite number. The quadratic polynomial $f(n) = n^2 - 79n + 1601$ yields primes for all $n < 80$. Polynomial functions can be obtained which will successively yield as many primes as desired, but no such function can be found which will always yield primes. It was about 1640 that Pierre de Fermat conjectured that $f(n) = 2^{2^n} + 1$ is prime for all nonnegative integral n, but this, as we have pointed out in Section 5-4, is incorrect. An interesting recent result along these lines is the proof, by W. H. Mills in 1947, of the existence of a real number A such that the largest integer not exceeding A^{3^n} is a prime for every positive integer n. Nothing was shown about the actual value, nor even the rough magnitude, of the real number A.

A remarkable generalization of Euclid's theorem on the infinitude of the primes was established by Lejeune-Dirichlet (1805–1859), who succeeded in showing that every arithmetic sequence,

$$a, \quad a + d, \quad a + 2d, \quad a + 3d, \quad \cdots,$$

in which a and d are relatively prime, contains an infinitude of primes. The proof of this result is far from simple.

Perhaps the most amazing result yet found concerning the distribution of the primes is the so-called *prime number theorem*. Suppose we let A_n denote the number of primes below n. The prime number theorem then says that $(A_n \log_e n)/n$ approaches 1 as n becomes larger and larger. In other words A_n/n, called the *density* of the primes among the first n integers, is approximated by $1/\log_e n$, the approximation improving as n increases. This theorem was conjectured by Gauss from an examination of a large table of primes, and was independently proved in 1896 by the French and Belgian mathematicians J. Hadamard and C. J. de la Vallée Poussin.

Extensive factor tables are valuable in researches on prime numbers. Such a table for all numbers up to 24,000 was published by J. H. Rahn in 1659, as an appendix to a book on algebra. In 1668, John Pell of England extended this table up to 100,000. As a result of appeals by the German mathematician J. H. Lambert, an extensive and ill-fated table was computed by a Viennese schoolmaster named Felkel. The first volume of Felkel's computations, giving factors of numbers up to 408,000, was published in 1776 at the expense of the Austrian imperial treasury. But there were very few subscribers to the volume, and so the treasury recalled almost the entire edition and converted the paper into cartridges to be used in a war for killing Turks! In the nineteenth century, the combined efforts of Chernac, Burckhardt, Crelle, Glaisher, and the lightning calculator Dase led to a table covering all numbers up to 10,000,000 and published in ten volumes. The greatest achievement of this sort, however, is the table calculated by J. P. Kulik (1773–1863), of the University of Prague. His as yet unpublished manuscript is the result of a 20-year hobby, and covers all numbers up to 100,000,000. The best available factor table is that of the American mathematician D. N. Lehmer* (1867–1938). It is a cleverly prepared one-volume table covering numbers up to 10,000,000.

There are many unproved conjectures regarding prime numbers. One of these is to the effect that there are infinitely many pairs of *twin primes*, or primes of the form p and $p + 2$, like 3 and 5, 11 and 13, and 29 and 31. Another is the conjecture made by C. Goldbach in 1742 in a letter to Euler.

*Father of D. H. Lehmer. D. N. Lehmer has pointed out that Kulik's table contains errors.

Goldbach had observed that every even integer, except 2, seemed representable as the sum of two primes. Thus $4=2+2$, $6=3+3$, $8=5+3$, $\cdots$, $16 = 13 + 3$, $18 = 11 + 7$, $\cdots$, $48 = 29 + 19$, $\cdots$, $100 = 97 + 3$, and so forth. Progress on this problem was not made until 1931 when the Russian mathematician Schnirelmann showed that every positive integer can be represented as the sum of not more than 300,000 primes! Somewhat later the Russian mathematician Vinogradoff showed that there exists a positive integer N such that any integer $n > N$ can be expressed as the sum of at most four primes, but the proof in no way permits us to appraise the size of N.

6-5 Apollonius

Euclid, Archimedes, and Apollonius are the three mathematical giants of the third century B.C. Apollonius, who was younger than Archimedes by some 25 years, was born about 262 B.C. in Perga in southern Asia Minor. The little that is known about the life of Apollonius is briefly told. As a young man he went to Alexandria, studied under the successors of Euclid, and remained there for a long time. Later he visited Pergamum in western Asia Minor, where there was a recently founded university and library patterned after that at Alexandria. He returned to Alexandria and died there somewhere around 200 B.C.

Although Apollonius was an astronomer of note and although he wrote on a variety of mathematical subjects, his chief bid to fame rests on his extraordinary *Conic Sections*, a work which earned him the name, among his contemporaries, of the "Great Geometer." Apollonius' *Conic Sections*, in eight books and containing about 400 propositions, is a thorough investigation of these curves, and completely superseded the earlier works on the subject by Menaechmus, Aristaeus, and Euclid. Only the first seven of the eight books have come down to us, the first four in Greek and the following three from a ninth-century Arabian translation. The first four books, of which I, II, and III are presumably founded on Euclid's previous work, deal with the general elementary theory of conics, while the later books are devoted to more specialized investigations.

Prior to Apollonius, the Greeks derived the conic sections from three types of cones of revolution, according as the vertex angle of the cone was less than, equal to, or greater than a right angle. By cutting each of three such cones with a plane perpendicular to an element of the cone an ellipse, parabola, and hyperbola respectively result. Only one branch of a hyperbola was considered. Apollonius, on the other hand, in Book I of his treatise, obtains all the conic sections in the now familiar way from *one* right or oblique circular *double* cone.

The names *ellipse, parabola,* and *hyperbola* were supplied by Apollonius, and were borrowed from the early Pythagorean terminology of application of areas. When the Pythagoreans applied a rectangle to a line segment (that is, placed the base of the rectangle along the line segment, with one end of the

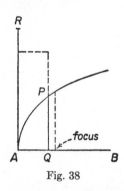

Fig. 38

base coinciding with one end of the segment) they said they had a case of "ellipsis," "parabole," or "hyperbole" according as the base of the applied rectangle fell short of the line segment, exactly coincided with it, or exceeded it. Now let AB (see Figure 38) be the principal axis of a conic, P any point on the conic, and Q the foot of the perpendicular from P on AB. At A, which is a vertex of the conic, draw a perpendicular to AB and mark off on it a distance AR equal to what we now call the *latus rectum,* or *parameter* p, of the conic. Apply, to the segment AR, a rectangle having AQ for one side and an area equal to $(PQ)^2$. According as the application falls short of, coincides with, or exceeds the segment AR, Apollonius calls the conic an *ellipse,* a *parabola,* or a *hyperbola.* In other words, if we consider the curve referred to a Cartesian coordinate system having its x and y axes along AB and AR respectively and if we designate the coordinates of P by x and y, then the curve is an ellipse, parabola, or hyperbola according as $y^2 \lesseqgtr px$. Actually, in the cases of the ellipse and hyperbola,

$$y^2 = px \mp px^2/d,$$

where d is the length of the diameter through vertex A. Apollonius derives the bulk of the geometry of the conic sections from the geometrical equivalents of these Cartesian equations. Facts like this cause some to defend the thesis that analytic geometry was an invention of the Greeks.

Book II of Apollonius' treatise on *Conic Sections* deals with properties of asymptotes and conjugate hyperbolas, and the drawing of tangents. Book III contains an assortment of theorems. Thus there are some area theorems like: *If the tangents at any two points A and B of a conic intersect in C and also intersect the diameters through B and A in D and E, then triangles CBD and ACE are equal in area.* One also finds the harmonic properties of poles and polars (a subject familiar to those who have had an elementary course in projective geometry), and theorems concerning the product of the segments of intersecting chords. As an example of the latter there is the theorem (sometimes today referred to as Newton's theorem): *If two chords PQ and MN, parallel to two given directions, intersect in O, then $(PO)(OQ)/(MO)(ON)$ is a constant independent of the position of O.* The well-known focal properties of the central conics occur toward the end of Book III. In the entire treatise there is no mention of the focus-directrix property of the conics, nor, for that matter, of the focus of the parabola. This is curious because, according to Pappus, Euclid was aware of these properties. Book IV of the treatise proves the converses of some of those propositions of Book III concerning harmonic properties of poles and polars. There also are some theorems about pairs of intersecting conics. Book V is the most remarkable and original of the extant books. It treats of normals considered as maximum and minimum line segments drawn from a point to the curve. The construction and enumeration of normals from a given point are dealt with. The subject is pushed to the point where one can write down the Cartesian equations of the evolutes (envelopes of normals) of the three conics! Book VI contains theorems and construction problems concerning equal and similar conics. Thus it is shown how in a given right cone to find a section equal to a given conic. Book VII contains a number of theorems involving conjugate diameters, such as the one about the constancy of the area of the parallelogram formed by the tangents to a central conic at the extremities of a pair of such diameters.

Conic Sections is a great treatise but, because of the extent and elaborateness of the exposition and the portentousness of the statements of many complicated propositions, is rather trying to read. Even from the above brief sketch of contents we see that the treatise is considerably more complete than the usual college course in the subject.

Pappus has given brief indications of the contents of six other works of

Apollonius. These are *On Proportional Section* (181 propositions), *On Spatial Section* (124 propositions), *On Determinate Section* (83 propositions), *Tangencies* (124 propositions), *Vergings* (125 propositions), and *Plane Loci* (147 propositions). Only the first of these has survived, and this is in Arabic. It deals with the general problem (see Figure 39): Given two lines a and b with the fixed points A on a and B on b, to draw through a given point O a line $OA'B'$ cutting a in A' and b in B' so that $AA'/BB' = k$, a given constant. The exhaustiveness of the treatment is indicated by the fact that

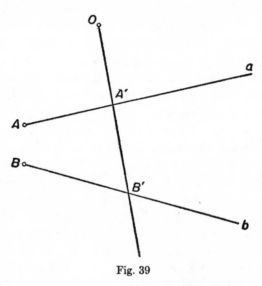

Fig. 39

Apollonius considers 77 separate cases. The second work dealt with a similar problem, only here we wish to have $(AA')(BB') = k$. The third work concerned itself with the problem: Given four points A, B, C, D on a line, to find a point P on the line such that we have $(AP)(CP)/(BP)(DP) = k$. The work on *Tangencies* dealt with the problem of constructing a circle tangent to three given circles, where the given circles are permitted to degenerate independently into straight lines or points. This problem, now known as the *problem of Apollonius*, has attracted many mathematicians, among them Viète, Euler, and Newton. One of the first solutions applying the new Cartesian geometry was given by Descartes' pupil, Princess Elizabeth, daughter of Frederick V of Bohemia. Probably the most elegant solution is that furnished by the French artillery officer and professor of mathematics, Joseph-Diez Gergonne (1771–1859). The general problem in *Vergings* was that of inserting a line segment between two given loci such that the line of the segment shall pass through a given point. In the last work, *Plane Loci*, appeared, among many others, the two theorems: (1) *If A and B are fixed points and k a given constant, then the locus of a point P, such that AP/BP = k, is either a circle (if k ≠ 1) or a straight line (if k = 1)*, and (2) *If A, B, · · · are fixed points and a, b, · · ·, k are given constants, then the locus of a point P, such that a(AP)² + b(BP)² + · · · = k, is a circle.* The circle of (1) is known, in modern college geometry texts, as a *circle of Apollonius*.

Attempts have been made to restore all six of the above works, the first two by Edmund Halley in 1706, the third by Robert Simson in 1749, the fourth by Viète in 1600, the fifth by Ghetaldi in 1607 and 1613, Alexander Anderson in 1612, and Samuel Horsley in 1770, and the last by Fermat in 1637 and, more completely, by Simson in 1746. In addition to these six works a number of other lost works by Apollonius are referred to by ancient writers.

6-6 Greek Trigonometry

The origins of trigonometry are obscure. There are some problems in the Rhind papyrus which seem to involve the cotangent of the dihedral angles at the base of a pyramid, and, as we have seen in Section 2-6, the Babylonian cuneiform tablet Plimpton 322 essentially contains a remarkable table of secants. It may be that modern investigations into the mathematics of ancient Mesopotamia will reveal an appreciable development of practical trigonometry. The Babylonian astronomers of the fourth and fifth centuries B.C. had accumulated a considerable mass of observational data and it is now known that much of this passed on to the Greeks. It was this early astronomy that gave birth to spherical trigonometry.

Probably the most eminent astronomer of antiquity was Hipparchus, who flourished about 140 B.C. Though there is an observation of the vernal equinox recorded by Hipparchus at Alexandria in 146 B.C., his most important observations were made at the famous observatory of the commercial center of Rhodes. Hipparchus was an extremely careful observer and is credited, in astronomy, with such feats as the determination of the length of the mean lunar month to within 1″ of the present accepted value, an accurate calculation of the inclination of the ecliptic, and the discovery and estimation of the annual precession of the equinoxes. He is said also to have computed the lunar parallax, to have determined the perigee and mean motion of the moon, and to have catalogued 850 fixed stars. It was Hipparchus, or perhaps Hipsicles (*ca.* 180 B.C.), who introduced into Greece the division of a circle into 360°, and he is known to have advocated location of positions on the earth by latitude and longitude. Knowledge of these achievements is secondhand, for almost nothing of Hipparchus' writings has reached us.

More important for us, though, than Hipparchus' achievements in astronomy, is the part he played in the development of trigonometry. The fourth-century commentator, Theon of Alexandria, has credited to Hipparchus a twelve-book treatise dealing with the construction of a *table of chords*. A subsequent table, given by Claudius Ptolemy and believed to have been adopted from Hipparchus' treatise, gives the lengths of the chords of all central angles of a given circle by half-degree intervals from $\frac{1}{2}$° to 180°. The radius of the circle is divided into 60 equal parts and the chord lengths then expressed sexagesimally in terms of one of these parts as a unit. Thus, using the symbol crd α to represent the length of the chord of a central angle α, one finds recordings like

$$\text{crd } 36° = 37^{p}4'55'',$$

meaning, of course, that the chord of a central angle of 36° is equal to 37/60 (or 37 small parts) of the radius, plus 4/60 of one of these small parts, plus

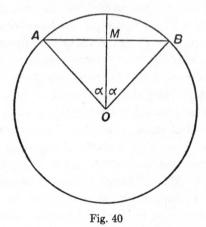

Fig. 40

55/3600 more of one of the small parts. It is seen from Figure 40 that a table of chords is equivalent to a table of trigonometric sines, for

$$\sin \alpha = AM/OA = AB/(\text{diameter of circle}) = (\text{crd } 2\alpha)/120.$$

Thus Ptolemy's table of chords gives, essentially, the sines of angles by $15'$ intervals, from $0°$ to $90°$. The mode of calculating these chord lengths, elegantly explained by Ptolemy, in all likelihood was known to Hipparchus. Evidence shows that Hipparchus made systematic use of his tables and was aware of the equivalents of several formulas now used in the solution of spherical right triangles.

Theon has also mentioned a six-book treatise on chords in a circle written by Menelaus of Alexandria, a contemporary of Plutarch (*ca.* 100 A.D.). This work, along with a variety of others by Menelaus, is lost to us. Fortunately, however, Menelaus' three-book treatise *Sphaerica* has been preserved in the Arabic. This work throws considerable light on the Greek development of trigonometry. In Book I there appears for the first time the definition of a *spherical triangle*. The book is devoted to establishing for spherical triangles many of the propositions Euclid established for plane triangles, such as the usual congruence theorems, theorems about isosceles triangles, and so on. In addition is established the congruence of two spherical triangles having the angles of one equal to the angles of the other (for which there is no analogue in the plane) and the fact that the sum of the angles of a spherical triangle is greater than two right angles. Symmetrical spherical triangles are regarded as congruent. Book II contains theorems of interest in astronomy. In Book III is developed the spherical trigonometry of the times, largely deduced from the spherical case of the powerful proposition known to students of college geometry as *Menelaus' theorem*: *If a transversal intersects the sides BC, CA, AB of a triangle ABC in the points L, M, N, respectively,* then

$$(AN/NB)(BL/LC)(CM/MA) = -1.$$

In the spherical analogue we have a great circle transversal intersecting the sides BC, CA, AB of a spherical triangle ABC in the points L, M, N, respectively. The corresponding conclusion is then equivalent to

$$(\sin \widehat{AN}/\sin \widehat{NB})(\sin \widehat{BL}/\sin \widehat{LC})(\sin \widehat{CM}/\sin \widehat{MA}) = -1.$$

The plane case is assumed by Menelaus as well known and is used by him to establish the spherical case. A great deal of spherical trigonometry can be deduced from this theorem by taking special triangles and special transversals. The converses of both the plane and spherical cases of the theorem are also true.

The definitive Greek work on astronomy was written by Claudius

Ptolemy of Alexandria about 150 A.D. This very influential treatise, called the *Syntaxis mathematica*, or "Mathematical Collection," was based on the writings of Hipparchus and is noted for its remarkable compactness and elegance. To distinguish it from other lesser works on astronomy, later commentators assigned to it the superlative *magiste*, or "greatest." Still later, the Arabian translators prefixed the Arabian article *al*, and the work has ever since been known as the *Almagest*. The treatise is in thirteen books. Book I contains, among some preliminary astronomical material, the table of chords referred to above, along with a succinct explanation of its derivation from the fertile geometrical proposition now known as *Ptolemy's theorem*: *In a cyclic quadrilateral the product of the diagonals is equal to the sum of the products of the two pairs of opposite sides* (see Problem Study 6.9). Book II considers phenomena depending on the sphericity of the earth. Books III, IV, V develop the geocentric system of astronomy by epicycles. In Book IV appears a solution of the three-point problem of surveying: To determine the point from which pairs of three given points are seen under given angles. This problem has had a long history and is sometimes referred to as the "Problem of Snell" (1617) or the "Problem of Pothenot" (1692). In Book VI, which gives the theory of eclipses, is found the four-place value of π alluded to in Section 4-8. Books VII and VIII are devoted to a catalogue of 1028 fixed stars. The remaining books are devoted to the planets. The *Almagest* remained the standard work on astronomy until the time of Copernicus and Kepler. Ptolemy also wrote on map projections (see Problem Study 6.10), optics, and music. He also attempted a proof of the parallel postulate.

6-7   Heron

Another worker in applied mathematics belonging to the period under consideration was Heron of Alexandria. His much disputed date, with contentions ranging from 150 B.C. to 250 A.D., has recently been plausibly placed in the second half of the first century A.D. His works on mathematical and physical subjects are so numerous and varied that it is customary to describe him as an encyclopedic writer in these fields. There are reasons to suppose he was an Egyptian with Greek training. At any rate his writings, which so often aim at practical utility rather than theoretical completeness, show a curious blend of the Greek and the Oriental. He did much to furnish a scientific foundation for engineering and land surveying. Fourteen or so treatises by Heron, some evidently considerably edited, have come down to us, and there are references to additional lost works.

Heron's works may be divided into two classes, the geometrical and the mechanical. The geometrical works deal largely with problems on mensuration and the mechanical ones with descriptions of ingenious mechanical devices.

The most important of Heron's geometrical works is his *Metrica*, written in three books and discovered in Constantinople by R. Schöne as recently as 1896. Book I deals with the area mensuration of squares, rectangles, triangles, trapezoids, various other specialized quadrilaterals, the regular polygons from the equilateral triangle to the regular dodecagon, circles and their segments, ellipses, parabolic segments, and the surfaces of cylinders, cones, spheres, and spherical zones. It is in this book that we find Heron's clever derivation of the famous formula for the area of a triangle in terms of its three sides (see Problem Study 6.11 (d)). Of particular interest, also, in this book, is Heron's method of approximating the square root of a nonsquare integer. It is a process frequently used by computers today, namely: if $n = ab$, then $\sqrt{n}$ is approximated by $(a + b)/2$, the approximation improving with the closeness of a to b. The method permits of successive approximations. Thus, if a_1 is a first approximation to $\sqrt{n}$, then

$$a_2 = (a_1 + n/a_1)/2$$

is a better approximation, and

$$a_3 = (a_2 + n/a_2)/2$$

is still better, and so on. Book II of the *Metrica* concerns itself with the volume mensuration of cones, cylinders, parallelepipeds, prisms, pyramids, frustums of cones and pyramids, spheres, spherical segments, tori (anchorrings), the five regular solids, and some prismatoids (see Problem Study 6.11 (g)). Book III deals with the division of certain areas and volumes into parts having given ratios to one another. We have seen such problems in Problem Study 3.11 (b) and (c).

In Heron's *Pneumatica* appear descriptions of about a hundred machines and toys, such as a siphon, a fire engine, a device for opening temple doors by a fire on the altar, and a wind organ. His work *Dioptra* concerns itself with the description and engineering applications of an ancient form of theodolite, or surveyor's transit. In *Catoptrica*, one finds the elementary properties of mirrors and problems concerning the construction of mirrors to satisfy certain requirements, such as for a person to see the back of his head or to appear upside down, and so on.

6-8   Diophantus

In 1842, G. H. F. Nesselmann conveniently characterized three stages in the historical development of algebraic notation. First we have *rhetorical algebra*, in which the solution of a problem is written, without abbreviations or symbols, as a pure prose argument. Then comes *syncopated algebra*, in which abbreviations are adopted for some of the more frequently recurring quantities and operations. Finally, as the last stage, we have *symbolic algebra*, in which solutions largely appear in a mathematical shorthand made up of symbols having little apparent connection with the entities they represent. It is fairly accurate to say that all algebra prior to the time of Diophantus was rhetorical. One of Diophantus' outstanding contributions to mathematics was the syncopation of Greek algebra. Rhetorical algebra, however, persisted pretty generally in the rest of the world, with the exception of India, for many hundreds of years. Specifically, in western Europe, most algebra remained rhetorical until the fifteenth century. Symbolic algebra made its first appearance in western Europe in the sixteenth century, but did not become prevalent until the middle of the seventeenth century. It is not often realized that much of the symbolism of our elementary algebra textbooks is only about three hundred years old.

One of our best sources of ancient Greek algebra problems is a collection known as the *Palatine*, or *Greek, Anthology*. This is a group of 46 number problems, in epigrammatic form, assembled about 500 A.D. by the grammarian Metrodorus. Although some of the problems may have originated with the author, there is every reason to believe that many of them are of considerably more ancient origin. The problems, apparently intended for mental recreation, are of a type alluded to by Plato, and closely resemble some of the problems in the Rhind papyrus. Half of them lead to simple linear equations in one unknown, a dozen more to easy simultaneous equations in two unknowns, one to three equations in three unknowns and one to four equations in four unknowns, and there are two cases of indeterminate equations of the first degree. A number of the problems are very much like many found in present-day elementary algebra textbooks. Some examples from the *Greek Anthology* are given in Problem Studies 6.13 and 6.14. Although these problems are easily solved with our modern algebraic symbolism, it must be conceded that a rhetorical solution would require pretty close mental attention. It has been remarked that many of these problems can be readily solved by geometrical algebra, but it is believed that they were actually solved arithmetically, perhaps by applying the *rule of false position* (see Section 2-8). Just when Greek algebra changed from a geometrical form to an arithmetical one is not known, but this probably occurred as early as the time of Euclid.

Of tremendous importance to the development of algebra, and of great influence on later European number theorists, was Diophantus of Alexandria. Diophantus is another mathematician, like Heron, of uncertain date and nationality. Recent historical research tends to place him in the first century of our era rather than the third, as had previously been thought more probable. Beyond the fact that he flourished at Alexandria nothing certain is known about him, although there is an epigram in the *Greek Anthology* which purports to give some details of his life (see Problem Study 6.15 (a)).

Diophantus wrote three works: *Arithmetica*, his most important one and of which six out of thirteen books are extant, *On Polygonal Numbers*, of which only a fragment is extant, and *Porisms*, which is lost. The *Arithmetica* had many commentators, but it was Regiomontanus who, in 1463, called for a Latin translation of the extant Greek text. A very meritorious translation, with commentary, was made in 1575 by Xylander (the Greek name assumed by Wilhelm Holzmann, a professor at the University of Heidelberg). This was used, in turn, by the Frenchman Bachet de Méziriac, who in 1621 published the first edition of the Greek text along with a Latin translation and notes. A second, carelessly printed, edition was brought out in 1670, and is historically important because it contained Fermat's famous marginal notes which stimulated such extensive number theory research. French, German, and English translations appeared later.

The *Arithmetica* is an analytical treatment of algebraic number theory and marks the author as a genius in this field. The extant portion of the work ·is devoted to the solution of about 130 problems, of considerable variety, leading to equations of the first and second degree. One very special cubic is solved. The first book concerns itself with determinate equations in one unknown, and the remaining books with indeterminate equations of the second, and sometimes higher, degree in two and three unknowns. Striking is the lack of general methods and the repeated application of ingenious devices designed for the needs of each individual problem. Diophantus recognized only positive rational answers and was, in most cases, satisfied with only one answer to a problem.

There are some penetrating number theorems stated in the *Arithmetica*. Thus we find, without proof but with an allusion to the *Porisms*, that *the difference of two rational cubes is also the sum of two rational cubes*, a matter which was later investigated by Viète, Bachet, and Fermat. There are many propositions concerning the representation of numbers as the sum of two, three, or four squares, a field of investigation later completed by Fermat, Euler, and Lagrange. Perhaps it might be interesting to list a few of the problems found in the *Arithmetica*; they are all alluring and some of them are challenging. It must be borne in mind that by "number" is meant "positive rational number."

> Problem 29, Book II: Find two square numbers such that when one forms their product and adds either of the numbers to it, the result is a square number.

> Problem 7, Book III: Find three numbers such that their sum is a square number and the sum of any two of them is a square number.

> Problem 9, Book III: Find three numbers in arithmetic progression such that the sum of any two of them is a square number.

> Problem 15, Book III: Find three numbers such that the product of any two of them minus the third is a square number. See Problem Study 6.16 (d).

> Problem 11, Book IV: Find two numbers such that their sum is equal to the sum of their cubes.

> Problem 22, Book IV: Find three numbers in geometric progression such that the difference between any two of them is a square number.

> Problem 18, Book VI: Find a Pythagorean triangle in which the length of the bisector of one of the acute angles is rational. See Problem Study 6.15 (c).

Indeterminate algebraic problems where one must find only the rational solutions have become known as *Diophantine problems*. In fact, modern usage of the terminology often implies the restriction of the solutions to integers. But Diophantus did not originate problems of this sort. Nor was he, as is sometimes stated, the first to work with indeterminate equations, nor the first to solve quadratic equations nongeometrically. He may have been, however, the first to take steps towards an algebraic notation. These steps were in the nature of stenographic abbreviations.

Diophantus had abbreviations for the unknown, powers of the unknown up through the sixth, subtraction, equality, and reciprocals. Our word "arithmetic" comes from the Greek word *arithmetike*, a compound of the words *arithmos* for "number" and *techne* for "science." It has been rather convincingly pointed out by Heath that Diophantus' symbol for the unknown was probably derived by merging the first two Greek letters, α and ρ, of the word *arithmos*. This came, in time, to look like the Greek final sigma *s*. While there is doubt about this, the meaning of the notation for powers of the unknown is quite clear. Thus "unknown squared" is denoted by Δ^T, the first two letters of the Greek word *dunamis* (ΔΥΝΑΜΙΣ) for "power." Again, "unknown cubed" is denoted by K^T, the first two letters of the Greek word *kubos* (ΚΥΒΟΣ) for "cube." Explanations are easily furnished for the succeeding powers of the unknown, $\Delta^T\Delta$ (square-square), ΔK^T (square-cube), and $K^T K$ (cube-cube). Diophantus' symbol for "minus" looks like an inverted V with the angle bisector drawn in. This has been explained as a compound of Λ and I, letters in the Greek word *leipis* (ΛΕΙΨΙΣ) for "lacking." All negative terms in an expression are gathered together and preceded by the minus symbol. Addition is indicated by juxtaposition, and the coefficient of any power of the unknown is represented by the alphabetic Greek numeral (see Section 1-6) following the power symbol. If there is a constant term then $\overset{o}{M}$, an abbreviation of the Greek word *monades* (ΜΟΝΑΔΕΣ) for "units," is used, with the appropriate number coefficient. Thus $x^3 + 13x^2 + 5x$ and $x^3 - 5x^2 + 8x - 1$ would appear as

$$K^T\alpha\Delta^T\iota\gamma s\epsilon \quad \text{and} \quad K^T\alpha s\eta \wedge \Delta^T\epsilon\overset{o}{M}\alpha,$$

which can be read literally as

unknown cubed 1, unknown squared 13, unknown 5

and

(unknown cubed 1, unknown 8) minus (unknown squared 5, units 1).

It is thus that rhetorical algebra became syncopated algebra.

6-9   Pappus

The immediate successors to Euclid, Archimedes, and Apollonius prolonged the great Greek geometrical tradition for a time, but then it began steadily to languish, and new developments were limited to astronomy, trigonometry, and algebra. Then toward the end of the third century A.D., 500 years after Apollonius, there lived the enthusiastic and competent Pappus of Alexandria, who strove to rekindle fresh interest in the subject.

Pappus wrote commentaries on Euclid's *Elements* and *Data*, and on Ptolemy's *Almagest* and *Planispherium*, but about all we know of these is through their influence on the writings of later commentators. Pappus' really great work is his *Mathematical Collection*, a combined commentary and guidebook of the existing geometrical works of his time, sown with numerous original propositions, improvements, extensions, and historical comments. Of the eight books the first, and part of the second, are lost.

Judging from what remains, Book II of the *Mathematical Collection* dealt with a method developed by Apollonius for writing and working with large numbers. Book III is in four sections, the first two dealing with the theory of means (see, for example, Problem Study 6.17 (a)) with attention given to the problem of inserting two mean proportionals between two given line segments, the third with some inequalities in a triangle, and the fourth with the inscription of the five regular polyhedra in a given sphere.

In Book IV is found Pappus' extension of the Pythagorean theorem (given in Problem Study 6.17 (c)), the "ancient proposition" on the arbelos (stated at the end of Problem Study 6.4), the description, genesis, and some properties of the spiral of Archimedes, the conchoid of Nicomedes, and the quadratrix of Dinostratus, with applications to the three famous problems, and a discussion of a special spiral drawn on a sphere.

Book V is largely devoted to *isoperimetry*, or the comparison of the areas of figures having equal bounding perimeters and of volumes of solids having equal bounding areas. This book also contains an interesting passage on bees and the maximum-minimum properties of the cells of their honeycombs. It is in this book that we find Pappus' reference, mentioned in Section 6-2, to the 13 semiregular polyhedra of Archimedes. Book VI is astronomical and deals with the treatises that were to be studied as an introduction to Ptolemy's *Almagest*.

Book VII is historically very important, for it gives an account of the works constituting *The Treasury of Analysis*, a collection which, after Euclid's *Elements*, purported to contain the material considered as essential equipment for the professional mathematician. The 12 treatises discussed are Euclid's *Data*, *Porisms*, and *Surface Loci*, Apollonius' *Conic Sections* and the six works considered toward the end of Section 6-5, Aristaeus' *Solid Loci*, and Eratosthenes' *On Means*. In this book we find an anticipation of the centroid theorem of P. Guldin (see Problem Study 6.18). Also a discussion is given of the famous "loci with respect to three or four lines": *If p_1, p_2, p_3, p_4 are the lengths of line segments drawn from a point P to four given lines, making given angles with these lines, and if $p_1p_2 = kp_3^2$, or $p_1p_2 = kp_3p_4$, where k is a constant, then the locus of P is a conic section.* This problem, solved by Apollonius, is historically important because it was in attempting to generalize it to n lines that Descartes was led in 1637 to formulate the method of coordinates; Pappus' contemporaries had unsuccessfully tried to generalize the problem. The linear case of the so-called *Stewart's theorem*, appearing in college geometry texts, is also found in this book, namely: *If A, B, C, D are any four points on a line, then*

$$(AD)^2(BC) + (BD)^2(CA) + (CD)^2(AB) + (BC)(CA)(AB) = 0,$$

where the segments involved are signed segments. Actually, Robert Simson anticipated Stewart in the discovery of this theorem for the more general case where D may be outside the line ABC. The *anharmonic*, or *cross*, *ratio* (AB,CD) of four collinear points A, B, C, D may be defined as $(AC/CB)/(AD/DB)$, that is, as the ratio of the ratios into which C and D divide the segment AB. In Book VII of the *Mathematical Collection*, Pappus proves that if four concurrent rays (see Figure 41) are cut by two transversals, giving the corresponding ranges A, B, C, D and A', B', C', D', then the two cross ratios (AB,CD) and $(A'B',C'D')$ are equal. In other words, the cross ratio of four collinear points is invariant under projection. This is a fundamental theorem of projective geometry. Book VII contains a solution of the problem: To inscribe in a given circle a triangle whose sides, produced if necessary, shall pass through three given collinear points. This has become known as the *Castillon-Cramer problem*, because in the eighteenth century the problem was generalized by Cramer to the case where the three points need not be collinear, and a solution of this generalization was published by Castillon in 1776. Solutions were also given by Lagrange, Euler, Lhuilier, Fuss, and Lexell in 1780. A few years later a gifted Italian lad of sixteen, named Giordano, generalized the problem to that of inscribing in a circle an n-gon whose sides shall pass through n given points, and he furnished an elegant solution. Poncelet extended the problem still further by replacing the circle with an arbitrary conic section. In Book VII also occurs the first recorded statement of the focus-directrix property of the three conic sections.

Book VIII, like Book VII, contains much that was probably original with Pappus. Here we find a solution of the problem of constructing a conic through five given points. An interesting proposition probably due to Pappus and found in this book is given in Problem Study 6.17 (d).

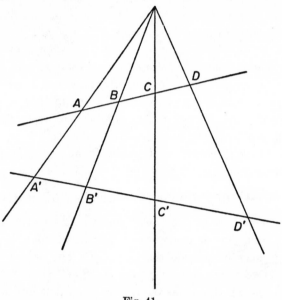

Fig. 41

Pappus' *Mathematical Collection* is a veritable mine of rich geometrical nuggets. Comparisons, where possible, have shown the historical comments contained in the work to be trustworthy. We owe much of our knowledge of Greek geometry to this great treatise, which cites from or refers to the works of over 30 different mathematicians of antiquity. It may be called the requiem of Greek geometry.

6-10 The Commentators

After Pappus, Greek mathematics ceased to be a living study and we find merely its memory perpetuated by minor writers and commentators. Among these were Theon of Alexandria, his daughter Hypatia, Proclus, Simplicius, and Eutocius.

Theon lived in the turbulent closing period of the fourth century A.D. and was the author of a commentary, in eleven books, on Ptolemy's *Almagest*. Also, it will be recalled, the modern editions of Euclid's *Elements* are based upon Theon's revision of the original work.

Theon's daughter, Hypatia, was distinguished in mathematics, medicine, and philosophy, and is reported to have written commentaries on Diophantus' *Arithmetica* and Apollonius' *Conic Sections*. She is the first woman mathematician to be mentioned in the history of mathematics. Her life and barbarous murder by a mob of fanatical Christians in March, 415, are reconstructed in Charles Kingsley's novel.*

Historians of mathematics are indebted to the Neoplatonic philosopher and mathematician Proclus for his *Commentary on Euclid, Book I*, one of our principal sources of information on the early history of elementary geometry. Proclus had access to historical and critical works now lost to us, chief of which were Eudemus' *History of Geometry* in four books and Geminus' apparently comprehensive *Theory of the Mathematical Sciences*. Proclus' commentary on Plato's *Republic* also contains passages of interest to the historian of mathematics. Proclus studied at Alexandria, became head of the Athenian school, and died in Athens in 485 when about 75 years old.

A debt is also owed to Simplicius, the commentator of Aristotle. He has given us accounts of Antiphon's attempt to square the circle, of the lunes of Hippocrates, and of a system of concentric spheres invented by Eudoxus to explain the apparent motions of the members of the solar system. He also wrote a commentary on the first book of Euclid's *Elements*, from which Arabian extracts were later made. Simplicius lived in the first half of the sixth century and studied at both Alexandria and Athens.

Probably contemporary with Simplicius was Eutocius, who wrote commentaries on Archimedes' *On the Sphere and Cylinder*, *Measurement of a Circle*, and *On Plane Equilibriums*, and on Apollonius' *Conic Sections*.

The Athenian school struggled on against the growing opposition of the Christians until the latter finally, in 529 A.D., obtained from Emperor Justinian a decree which closed the doors of the school forever. The school at Alexandria fared little better at the hands of the Christians but at least remained open until the fall of Alexandria to the Arabs in 641, who then put the torch to what the Christians had left. The long and glorious history of Greek mathematics came to an end.

Hypatia, or New Foes with an Old Face (New York: E. P. Dutton & Co., Inc., 1907).

Problem Studies

6.1 Measurements by Aristarchus and Eratosthenes

Aristarchus of Samos (*ca.* 287 B.C.) applied mathematics to astronomy. Since he put forward the heliocentric hypothesis of the solar system, he has become known as the Copernicus of antiquity.

(a) Using crude instruments Aristarchus observed that the angular distance between the moon, when at first quadrant, and the sun is 29/30 of a right angle. On the basis of this measurement he showed (without benefit of trigonometry) that the distance from the earth to the sun is between 18 and 20 times the distance from the earth to the moon. Verify this, using the result of Aristarchus' observation. (The angle concerned is actually about 89° 50′.)

(b) Aristarchus, in his tract *On Sizes and Distances of the Sun and Moon,* used the equivalent of the fact that

$$\sin a/\sin b < a/b < \tan a/\tan b,$$

where $0 < b < a < \pi/2$. From a knowledge of the graphs of the functions $\sin x$ and $\tan x$ show that $(\sin x)/x$ *decreases,* and $(\tan x)/x$ *increases,* as x increases from 0 to $\pi/2$, and thus establish the above inequalities.

(c) Eratosthenes (*ca.* 230 B.C.) made a famous measurement of the earth. He observed at Syene, at noon and at the summer solstice, that a vertical stick had no shadow, while at Alexandria (on the same meridian with Syene) the sun's rays were inclined 1/50 of a complete circle to the vertical. He then calculated the circumference of the earth from the known distance of 5000 stades between Alexandria and Syene. Obtain Eratosthenes' result of 250,000 stades for the circumference of the earth. There is reason to suppose that a stade is about equal to 516.7 feet. Assuming this, calculate from the above result the polar diameter of the earth in miles. (The actual polar diameter of the earth, to the nearest mile, is 7900 miles.)

6.2   On the Sphere and Cylinder

Cicero has related that when serving as Roman quaestor in Sicily he found and repaired Archimedes' then neglected (but now vanished) tomb, upon which was engraved a sphere inscribed in a cylinder. This device commemorates Archimedes' favorite work, *On the Sphere and Cylinder*. Verify the following two results established by Archimedes in this work:

(a) The volume of the sphere is 2/3 that of the circumscribed cylinder.

(b) The area of the sphere is 2/3 of the total area of the circumscribed cylinder.

(c) Define *spherical zone* (of one and two bases), *spherical segment* (of one and two bases), and *spherical sector*.

(d) Assuming the theorem: *The area of a spherical zone is equal to the product of the circumference of a great circle by the altitude of the zone*, obtain the familiar formula for the area of a sphere and establish the theorem: *The area of a spherical zone of one base is equal to that of a circle whose radius is the chord of the generating arc.*

Assuming that the volume of a spherical sector is given by one-third the product of the area of its base and the radius of the sphere, obtain the following results:

(e) The volume of a spherical segment of one base, cut from a sphere of radius R, having h as altitude and a as the radius of its base, is given by

$$V = \pi h^2 (R - h/3) = \pi h(3a^2 + h^2)/6.$$

(f) The volume of a spherical segment of two bases, having h as altitude and a and b as the radii of its bases, is given by

$$V = \pi h(3a^2 + 3b^2 + h^2)/6.$$

(g) The spherical segment of part (f) is equivalent to the sum of a sphere of radius $h/2$ and two cylinders whose altitudes are each $h/2$ and whose radii are a and b, respectively.

6.3   The Problem of the Crown

Proposition 7 of the first book of Archimedes' work *On Floating Bodies* is the famous law of hydrostatics: *A body immersed in a fluid is buoyed up by a force equal to the weight of the displaced fluid.*

(a) Let a crown of weight w pounds be made up of w_1 pounds of gold and w_2 pounds of silver. Suppose that w pounds of pure gold loses f_1 pounds when weighed in water, that w pounds of pure silver loses f_2 pounds when weighed in water, and that the crown loses f pounds when weighed in water. Show that

$$w_1/w_2 = (f_2 - f)/(f - f_1).$$

(b) Suppose the crown of part (a) displaces a volume of v cubic inches when immersed in water, and that lumps, of the same weight as the crown, of pure gold and pure silver displace, respectively, v_1 and v_2 cubic inches when immersed in water. Show that

$$w_1/w_2 = (v_2 - v)/(v - v_1).$$

6.4   The Arbelos

The *Liber assumptorum*, or *Book of Lemmas*, which has been pre-served in an Arabic version, contains some elegant geometrical theorems credited to Archimedes. Among them are some properties of the "arbelos" or "shoemaker's knife." Let A, C, B be three points on a straight line, C lying between A and B. Semicircles are drawn on the same side of the line and having AC, CB, AB as diameters. The "arbelos" is the figure bounded by these three semicircles. At C erect a perpendicular to AB to cut the largest semicircle in G. Let the common external tangent to the two smaller semi-circles touch these curves at T and W. Denote AC, CB, AB by $2r_1$, $2r_2$, $2r$. Establish the following elementary properties of the arbelos:

(a) GC and TW are equal and bisect each other.

(b) The area of the arbelos equals the area of the circle on GC as diameter.

(c) The lines GA and GB pass, respectively, though T and W.

The arbelos has many properties not so easily established. For example it is alleged that Archimedes showed that the circles inscribed in the curvi-linear triangles ACG and BCG are equal, the diameter of each being r_1r_2/r. The smallest circle that is tangent to and circumscribes these two circles is equal to the circle on GC, and therefore equal in area to the arbelos. Con-sider, in the arbelos, a chain of circles c_1, c_2, $\cdots$, all tangent to the semicircles on AB and AC, where c_1 is also tangent to the semicircle on BC, c_2 to c_1, and so on. Then, if r_n represents the radius of c_n and h_n the distance of its center from ACB, we have $h_n = 2nr_n$. This last proposition is found in Book IV of

Pappus' *Mathematical Collection* and is there referred to as an "ancient proposition."

6.5 Prime Numbers

(a) Find, by the sieve of Eratosthenes, all the primes below 500.

(b) Prove that a positive integer p is prime if it has no prime factor not exceeding the greatest integer whose square does not exceed p. This theorem says that, in the elimination process of the sieve of Eratosthenes, we may stop as soon as we reach a prime $p > \sqrt{n}$, for the cancellation of every pth number from p will merely be a repetition of cancellations already effected. Thus, in finding the primes less than 500, we may stop after crossing off every nineteenth number from 19, since the next prime, 23, is greater than $\sqrt{500}$.

(c) Compute $(A_n \log_e n)/n$ for $n = 500$, 10^8, and 10^9.

(d) Prove that there can always be found n consecutive composite integers, however great n may be.

(e) How many pairs of twin primes are there less than 100?

(f) Express each even positive integer less than 100, other than 2, as the sum of two primes.

6.6 The Focus-Directrix Property

Although the Greeks defined the conic sections as sections of cones, it is customary, in college courses in analytic geometry, to define them by the focus-directrix property. Establish the following lemma (a) and then complete the simple proof in (b) that any section of a right circular cone possesses the focus-directrix property.

(a) The lengths of any two line segments from a point to a plane are inversely proportional to the sines of the angles which the line segments make with the plane.

(b) Denote the plane of the section of the right circular cone by p. Let a sphere touch the cone along a circle whose plane we shall call q and also touch plane p at point F (see Figure 42). Let planes p and q intersect in line d. From P, any point on the conic section, drop a perpendicular PR on line d. Let the element of the cone through P cut plane q in point E. Finally, let α be the angle between planes p and q and β the angle an element of the cone makes with plane q. Show that $PF/PR = PE/PR = (\sin \alpha)/(\sin \beta) = e$, a constant. Thus F is a focus, d the corresponding directrix, and e the eccentricity of the conic section. (This simple and elegant approach was discovered around the first quarter-mark of the nineteenth century by the two Belgian mathematicians Adolphe Quetelet (1796–1874) and Germinal Dandelin (1794–1847).)

(c) Show that if p cuts every element of one nappe of the cone then $e < 1$; if p is parallel to one and only one element of the cone then $e = 1$; if p cuts both nappes of the cone then $e > 1$.

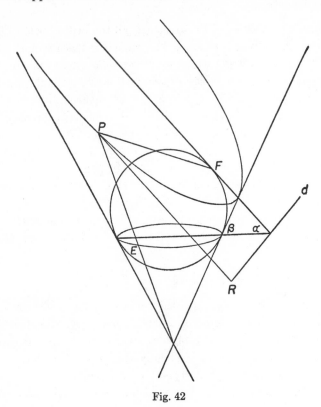

Fig. 42

6.7　Tangencies

In his lost treatise on *Tangencies*, Apollonius considered the problem of drawing a circle tangent to three given circles A, B, C, where each of A, B, C may independently assume either of the degenerate forms of point or straight line. This problem has become known as the *problem of Apollonius*.

(a) Show that there are ten cases of the problem of Apollonius, depending on whether each of A, B, C is a point, a line, or a circle. What is the number of solutions for each general case?

(b) Solve the problem where A, B, C are two points and a line.

(c) Reduce the problem where A, B, C are two lines and a point to the case of part (b).

(d) Given the focus and directrix of a parabola p, and a line m. With Euclidean tools find the points of intersection of p and m.

6.8 Problems from Apollonius

(a) Solve the following easy verging problem considered by Apollonius in his work *Vergings*: In a given circle to insert a chord of given length and verging to a given point.

A more difficult verging problem considered by Apollonius is: Given a rhombus with one side produced, to insert a line segment of given length in the exterior angle so that it verges to the opposite vertex. Several solutions to this problem were furnished by Huygens (1629–1695).

(b) Establish, by analytical geometry, the two problems (1) and (2) stated in Section 6-5 in connection with Apollonius' work *Plane Loci*.

(c) Establish synthetically the first problem in part (b) and also the following special case of the second problem in part (b): The locus of a point, the sum of the squares of whose distances from two fixed points is constant, is a circle whose center is the mid-point of the segment joining the two points.

6.9 Ptolemy's Table of Chords

(a) Prove *Ptolemy's theorem*: *In a cyclic quadrilateral the product of the diagonals is equal to the sum of the products of the pairs of opposite sides.*

Derive, from Ptolemy's theorem, the following relations:

(b) If a and b are the chords of two arcs of a circle of unit radius, then

$$s = (a/2)(4 - b^2)^{1/2} + (b/2)(4 - a^2)^{1/2}$$

is the chord of the sum of the two arcs.

(c) If a and b, $a \geqq b$, are the chords of two arcs of a circle of unit radius, then

$$d = (a/2)(4 - b^2)^{1/2} - (b/2)(4 - a^2)^{1/2}$$

is the chord of the difference of the two arcs.

(d) If t is the chord of an arc of a circle of unit radius, then

$$s = \{2 - (4 - t^2)^{1/2}\}^{1/2}$$

is the chord of half the arc.

In a circle of unit radius crd $60° = 1$, and it may be shown that crd $36° =$

smaller segment of the radius when divided in extreme and mean ratio (see Problem Study 3.10 (d)) = 0.3820. By part (c), crd 24° = crd (60° − 36°) = 0.4158. By part (d) we may calculate the chords of 12°, 6°, 3°, 90′, and 45′, obtaining crd 90′ = 0.0524 and crd 45′ = 0.0262. By Problem Study 6.1 (b), crd 60′/crd 45′ < 60/45 = 4/3, or crd 1° < (4/3)(0.0262) = 0.0349. Also, crd 90′/crd 60′ < 90/60 = 3/2, or crd 1° > (2/3)(0.0524) = 0.0349. Therefore crd 1° = 0.0349. By part (d) we may find crd $\frac{1}{2}$°. Now one can construct a table of chords for $\frac{1}{2}$° intervals. This is the gist of Ptolemy's method of constructing his table of chords.

(e) Show that the relations of parts (b), (c), (d) are equivalent to the trigonometrical formulas for sin $(\alpha + \beta)$, sin $(\alpha - \beta)$, and sin $(\theta/2)$.

(f) Establish the following interesting results as consequences of Ptolemy's theorem: If P lies on the arc AB of the circumcircle of

     (1) an equilateral triangle ABC, then $PC = PA + PB$;

     (2) a square $ABCD$, then $(PA + PC)PC = (PB + PD)PD$;

     (3) a regular pentagon $ABCDE$, then $PC + PE = PA + PB + PD$;

     (4) a regular hexagon $ABCDEF$, then $PD + PE = PA + PB + PC + PF$.

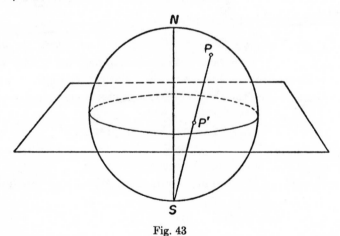

Fig. 43

6.10   Stereographic Projection

In his *Planisphaerium*, Ptolemy developed *stereographic projection* as a mapping by which the points on a sphere are represented on the plane of its equator by projection from the south pole. Under this mapping (see Figure 43), into what do

     (a) The circles of latitude go?

(b) The meridian circles?

(c) A small circle, on the sphere, passing through the south pole?

It can be shown that any circle on the sphere, not through the south pole, maps into a circle on the plane. Very important is the property that stereographic projection is a *conformal* mapping, that is, a mapping which preserves angles between curves. Why is this property important in mapping a small part of the earth's surface onto a plane? (An interesting development of spherical trigonometry from plane trigonometry by stereographic projection is given in J. D. H. Donnay, *Spherical Trigonometry after the Cesàro Method* (New York: Interscience Publishers, Inc., 1945).)

6.11 Problems from Heron

(a) A regular heptagon (seven-sided polygon) cannot be constructed with Euclidean tools. In his work *Metrica*, Heron takes, for an approximate construction, the side of the heptagon equal to the apothem of a regular hexagon having the same circumcircle. How good an approximation is this?

(b) In *Catoptrica*, Heron proves, on the assumption that light travels by the shortest path, that the angles of incidence and reflection in a mirror are equal. Prove this.

(c) A man wishes to go from his house to the bank of a straight river for a pail of water, which he will then carry to his barn, on the same side of the river as his house. Find the point on the riverbank which will minimize the distance the man must travel.

(d) Complete the details of the following indication of Heron's derivation of the formula for the area Δ of a triangle ABC in terms of its sides a, b, c. (1) Let the incircle, with center I and radius r, touch the sides BC, CA, AB in D, E, F, as in Figure 44. On BC produced take G such that $CG = AE$. Draw IH perpendicular to BI to cut BC in J and to meet the perpendicular to BC at C in H. (2) If $s = (a + b + c)/2$, then $\Delta = rs = (BG)(ID)$. (3) B, I, C, H are concyclic, whence $\angle CHB$ is the supplement of $\angle BIC$ and hence equal to $\angle EIA$. (4) $BC/CG = BC/AE = CH/IE = CH/ID = CJ/JD$. (5) $BG/CG = CD/JD$. (6) $(BG)^2/(CG)(BG) = (CD)(BD)/(JD)(BD) = (CD)(BD)/(ID)^2$. (7) $\Delta = (BG)(ID) = \{(BG)(CG)(BD)(CD)\}^{1/2} = \{s(s - a)(s - b)(s - c)\}^{1/2}$.

(e) Derive the formula of part (d) by the following process: Let h be the altitude on side c and let m be the projection of side b on side c. (1) Show that $m = (b^2 + c^2 - a^2)/2c$. (2) Substitute this value for m in $h = (b^2 - m^2)^{1/2}$. (3) Substitute this value for h in $\Delta = (ch)/2$.

(f) Approximate successively, by Heron's method, $\sqrt{3}$ and $\sqrt{720}$.

(g) A *prismatoid* is a polyhedron all of whose vertices lie in two parallel planes. The two faces in these parallel planes are called the *bases* of the prismatoid, the perpendicular distance between the two planes is called the *altitude* of the prismatoid, and the section parallel to the bases and midway

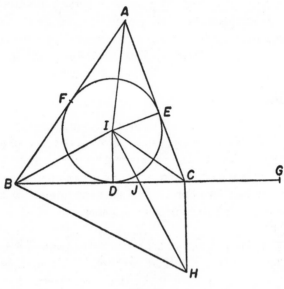

Fig. 44

between them is called the *mid-section* of the prismatoid. Let us denote the volume of the prismatoid by V, the areas of the upper base, lower base, and mid-section by U, L, M, and the altitude by h, as indicated in Figure 45. In books on solid geometry it is shown that

$$V = h(U + L + 4M)/6.$$

In Book II of the *Metrica*, Heron gives, as the volume of a prismatoid having similarly oriented rectangular bases with corresponding pairs of dimensions a, b and c, d,

$$V = h[(a + c)(b + d)/4 + (a - c)(b - d)/12].$$

Show that this result is equivalent to that given by the prismatoid formula above.

6.12   Simultaneous Equations

(a) Thymaridas, a lesser mathematician of the fourth century B.C., gave the following rule for solving a certain set of n simultaneous linear equations connecting n unknowns. The rule became so well known that it

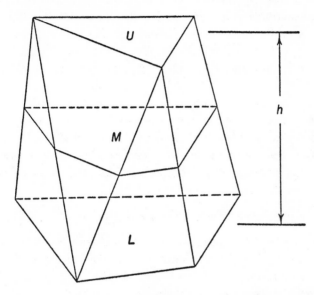

Fig. 45

went by the title of the "bloom" of Thymaridas. *If the sum of* n *quantities be given, and also the sum of every pair which contains a particular one of them, then this particular quantity is equal to* $1/(n - 2)$ *of the difference between the sums of these pairs and the first given sum.* Prove this rule.

(b) In some problems given in the Heronian collections appear the formulas

$$a, b = [(r + s) \pm \{(r + s)^2 - 8rs\}^{1/2}]/2,$$

for the legs a and b of a right triangle of perimeter $2s$ and inradius r. Obtain these formulas.

6.13   Problems from the *Greek Anthology*

(a) How many apples are needed if four persons of six receive one third, one eighth, one fourth, and one fifth, respectively, of the total number, while the fifth receives ten apples, and one apple remains left for the sixth person?

(b) Demochares has lived a fourth of his life as a boy, a fifth as a youth, a third as a man, and has spent 13 years in his dotage. How old is he?

(c) After staining the holy chaplet of fair-eyed Justice that I might see thee, all-subduing gold, grow so much, I have nothing; for I gave 40 talents under evil auspices to my friends in vain, while O ye varied mischances of men, I see my enemies in possession of the half, the third, and the eighth of my fortune. (How many "talents" did the unfortunate man once possess?)

(d) The Graces were carrying baskets of apples, and in each was the same number. The nine Muses met them and asked each for apples and they gave the same number to each Muse and the nine and the three each had the same number. Tell me how many they gave and how they all had the same number. (This problem is indeterminate. Find the smallest permissible solution.)

6.14   Type Problems from the *Greek Anthology*

There are certain standard types of problems found in present-day elementary algebra texts which date back to ancient times. Consider, for example, the following "work" problem, "cistern" problem, and "mixture" problem found in the *Greek Anthology*.

(a) Brickmaker, I am in a hurry to erect this house. Today is cloudless, and I do not require many more bricks, for I have all I want but three hundred. Thou alone in one day couldst make as many, but thy son left off working when he had finished two hundred, and thy son-in-law when he had made two hundred and fifty. Working all together, in how many days can you make these?

(b) I am a brazen lion; my spouts are my two eyes, my mouth, and the flat of my right foot. My right eye fills a jar in two days [1 day = 12 hours], my left eye in three, and my foot in four. My mouth is capable of filling it in six hours. Tell me how long all four together will take to fill it.

(c) Make a crown of gold, copper, tin, and iron weighing 60 minae: gold and copper shall be two thirds of it; gold and tin three fourths of it; and gold and iron three fifths of it: find the weights of gold, copper, tin, and iron required. (This is a numerical illustration of the "bloom" of Thymaridas. See Problem Study 6.12 (a).)

6.15 Diophantus

(a) About all we know of Diophantus' personal life is that contained in the following summary of an epitaph given in the *Greek Anthology:* "Diophantus passed one sixth of his life in childhood, one twelfth in youth, and one seventh more as a bachelor. Five years after his marriage was born a son who died four years before his father, at half his father's [final] age." How old was Diophantus when he died?

(b) Solve the following problem, which appears in Diophantus' *Arithmetica* (Problem 17, Book I): Find four numbers, the sum of every arrangement three at a time being given; say, 22, 24, 27, and 20.

(c) Solve the following problem, also found in the *Arithmetica* (Problem 18, Book VI): In the right triangle ABC, right angled at C, AD bisects angle A. Find the set of smallest integers for AB, AD, AC, BD, DC such that $DC : CA : AD = 3 : 4 : 5$.

(d) Augustus De Morgan, who lived in the nineteenth century, proposed the conundrum: "I was x years old in the year x^2." When was he born?

6.16 Some Number Theory in the *Arithmetica*

(a) Establish the identities

$$(a^2 + b^2)(c^2 + d^2) = (ac \pm bd)^2 + (ad \mp bc)^2$$

and use them to express $481 = (13)(37)$ as the sum of two squares in two different ways.

These identities were given later, in 1202, by Fibonacci in his *Liber abaci*. They show that the product of two numbers each expressible as the sum of two squares is also expressible as the sum of two squares. It can be shown that these identities include the addition formulas for the sine and cosine. The identities later became the germ of the Gaussian theory of arithmetical quadratic forms and of certain developments in modern algebra.

(b) Express $1105 = (5)(13)(17)$ as the sum of two squares in four different ways.

In the following two problems "number" means "positive rational number."

(c) If m and n are numbers differing by 1, and if x, y, a are numbers such that $x + a = m^2$, $y + a = n^2$, show that $xy + a$ is a square number.

(d) If m is any number and $x = m^2$, $y = (m + 1)^2$, $z = 2(x + y + 1)$, show that the six numbers $xy + x + y$, $yz + y + z$, $zx + z + x$, $xy + z$, $yz + x$, $zx + y$ are all square numbers.

6.17   Problems from Pappus

(a) In Book III of Pappus' *Mathematical Collection* we find the following interesting geometrical representation of some means. Take B on segment AC, B not being the mid-point O of AC. Erect the perpendicular to AC at B to cut the semicircle on AC in D, and let F be the foot of the perpendicular from B on OD. Show that OD, BD, FD represent the arithmetic mean, the geometric mean, and the harmonic mean of the segments AB and BC, and show that, if $AB \neq BC$,

$$\text{arith. mean} > \text{geom. mean} > \text{harm. mean.}$$

(b) In Book III of *Mathematical Collection*, Pappus gives the following neat construction for the harmonic mean of the two given segments OA and OB in Figure 46. On the perpendicular to OB at B mark off $BD = BE$, and let the perpendicular to OB at A cut OD in F. Draw FE to cut OB in C. Then OC is the sought harmonic mean. Prove this.

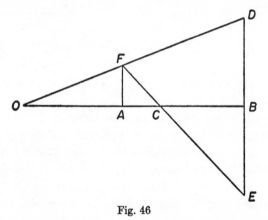

Fig. 46

(c) Prove the following extension of the Pythagorean theorem given by Pappus in Book IV of *Mathematical Collection*. Let ABC (see Figure 47) be any *triangle and* $ABDE$, $ACFG$ *any parallelograms described externally on* AB *and* AC. *Let* DE *and* FG *meet in* H *and draw* BL *and* CM *equal and parallel to* HA. *Then*

$$\square BCML = \square ABDE + \square ACFG.$$

(d) In Book VIII of *Mathematical Collection* Pappus establishes the following theorem: *If D, E, F are points on the sides BC, CA, AB of triangle ABC such that BD/DC = CE/EA = AF/FB, then triangles DEF and ABC have a common centroid.* Prove this either synthetically or analytically.

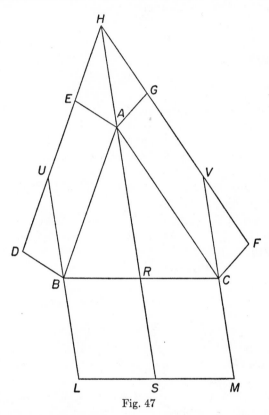

Fig. 47

6.18   The Centroid Theorems

In Book VII of *Mathematical Collection,* Pappus anticipated one of the centroid theorems sometimes credited to P. Guldin (1577–1642). These theorems may be stated as follows: (1) *If a plane arc be revolved about an axis in its plane, but not cutting the arc, the area of the surface of revolution so formed is equal to the product of the length of the arc and the length of the path traced by the centroid of the arc.* (2) *If a plane area be revolved about an axis in its plane, but not intersecting the area, the volume of the solid of revolution so formed is equal to the product of the area and the length of the path traced by the centroid of the area.* Using these theorems find

(a) The volume and surface area of the torus formed by revolving a circle of radius r about an axis, in the plane of the circle, at distance $R > r$ from the center of the circle.

(b) The centroid of a semicircular arc.

(c) The centroid of a semicircular area.

Bibliography

APOLLONIUS OF PERGA, *Conics*, 3 vols., tr. by R. Catesby Taliaferro. Classics of the St. John's program. Annapolis, Md.: R. C. Taliaferro, 1939.

DIJKSTERHUIS, E. J., *Archimedes*. New York: Humanities Press, Inc., 1957.

GOW, JAMES, *A Short History of Greek Mathematics*. New York: Hafner Publishing Company, 1923.

HEATH, T. L., *Apollonius of Perga, Treatise on Conic Sections*. New York: Barnes and Noble, Inc., 1961.

————, *Aristarchus of Samos*. New York: Oxford University Press, 1913.

————, *Diophantus of Alexandria*. New York: Cambridge University Press, 1910.

————, *History of Greek Mathematics*, Vol. 2. New York: Oxford University Press, 1921.

————, *A Manual of Greek Mathematics*. New York: Oxford University Press, 1931.

————, *The Works of Archimedes*. New York: Cambridge University Press, 1897. Reprinted by Dover Publications, Inc.

ORE, OYSTEIN, *Number Theory and Its History*. New York: McGraw-Hill Book Company, Inc., 1948.

SARTON, GEORGE, *Ancient Science and Modern Civilization*. Lincoln, Neb.: The University of Nebraska Press, 1954.

VAN DER WAERDEN, B. L., *Science Awakening*, tr. by Arnold Dresden. Groningen, Netherlands: P. Noordhoff, N.V., 1954.

7

Hindu and Arabian
Mathematics*

INDIA

7-1   General Survey

Because of the lack of authentic records very little is known of the development of ancient Hindu mathematics. The earliest history is preserved in the 5000-year-old ruins of a city at Mohenjo Daro. Evidence of wide streets, brick dwellings and apartment houses with tiled bathrooms, covered city drains, and community swimming pools, indicates a civilization as advanced as that found anywhere else in the ancient orient. These early peoples had systems of writing, counting, weighing, and measuring, and they dug canals for irrigation. All this required considerable basic mathematics and engineering. It is not known what became of these peoples.

It was about 4000 years ago that wandering bands crossed the Himalaya passes into India from the great plains of central Asia. These people were called *Aryans*, from a Sanskrit word meaning "noblemen" or "owners of land." Many of these remained and others wandered into Europe and formed the root of the Indo-European stock. The influence of the Aryans gradually extended over all India. During their first 1000 years they perfected both written and spoken Sanskrit. To them is also due the introduction of the caste system. In the sixth century B.C., the Persian armies under Darius entered India but made no permanent conquests. To this period belong two great early Indians, the grammarian Panini and the religious teacher Buddha. This probably is also the approximate time of the *Śulvasūtras*

---

*For the pronunciation of Hindu and Arabian names, see Introduction, p. 2.

("the rules of the cord"), some religious writings of interest in the history of mathematics because they embody geometrical rules for the construction of altars by rope stretching and show an acquaintance with Pythagorean numbers.

After the temporary conquest of northwest India by Alexander the Great in 326 B.C., the Maurya Empire was established and in time spread over all India and parts of central Asia. The most famous Maurya ruler was King Ásoka (272–232 B.C.), some of whose great stone pillars, erected in every important city in India of his day, still stand. These pillars are of interest to us because, as stated in Section 1-9, some of them contain the earliest preserved specimens of our present number symbols.

After Ásoka, India underwent a series of invasions, which were finally followed by the Gupta dynasty under the rule of native Indian emperors. The Gupta period proved to be the golden age of the Sanskrit renaissance and India became a center of learning, art, and medicine. Rich cities grew up and universities were founded. The first important astronomical work, the anonymous *Sūrya Siddhānta* ("knowledge from the sun"), dates from this period, probably about the beginning of the fifth century. Hindu mathematics from here on becomes subservient to astronomy rather than religion. In the sixth century work, *Pañca Siddhāntikā*, of the astronomer Varāhamihira and based on the earlier *Sūrya Siddhānta*, is found a good summary of early Hindu trigonometry and a table of sines apparently derived from Ptolemy's table of chords.

The degree of influence of Greek, Babylonian, and Chinese mathematics on Hindu mathematics, and vice versa, is still an unsettled matter, but there is ample evidence that it was quite appreciable. One of the pronounced benefits of the Pax Romana was the diffusion of knowledge between East and West, and from a very early date India exchanged diplomats with both the West and the Far East.

From about 450 A.D. until near the end of the 1400's, India was again subjected to numerous foreign invasions. First came the Huns, then the Arabs in the eighth century, and the Persians in the eleventh. During this period there were several Hindu mathematicians of prominence. Among these were the two Āryabhatas, Brahmagupta, Mahāvīra, and Bhāskara. The elder Āryabhata flourished in the sixth century and was born near present-day Patna on the Ganges. He wrote a work on astronomy of which the third chapter is devoted to mathematics. There is some confusion between the two Āryabhatas and it may be that their work is not correctly differentiated. Brahmagupta was the most prominent Hindu mathematician of the seventh century. He lived and worked in the astronomical center of Ujjain, in central India. In 628, he wrote his *Brahma-sphuta-sidd'hānta* ("the revised system of Brahma"), an astronomical work of 21 chapters of which Chapters 12 and 18 deal with mathematics. Mahāvīra, who flourished about 850, was from southern India, and wrote on elementary mathematics. Bhāskara's work, *Siddhānta Śiromani* ("diadem of an astronomical system"), was written in 1150 and shows little advancement over the work of Brahmagupta of more than 500 years earlier. The important mathematical parts of Bhāskara's work are the *Lilāvati* ("the beautiful") and *Vījaganita* ("seed arithmetic")*, which respectively deal with arithmetic and algebra. The mathematical parts of Brahmagupta's and Bhāskara's works were translated into English in 1817 by H. T. Colebrooke. The *Sūrya Siddhānta* was translated by E. Burgess in 1860, and Mahāvīra's work was published in 1912 by M. Rangācārya.

Hindu mathematics after Bhāskara actually retrograded until modern times. In 1907, the Indian Mathematical Society was founded, and two years later the *Journal of the Indian Mathematical Society* started in Madras. The Indian statistics journal, *Sankhyā*, began publication in 1933.

Texts on the history of mathematics show some contradictions and confusion when dealing with the Hindus. This is probably due, in no small measure, to the obscure and at times nearly unintelligible writing of the Hindu authors. The history of Hindu mathematics still awaits a more reliable and scholarly treatment.

7-2   Number Computing

In Section 1-9 we considered briefly the little that is known concerning the part played by the Hindus in the development of our present positional numeral system. We shall now give some account of Hindu methods of computing with this system. The key to an understanding of the algorisms that were elaborated lies in a realization of the writing materials which were at the disposal of the calculators. According to the German historian H. Hankel, they generally wrote either upon a small blackboard with a cane pen dipped in a thin white paint which could easily be rubbed

---

*It is not certain that the *Lilāvati* and *Vījaganita* are parts of the *Siddhānta Śiromani;* they may be separate works.

off, or, with a stick, upon a white tablet less than a foot square and coated with a sprinkling of red flour. In either case the writing space was small and legibility demanded fairly large figures, but erasures and corrections were very easily effected. The calculation processes accordingly were schemed to conserve the writing space by erasing a digit as soon as it had served its purpose.

Early Hindu addition was perhaps done from left to right, instead of from right to left as we prefer to do it today. As an example consider the addition of 345 and 488. These would probably be written, one under the other, a little below the top of the computing tablet, as shown in the accompanying illustration. The computer would say 3 + 4 = 7, and write the 7 at the head of the left column. Next, 4 + 8 = 12, which changes the 7 to an 8 followed by a 2. The 7 is accordingly rubbed off and 82 written down. In our illustration we have, instead, crossed out the 7 and written the 8 above it.

8 3

7 2 3

3 4 5

4 8 8

Then 5 + 8 = 13, which changes the 2 to a 3 followed by another 3. Again things are corrected with a quick rub of the finger, and the final answer 833 appears at the top of the tablet. Now the 345 and 488 can be rubbed off and we have the rest of the tablet clear for further work.

In an undated commentary of Bhāskara's *Lilāvati* we find another method, by which the addition of 345 and 488 would be effected by the following process:

$$\text{sum of units} \quad 5 + 8 = \quad 13$$
$$\text{sum of tens} \quad 4 + 8 = 12\cdot$$
$$\text{sum of hundreds } 3 + 4 = 7\cdot\cdot$$
$$\text{sum of sums} \quad\quad = 833$$

Several methods were used for multiplication. The written work for the simple multiplication of, say, 569 by 5, might appear as follows, again working from left to right. On the tablet, a little below the top, write down 569 followed, on the same line, by the multiplier 5. Then, since 5 × 5 = 25, 25 is written above the 569 as shown in the accompanying illustration. Next, 5 × 6 = 30, which changes the 5 in 25 to an 8 followed by a 0. A quick erasure fixes this. Again, in the illustration we have, instead, crossed out the 5 and written the 8 above it. Then 5 × 9 = 45, which changes the 0 to a 4 followed by a 5. The final product, 2845, now appears at the top of the computing tablet.

8 4

2 5 0 5

5 6 9   5

A more complicated multiplication, like 135 × 12 say, might be accomplished by first finding, as above, 135 × 4 = 540, then 540 × 3 = 1620, or by adding 135 × 10 = 1350 and 135 × 2 = 270 to get 1620. Or again, according to Hankel, it might be accomplished as follows. A little below the top of the tablet write the multiplicand 135 and the multiplier 12 so that the units digit in the multiplicand falls beneath the extreme left digit in the multiplier. Now 135 × 1 = 135, which is written at the top of the tablet. Next, by erasing, shift the multiplicand 135 one place to the right and multiply by the 2 of the 12. In doing this we find 2 × 1 = 2, which changes the 3 in our partial product to a 5. Then 2 × 3 = 6, which changes the two 5's in our new partial product to 61. Finally, 2 × 5 = 10, which changes the final 1 in our partial product to 2 followed by a 0. The finished product, 1620, now appears at the top of the tablet.

$$\begin{array}{cccc} & & 6 & 2 \\ & & \cancel{5} & \cancel{1} \\ 1 & \cancel{3} & \cancel{5} & 0 \\ & & 1 & 2 \\ \cancel{1} & \cancel{3} & \cancel{5} & \\ 1 & 3 & 5 & \end{array}$$

Another method of multiplication, known to the Arabians and probably obtained from the Hindus, which closely resembles our present process is indicated in the accompanying illustration, where we again find the prod-

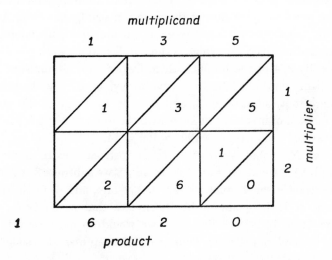

multiplicand

product

uct of 135 by 12. The lattice diagram is actually drawn and the additions performed diagonally. Note, because of the way each cell is divided in two by a diagonal, no carrying over is required in the multiplication.

The Arabians, who later borrowed some of the Hindu processes, were unable to improve on them and accordingly adapted them to "paper" work, where erasures were not easily effected, by crossing off undesired digits and writing the new ones above or below the old ones, as we have done in the illustrations above.

The development of algorisms for our elementary arithmetic operations started in India, perhaps about the tenth or eleventh century, were adopted by the Arabians, and later carried to western Europe, where they were modified into their present forms. This work received considerable attention from the fifteenth-century European writers on arithmetic.

7-3   Arithmetic and Algebra

The Hindus were gifted arithmeticians and made significant contributions to algebra.

Many of the arithmetical problems were solved by *false position*. Another favorite method of solution was that of *inversion*, where one works backward from a given piece of information. Consider, for example, the following problem given during the sixth century by the elder Āryabhata: "Beautiful maiden with beaming eyes, tell me, as thou understandst the right method of inversion, which is the number which multiplied by 3, then increased by 3/4 of the product, then divided by 7, diminished by 1/3 of the quotient, multiplied by itself, diminished by 52, by the extraction of the square root, addition of 8, and division by 10 gives the number 2?" By the method of inversion we start with the number 2 and work backward. Thus $[(2)(10) - 8]^2 + 52 = 196$, $\sqrt{196} = 14$, $(14)(3/2)(7)(4/7)/3 = 28$, the answer. Note that where the problem instructed us to divide by 10 we multiply by 10, where we were told to add 8 we subtract 8, where we were told to extract a square root we take the square, and so forth. It is the replacement of each operation by its inverse that accounts for the name *inversion*. It is, of course, just what we would do if we were to solve the problem by modern methods. Thus, if we let x represent the sought number, we have

$$\{\sqrt{[(2/3)(7/4)(3x)/7]^2 - 52} + 8\}/10 = 2.$$

To solve this we *multiply* both sides by 10, then *subtract* 8 from each side, then *square* both sides, and so forth. This problem also illustrates the Hindu practice of clothing arithmetical problems in poetic garb. This was because school texts were written in verse and because the problems were frequently used for social amusement.

The Hindus summed arithmetic and geometric progressions and solved commercial problems in simple and compound interest, discount, and partnership. They also solved "mixture" and "cistern" problems, similar to those found in modern texts. Several specimens of Hindu arithmetical problems may be found in Problem Studies 7.1, 7.2, and 7.3.

Much of our knowledge of Hindu arithmetic stems from Bhāskara's *Lilāvati*. A romantic story is told about this work. According to the tale, the stars foretold dire misfortune if Bhāskara's daughter Lilāvati should marry other than at a certain hour on a certain propitious day. On that day, as the bride was watching the sinking water level of the hour cup, a pearl fell from her headdress and, stopping the hole in the cup, arrested the outflow of water, and so the lucky moment passed unnoticed. To console the unhappy girl, Bhāskara gave her name to his book.

The Hindus syncopated their algebra. Like Diophantus, addition was usually indicated by juxtaposition. Subtraction was indicated by placing a dot over the subtrahend, multiplication by writing *bha* (the first syllable of the word *bhavita*, "the product") after the factors, division by writing the divisor beneath the dividend, square root by writing *ka* (from the word *karana*, "irrational") before the quantity. Brahmagupta indicated the unknown by *yā* (from *yāvattāvat*, "so much as"). Known integers were prefixed by *rū* (from *rūpa*, "the absolute number"). Additional unknowns were indicated by the initial syllables of words for different colors. Thus a second unknown might be denoted by *kā* (from *kālaka*, "black"), and $8xy + \sqrt{10} - 7$ might appear as

$$yā\ kā\ 8\ bha\ ka\ 10\ rū\ \dot{7}.$$

The Hindus admitted negative and irrational numbers, and recognized that a quadratic (having real answers) has two formal roots. They unified the algebraic solution of quadratic equations by the familiar method of completing the square. This method is today often referred to as *the Hindu method*. Bhāskara gave the two remarkable identities

$$\sqrt{a \pm \sqrt{b}} = \sqrt{(a + \sqrt{a^2 - b})/2} \pm \sqrt{(a - \sqrt{a^2 - b})/2},$$

which are sometimes employed in our algebra texts for finding the square root of a binomial surd. These identities are also found in Book X of Euclid's

Elements, but are there given in an involved language which is difficult to comprehend.

The Hindus showed remarkable ability in indeterminate analysis and were perhaps the first to devise general methods in this branch of mathematics. Unlike Diophantus, who sought *any one rational* solution to an indeterminate equation, the Hindus endeavored to find *all possible integral* solutions. Āryabhata and Brahmagupta found the integral solutions of the linear indeterminate equation $ax + by = c$, where a, b, c are integers. The indeterminate quadratic equation $xy = ax + by + c$ was solved by a method later reinvented by Euler. The work of Brahmagupta and Bhāskara on the so-called Pell equation, $y^2 = ax^2 + 1$, where a is a nonsquare integer, is highly regarded by some. They showed how, from one solution x, y, $xy \neq 0$, infinitely many others could be found. The complete theory of the Pell equation was finally worked out by Lagrange in 1766–1769. The Hindu work on indeterminate equations reached western Europe too late to exert any beneficial influence.

7-4　Geometry and Trigonometry

The Hindus were not proficient in geometry. Rigid demonstrations were unusual and postulational developments were nonexistent. Their geometry was largely empirical and generally connected with mensuration.

The ancient *Śulvasūtras* show that the early Hindus applied geometry to the construction of altars and in doing so made use of the Pythagorean relation. The rules furnish instructions for finding a square equal to the sum or difference of two given squares and of a square equal to a given rectangle. There also appear solutions of the circle-squaring problem which are equivalent to taking $d = (2 + \sqrt{2})s/3$ and $s = 13d/15$, where d is the diameter of the circle and s the side of the equal square. There also appears the expression

$$\sqrt{2} = 1 + 1/3 + 1/(3)(4) - 1/(3)(4)(34),$$

which is interesting in that all the fractions are unit fractions and that the expression is correct to five decimal places.

Both Brahmagupta and Mahāvīra not only gave Heron's formula for the area of a triangle in terms of the three sides but also the remarkable extension,*

$$K = [(s - a)(s - b)(s - c)(s - d)]^{1/2},$$

for the area of a cyclic quadrilateral having sides a, b, c, d and semiperimeter s. It seems that later commentators failed to realize the limitation on the quadrilateral. The formula for the general case is

$$K^2 = (s - a)(s - b)(s - c)(s - d) - abcd \cos^2[(A + C)/2],$$

where A and C are a pair of opposite vertex angles of the quadrilateral.

Most remarkable in Hindu geometry, and solitary in its excellence, are Brahmagupta's theorems that the diagonals m and n of a cyclic quadrilateral having consecutive sides a, b, c, d are given by

$$m^2 = (ab + cd)(ac + bd)/(ad + bc),$$

$$n^2 = (ac + bd)(ad + bc)/(ab + cd),$$

and that if a, b, c, A, B, C are positive integers such that $a^2 + b^2 = c^2$ and $A^2 + B^2 = C^2$, then the cyclic quadrilateral having consecutive sides aC, cB, bC, cA (called a *Brahmagupta trapezium*) has rational area and diagonals, and the diagonals are perpendicular to each other (see Problem Studies 7.6 and 7.7). Brahmagupta knew Ptolemy's theorem on the cyclic quadrilateral.

Many inaccuracies appear in Hindu mensuration formulas. Thus Āryabhata gives the volume of a pyramid as *half* the product of the base and altitude, and the volume of a sphere as $\pi^{3/2}r^3$. The Hindus gave some accurate values for π but also frequently used $\pi = 3$ and $\pi = \sqrt{10}$.

Most students of high school geometry have seen Bhāskara's dissection proof of the Pythagorean theorem, in which the square on the hypotenuse is cut up, as indicated in Figure 48, into four triangles each congruent to the given triangle plus a square with side equal to the difference of the legs of the given triangle. The pieces are easily rearranged to give the sum of the squares on the two legs. Bhāskara drew the figure and offered no further explanation than the word "Behold!" A little algebra, however, supplies a proof. For, if c is the hypotenuse and a and b are the legs of the triangle,

$$c^2 = 4(ab/2) + (b - a)^2 = a^2 + b^2.$$

*For a derivation of this formula see, for example, E. W. Hobson, *A Treatise on Plane Trigonometry*, 4th ed. (New York: The Macmillan Company, 1902), p. 204, or R. A. Johnson, *Modern Geometry* (Boston: Houghton Mifflin Company, 1929), p. 81. Each of these works has been reprinted by Dover Publications, Inc.

This dissection proof is found much earlier in China. Bhāskara also gave a second demonstration of the Pythagorean theorem by drawing the altitude on the hypotenuse. From similar right triangles in Figure 49 we have

$$c/b = b/m, \quad c/a = a/n,$$

or

$$cm = b^2, \qquad cn = a^2.$$

Adding we get

$$a^2 + b^2 = c(m + n) = c^2.$$

This proof was rediscovered by John Wallis in the seventeenth century.

The Hindus, as the Greeks, regarded trigonometry as a tool for their astronomy. They used our familiar degree, minute, and second divisions

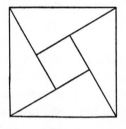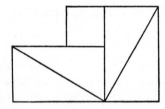

Fig. 48

and constructed tables of *sines* (that is, tables of half chords, and not chords, as did the Greeks). The Hindus employed the equivalents of sines, cosines, and versed sines (versin $A = 1 - \cos A$). They computed the sines of halves

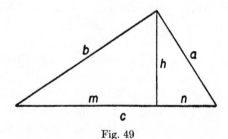

Fig. 49

of angles by the relation versin $2A = 2 \sin^2 A$. In their astronomy they solved plane and spherical triangles. The astronomy itself is of poor quality and shows an inaptness in observing, collecting and collating facts, and inducing laws. The trigonometry may be described as arithmetical rather than geometrical.

7-5   Contrast between Greek and Hindu Mathematics

There are many differences between Greek and Hindu mathematics. In the first place, the Hindus who worked in mathematics regarded themselves primarily as astronomers, and thus Hindu mathematics remained largely a handmaiden to astronomy; with the Greeks, mathematics attained an independent existence and was studied for its own sake. Also, due to the caste system, mathematics in India was cultivated almost entirely by the priests; in Greece, mathematics was open to any one who cared to study the subject. Again, the Hindus were accomplished computers but mediocre geometers; the Greeks excelled in geometry but cared little for computational work. Even Hindu trigonometry, which was meritorious, was arithmetical in nature; Greek trigonometry was geometrical in character. The Hindus wrote in verse and often clothed their works in obscure and mystic language; the Greeks strove for clarity and logic in presentation. Hindu mathematics is largely empirical with proofs or derivations seldom offered; an outstanding characteristic of Greek mathematics is its insistence on rigorous demonstration. Hindu mathematics is of very uneven quality, good and poor mathematics often appearing side by side; the Greeks seemed to have an instinct which led them to distinguish good from poor quality and to preserve the former while abandoning the latter.

Some of the contrast between Greek and Hindu mathematics is perpetuated today in the differences between our elementary geometry and algebra textbooks.

ARABIA

7-6   The Rise of Moslem Culture

The rise and decline of the Arabian empire is one of the most spectacular episodes in history. Within the decade following Mohammed's flight from Mecca to Medina in 622 A.D., the scattered and disunited tribes of the Arabian peninsula were consolidated by a strong religious fervor into a powerful nation. Within a century, force of arms had extended the Moslem rule and influence over a territory reaching from India, through Persia,

Mesopotamia, and northern Africa, clear into Spain. Opposing contenders for the caliphate caused an east-west split in the empire in 755, resulting in one caliph reigning in Bagdad and another in Córdoba. Until about the year 1000, the eastern empire enjoyed spiritual supremacy. At that time, however, much of the eastern territory became overrun by the ruthless Seljuk Turks. Between 1100 and 1300, the Christian Crusades were launched to dislodge the Moslems from the Holy Land. In 1258, Bagdad was taken by the Mongols, the eastern caliph fell from power, and the Arabian empire began to decline. In the 1400's, Spain overthrew the last of its Moorish rulers, and the Arabs lost their European foothold.

Of considerable importance for the preservation of much of world culture was the manner in which the Arabs seized upon Greek and Hindu erudition. The Bagdad caliphs not only governed well but many became patrons of learning and invited distinguished scholars to their courts. Numerous Hindu and Greek works in astronomy, medicine, and mathematics were industriously translated into the Arabic tongue and thus were saved until later European scholars were able to retranslate them into Latin and other languages. But for the work of the Arabian scholars much of Greek and Hindu science would have been irretrievably lost over the long period of the Dark Ages.

During the reign of the caliph al-Mansûr, Brahmagupta's works were brought to Bagdad (ca. 766) and, under royal patronage, translated into Arabic. It has been said that this was the means by which the Hindu numerals were brought into Arabic mathematics. The next caliph was Harun al-Rashid (Aaron the Just), who reigned from 786 to 808 and is known to us in connection with *The Arabian Nights*. Under his patronage, several Greek classics in science were translated into Arabic, among them part of Euclid's *Elements*. There was also a further influx of Hindu learning into Bagdad during his reign. Harun al-Rashid's son, al-Mâmûn, who reigned from 809 to 833, also was a patron of learning and was himself an astronomer. He built an observatory at Bagdad and undertook the measurement of the earth's meridian. The difficult task of obtaining satisfactory translations of Greek classics continued under his orders; the *Almagest* was put into Arabic and the translation of the *Elements* completed. Greek manuscripts were secured, as a condition in a peace treaty, from the emperor of the Byzantine Empire and were then translated by Syrian Christian scholars invited to al-Mâmûn's court. Many scholars wrote on mathematics and astronomy during this reign, the most famous being Mohammed ibn Mûsâ al-Khowârizmî (Mohammed, the son of Moses of Khwarezm). He wrote a treatise on algebra and a book on the Hindu numerals, both of which later exerted tremendous influence in Europe when translated into Latin in the twelfth century. A somewhat later scholar was Ṭâbit ibn Qorra (826–901), famed as a physician, philosopher, linguist, and mathematician. He produced the first really satisfying Arabic translation of the *Elements*. His translations of Apollonius, Archimedes, Ptolemy, and Theodosius are said to rank among the best made. He also wrote on astronomy, the conics, elementary algebra, magic squares, and amicable numbers (see Problem Study 7.8).

Probably the most celebrated Moslem mathematician of the tenth century was Abû'l-Wefâ (940–998), born in the Persian mountain region of Khorâsân. He is known for his translation of Diophantus, his introduction of the *tangent* function into trigonometry, and his computation of a table of sines and tangents for 15' intervals. He wrote on a number of mathematical topics. Abû Kâmil and al-Karkhî, who wrote in the tenth and eleventh centuries, should be mentioned for their work in algebra. The former drew upon al-Khowârizmî, and was in turn drawn upon by the European mathematician Fibonacci (1202). Al-Karkhî, who was a disciple of Diophantus, produced a work called the *Fakhrî*, one of the most scholarly of the Moslem works on algebra. But perhaps the deepest and most original algebraic contribution was the geometrical solution of cubic equations by Omar Khayyam (*ca.* 1100), another native of Khorâsân, and known to the western world as the author of the exquisite *Rubaiyat*. Khayyam is also noted for his very accurate proposed calendar reform.

A considerably later writer was Nasîr ed-dîn (*ca.* 1250), also of Khorâsân. He wrote the first work on plane and spherical trigonometry considered independently of astronomy. Saccheri started his work on non-Euclidean geometry through a knowledge of Nasîr ed-dîn's writings on Euclid's parallel postulate. These writings were translated into Latin by John Wallis in the seventeenth century and used by him in his geometrical lectures at Oxford. Finally, there was Ulugh Beg, a fifteenth-century Persian astronomer of royal blood, who compiled remarkable tables of sines and tangents for 1' intervals correct to eight or more decimal places.

7-7   Arithmetic and Algebra

Before Mohammed, the Arabians wrote out all numbers in words. The subsequent extensive administration of conquered lands was partly responsible for the introduction of a short symbolism. Sometimes local numeral systems were adopted and at one time it was rather common practice to use a ciphered numeral system, like the Ionic Greek, employing the 28 Arabic letters. This notation was, in turn, superseded by the Hindu notation, which was first adopted by merchants and writers on arithmetic. Strangely enough, the Hindu numerals are excluded from some of the later arithmetics of the eastern empire. Thus Abû'l-Wefâ and al-Karkhî, of the tenth and eleventh centuries, wrote arithmetics in which all numbers are again written out in words. These later Arabian writers departed from Hindu teachings and became influenced by Greek methods. No trace of the use of an abacus has been discovered among the early Arabs.

The first Arabic arithmetic known to us is that of al-Khowârizmî; it was followed by a host of other Arabic arithmetics by later authors. These arithmetics generally explained the rules for computing, modeled after the Hindu algorisms. They also gave the process known as *casting out 9's*, used for checking arithmetical computations, and the rules of *false position* and *double false position*, by which certain algebra problems can be solved non-algebraically (see Problem Studies 7.9 and 7.10). Square and cube roots, fractions, and the *rule of three* were also frequently explained.

The *rule of three*, like much else in elementary arithmetic, seems to have originated with the Hindus, and was actually called by this name by Brahmagupta and Bhāskara. For centuries, the rule was very highly regarded by merchants. It was mechanically stated without reason, and its connection with proportion was not recognized until the end of the fourteenth century. Here is how Bramagupta stated the rule: *In the rule of three, Argument, Fruit, and Requisition are the names of the terms. The first and last terms must be similar. Requisition multiplied by Fruit, and divided by Argument, is the Produce.* For clarification consider the following problem given by Bhāskara: If two and a half palas of saffron are purchased for three sevenths of a niska, how many palas will be purchased for nine niskas? Here 3/7 and 9, which are of the same denomination, are the Argument and the Requisition, and 5/2 is the Fruit. The answer, or Produce, is then given by $(9)(5/2)/(3/7) = 52\frac{1}{2}$. Today we would regard the problem as a simple application of proportion,

$$x : 9 = 5/2 : 3/7.$$

Much space was devoted to the *rule of three* by the early European writers on arithmetic, the mechanical nature of the rule being observable in the doggerel verse and the schematic diagrams often used to explain it.

Al-Khowârizmî's algebra shows little originality. The four elementary operations are explained and linear and quadratic equations are solved, the latter both arithmetically and geometrically. The work contains some geometrical mensuration and some problems on inheritance.

The Moslem mathematicians made their best contributions in the field of geometrical algebra, the peak being reached in Omar Khayyam's geometrical solution of cubic equations. Here cubic equations are systematically classified and a root obtained as the abscissa of a point of intersection of a circle and a rectangular hyperbola or of two rectangular hyperbolas (see Problem Study 7.11). Khayyam rejected negative roots and frequently failed to discover all the positive ones. Cubic equations arose from the consideration of such problems as the construction of a regular heptagon and the Archimedean problem of cutting a sphere into two segments having a prescribed ratio. Abû'l-Wefâ gave geometrical solutions to some special quartic equations.

Some of the Moslem mathematicians showed interest in indeterminate analysis. Thus a proof (probably defective and now lost) was given to the theorem that it is impossible to find two positive integers the sum of whose cubes is the cube of a third integer. This is a special case of Fermat's famous *last theorem*, to which we will return in Chapter X. Mention has already been made of Tâbit ibn Qorra's rule for finding amicable numbers. This is said to be the first piece of original mathematical work done by an Arabian. Al-Karkhî was the first Arabian writer to give and prove theorems furnishing the sums of the squares and cubes of the first n natural numbers.

Arabian algebra, except for that of the later western Arabs, was rhetorical.

7-8   Geometry and Trigonometry

The important role played by the Arabs in geometry was not one of discovery but one of preservation. The world owes them a large debt for their persevering efforts to translate satisfactorily the great Greek classics. There was a nice piece of geometrical research done by Abû'l-Wefâ, in which he showed how to locate the vertices of the regular polyhedra on their circumscribed spheres, using compasses of fixed opening. We have already mentioned Omar Khayyam's geometrical solution of cubic equations and Nasîr ed-dîn's influential work on the parallel postulate. This last scholar is also credited with an original proof of the Pythagorean theorem. The proof is essentially the one we have suggested, in the notes to Problem Study 6.17 (c), for Pappus' extension of the Pythagorean theorem. The name al-Haitam (Alhazen, *ca.* 965–1039) has been preserved in mathematics in connection with the so-called *problem of Alhazen:* To draw from two given points in the plane of a given circle lines which intersect on the circle and make equal angles with the circle at that point. The problem leads to a quartic equation which was solved in Greek fashion by an intersecting hyperbola and circle. Al-Haitam lived in Egypt and was perhaps the greatest of the Moslem physicists. The above problem arose in connection with his *Optics,* a treatise which later had great influence in Europe.

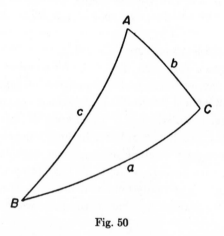

Fig. 50

Like the Hindus, the Arabian mathematicians generally regarded themselves primarily as astronomers and thus showed considerable interest in trigonometry. We have already mentioned some of the Moslem accomplishments in the construction of trigonometric tables. They may also be credited with using all six of the trigonometric functions and with improving upon the derivation of the formulas of spherical trigonometry. The law of cosines for an oblique spherical triangle,

$$\cos a = \cos b \cos c + \sin b \sin c \cos A,$$

was given by al-Battânî (Latinized as Albategnius, *ca.* 920), and the formula

$$\cos B = \cos b \sin A,$$

for a spherical triangle ABC with a right angle at C (see Figure 50), is sometimes called *Geber's theorem*, after the western Moslem astronomer Jabir ibn Aflah (frequently called Geber, *ca.* 1130) who flourished at Seville.

7-9   Some Etymology

Many names and words used today may be traced back to the Arabian period. Thus anyone interested in observational astronomy probably is aware that a large number of star names, particularly those of the fainter stars, are Arabic. As well-known examples we have Aldebaran, Vega, and Rigel among the brighter stars, and Algol, Alcor, and Mizar among the fainter ones. Many of the star names were originally expressions locating the stars in the constellations. These descriptive expressions, when transcribed from Ptolemy's catalogue into the Arabic, later degenerated into single words. Thus we have Betelgeuse (armpit of the Central One), Fomalhaut (mouth of the Fish), Deneb (tail of the Bird), Rigel (leg of the Giant), and so forth. Earlier, in Section 6-6, we traced the derivation of *Almagest*, the Arabic name by which Ptolemy's great work is commonly referred.

Very interesting is the origin of our word *algebra* from the title, *Hisâb al-jabr w'al-muqâbalah*, of al-Khowârizmî's treatise on the subject. This title has been literally translated as "science of the reunion and the opposition," or more freely as "science of reduction and cancellation." The text, which is extant, became known in Europe through Latin translations, and made the word *al-jabr*, or *algebra*, synonymous with the science of equations. Since the middle of the nineteenth century, *algebra* has come, of course, to mean a great deal more.

The Arabic word *al-jabr*, used in a nonmathematical sense, found its way into Europe through the Moors of Spain. There an *algebrista* was a bonesetter (reuniter of broken bones), and it was usual for a barber of the times to call himself an *algebrista*, for bonesetting, and bloodletting, were sidelines of the medieval barber.

Al-Khowârizmî's book on the use of the Hindu numerals also introduced a word into the vocabulary of mathematics. This book is not extant in the original, but in 1857 a Latin translation was found which begins, "Spoken has Algoritmi, ..." Here the name *al-Khowârizmî* had become *Algoritmi*, from which, in turn, was derived our present word "algorism," meaning the art of calculating in any particular way.

The meanings of the present names of the trigonometric functions, with the exception of *sine*, are clear from their geometrical interpretations when the angle is placed at the center of a circle of unit radius. Thus, in Figure 51, if the radius of the circle is one unit, the measures of tan θ and sec θ are given

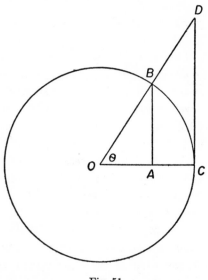

Fig. 51

by the lengths of the *tangent* segment CD and the *secant* segment OD. And, of course, *co*tangent merely means *co*mplement's tangent, and so on. The functions tangent, cotangent, secant, and cosecant have been known by various other names, these present ones appearing as late as the end of the sixteenth century.

The origin of the word *sine* is curious. Āryabhata called it *ardhā-jyā* ("half chord") and also jyā-ardhā ("chord half"), and then abbreviated the term by simply using *jyā* ("chord"). From *jyā* the Arabs phonetically derived *jība*, which, following the Arabian practice of omitting vowels, was written as *jb*. Now *jība*, aside from its technical significance, is a meaningless word in Arabic. Later writers, coming across *jb* as an abbreviation for the meaningless *jība* decided to substitute *jaib* instead, which contains the same letters and is a good Arabian word meaning "bosom." Still later, Gherardo of Cremona (*ca.* 1150), when he made his translations from the Arabic, replaced the Arabian *jaib* by its Latin equivalent, *sinus*, whence came our present word *sine*.

7-10 The Arabian Contribution

Estimates of the Arabian contribution to the development of mathematics are by no means in agreement. Some have assigned to the Moslem writers very high originality and genius, particularly in their work in algebra and trigonometry. Others see these writers as perhaps learned, but scarcely creative, and point out that their work is quite secondary both in quantity and quality to that of either the Greek or the modern writers. On the one hand it must be admitted that they made at least small advances, and on the other hand it may be that to some their achievements, when viewed against the scientifically sterile backdrop of the rest of the world of the time, seem greater than they really were. There is still, in the balance in their favor, the outstanding fact that they served admirably as custodians of much of the world's intellectual possessions, which were transmitted to the later Europeans after the Dark Ages had passed.

Problem Studies

7.1 Some Early Hindu Problems

(a) Solve the following problem generalized from one given by Brahmagupta (*ca.* 630): "Two ascetics lived at the top of a cliff of height h, whose base was distant d from a neighboring village. One descended the cliff and walked to the village. The other, being a wizard, flew up a height x and then flew in a straight line to the village. The distance traversed by each was the same. Find x." In the original problem $h = 100$ and $d = 200$.

(b) Solve the following problem, also given by Brahmagupta: "A bamboo 18 cubits high was broken by the wind. Its top touched the ground 6 cubits from the root. Tell the lengths of the segments of the bamboo." There is an older Chinese version of this problem, given about 176 B.C. by Ch'ang Ts'ang in his *K'iu-ch'ang Suan-shu* (Arithmetic in Nine Sections).

(c) An anonymous arithmetic, known as the *Bakhshālī manuscript*, was unearthed in 1881 at Bakhshālī, in northwest India. It consists of 70 pages of birch bast. Its origin and date have been the subject of much conjecture, estimates of the date ranging from the third to the twelfth century A.D. Solve the following problem found in this manuscript: "A merchant pays duty on certain goods at three different places. At the first he gives 1/3 of the goods, at the second 1/4 [of the remainder], and at the third 1/5 [of the remainder]. The total duty is 24. What was the original amount of goods?"

7.2 Problems from Mahāvīra

The nature of many of the Hindu arithmetical problems may be judged from the following, adapted from Mahāvīra (*ca.* 850). Solve these problems.

(a) A powerful unvanquished excellent black snake which is 80 angulas in length, enters into a hole at the rate of 7 1/2 angulas in 5/14 of a day, and in the course of 1/4 of a day its tail grows 11/4 of an angula. O ornament of arithmeticians, tell me by what time this serpent enters fully into the hole?

(b) Of a collection of mango fruits, the king took 1/6, the queen 1/5 of the remainder, and the three chief princes 1/4, 1/3, and 1/2 of the successive remainders, and the youngest child took the remaining three mangoes. O you who are clever in miscellaneous problems on fractions, give out the measure of that collection of mangoes.

(c) The mixed price of 9 citrons and 7 fragrant wood apples is 107; again, the mixed price of 7 citrons and 9 fragrant wood apples is 101. O you arithmetician, tell me quickly the price of a citron and of a wood apple here, having distinctly separated those prices well.

(d) One fourth of a herd of camels was seen in the forest; twice the square root of that herd had gone on to the mountain slopes; and three times five camels remained on the riverbank. What is the numerical measure of that herd of camels?

7.3 Problems from Bhāskara

Hindu arithmetical problems usually involved quadratics, the Pythagorean theorem, arithmetic progressions, and permutations. Consider the following problems adapted from Bhāskara (*ca.* 1150).

(a) The square root of half the number of bees in a swarm has flown out upon a jessamine bush, 8/9 of the swarm has remained behind; one female bee flies about a male that is buzzing within a lotus flower into which he was allured in the night by its sweet odor, but is now imprisoned in it. Tell me, most enchanting lady, the number of bees.

(b) A snake's hole is at the foot of a pillar which is 15 cubits high, and a peacock is perched on its summit. Seeing a snake, at a distance of thrice the pillar's height, gliding toward his hole, he pounces obliquely upon him. Say quickly at how many cubits from the snake's hole do they meet, both proceeding an equal distance?

(c) In an expedition to seize his enemy's elephants, a king marched 2 yojanas the first day. Say, intelligent calculator, with what increasing rate of daily march did he proceed, since he reached his foe's city, a distance of 80 yojanas, in a week?

(d) How many are the variations in the form of the god Sambu (Siva) by the exchange of his ten attributes held reciprocally in his several hands: namely, the rope, the elephant's hook, the serpent, the tabor, the skull, the trident, the bedstead, the dagger, the arrow, the bow: as those of Hari by the exchange of the mace, the discus, the lotus, and the conch?

(e) Arjuna, exasperated in combat, shot a quiver of arrows to slay Carna. With half his arrows he parried those of his antagonist; with four times the square root of the quiverful he killed his horse; with six arrows he slew Salya (Carna's charioteer); with three he demolished the umbrella, standard, and bow; and with one he cut off the head of the foe. How many were the arrows which Arjuna let fly?

7.4 Quadratic Surds

A numerical radical in which the radicand is rational but the radical itself is irrational, is called a *surd*. A surd is called *quadratic, cubic*, and so on, according as its index is two, three, and so on.

(a) Show that a quadratic surd cannot be equal to the sum of a non-zero rational number and a quadratic surd.

(b) Show that if $a + \sqrt{b} = c + \sqrt{d}$, where $\sqrt{b}$ and $\sqrt{d}$ are surds and a and c are rational, then $a = c$ and $b = d$.

(c) Establish Bhāskara's identities given in Section 7-3 and use one of them to express $\sqrt{17 + \sqrt{240}}$ as the sum of two quadratic surds.

7.5 Indeterminate Equations of the First Degree

The Hindus solved the problem of finding all integral solutions of the linear indeterminate equation $ax + by = c$, where a, b, c are integers.

(a) If $ax + by = c$ has an integral solution, show that the greatest common divisor of a and b is a divisor of c. (This theorem says that there is no loss in generality if we consider a and b to be relatively prime.)

(b) If x_1 and y_1 constitute an integral solution of $ax + by = c$, where

a and b are relatively prime, show that all integral solutions are given by $x = x_1 + mb$, $y = y_1 - ma$, where m is an arbitrary integer. (This theorem says that all integral solutions are known if just one integral solution can be found. A simple way of finding one integral solution is illustrated in the suggestions for Problem Study, 7.5 (c).)

(c) Solve $7x + 16y = 209$ for positive integral solutions.

(d) Solve $23x + 37y = 3000$ for positive integral solutions.

(e) In how many ways can the sum of five dollars be paid in dimes and quarters?

(f) Find the smallest permissible answer to the following indeterminate problem of Mahāvīra: "Into the bright and refreshing outskirts of a forest, which were full of numerous trees with their branches bent down with the weight of flowers and fruits, trees such as jambu trees, lime trees, plantains, areca palms, jack trees, date palms, hintala trees, palmyras, punnāga trees, and mango trees—outskirts, the various quarters whereof were filled with the many sounds of crowds of parrots and cuckoos found near springs containing lotuses with bees roaming about them—into such forest outskirts a number of weary travelers entered with joy. There were 63 numerically equal heaps of plantain fruits put together and combined with 7 more of those same fruits, and these were equally distributed among 23 travelers so as to have no remainder. You tell me now the numerical measure of a heap of plantains."

7.6   The Diagonals of a Cyclic Quadrilateral

Establish the following chain of theorems:

(a) The product of two sides of a triangle is equal to the product of the altitude on the third side by the diameter of the circumscribed circle.

(b) Let $ABCD$ be a cyclic quadrilateral of diameter δ. Denote the lengths of the sides AB, BC, CD, DA by a, b, c, d, the diagonals BD and AC by m and n, and the angle between either diagonal and the perpendicular upon the other by θ. Show that

$$m\delta \cos \theta = ab + cd, \qquad n\delta \cos \theta = ad + bc.$$

(c) Show, for the above quadrilateral, that

$$m^2 = (ac + bd)(ab + cd)/(ad + bc),$$
$$n^2 = (ac + bd)(ad + bc)/(ab + cd).$$

(d) If, in the above quadrilateral, the diagonals are perpendicular to each other, then

$$\delta^2 = (ad + bc)(ab + cd)/(ac + bd).$$

7.7 Brahmagupta's Quadrilaterals

(a) Brahmagupta gave the formula $K^2 = (s-a)(s-b)(s-c)(s-d)$ for the area K of a cyclic quadrilateral of sides a, b, c, d and semiperimeter s. Show that Heron's formula for the area of a triangle is a special case of this formula.

(b) Using Brahmagupta's formula of part (a) show that the area of a quadrilateral possessing both an inscribed and a circumscribed circle is equal to the square root of the product of its sides.

(c) Show that a quadrilateral has perpendicular diagonals if and only if the sum of the squares of one pair of opposite sides is equal to the sum of the squares of the other pair of opposite sides.

(d) Brahmagupta showed that if $a^2 + b^2 = c^2$ and $A^2 + B^2 = C^2$, then any quadrilateral having aC, cB, bC, cA for consecutive sides has perpendicular diagonals. Prove this.

(e) Find the sides, diagonals, circumdiameter, and area of the Brahmagupta trapezium (see Section 7-4) determined by the two Pythagorean triples (3,4,5) and (5,12,13).

7.8 Tâbit ibn Qorra and al-Karkhî

(a) Tâbit ibn Qorra (826–901) invented the following rule for finding amicable numbers: *If $p = 3 \cdot 2^n - 1$, $q = 3 \cdot 2^{n-1} - 1$, $r = 9 \cdot 2^{2n-1} - 1$ are three primes, then $2^n pq$ and $2^n r$ are a pair of amicable numbers.* Verify this for $n = 2$ and $n = 4$ (see Section 3-3).

(b) Al-Karkhî (*ca.* 1020) wrote a work on algebra called the *Fakhrî*, named after his patron Fakhr al-Mulk, the grand vizier of Bagdad at the time. Problem 1 of Section 5 of the *Fakhrî* requests one to find two rational numbers such that the sum of their cubes is the square of a rational number. In other words, find rational numbers x, y, z such that

$$x^3 + y^3 = z^2.$$

Al-Karkhî essentially takes

$$x = n^2/(1 + m^3), \qquad y = mx, \qquad z = nx,$$

where m and n are arbitrary rational numbers. Verify this, and find x, y, z for $m = 2$ and $n = 3$.

7.9 Casting Out 9's

(a) Show that when the sum of the digits of a natural number is divided by 9, one obtains the same remainder as when the number itself is divided by 9.

The act of obtaining the remainder when a given natural number is divided by an integer n is known as *casting out* n's. The above theorem shows that it is particularly easy to cast out 9's.

(b) Let us call the remainder obtained when a given natural number is divided by 9, the *excess* for that number. Prove the following two theorems: (1) *The excess for a sum is equal to the excess for the sum of the excesses of the addends.* (2) *The excess for the product of two numbers is equal to the excess for the product of the excesses of the two numbers.*

These two theorems furnish the basis for checking addition and multiplication by casting out 9's.

(c) Add and then multiply 478 and 993, and check by casting out 9's.

(d) Show that if the order of the digits of a natural number are permuted in any way to form a new number, then the difference between the old and the new numbers is divisible by 9.

This furnishes the basis for the *bookkeeper's check*. If the sums of the debit and credit entries in double entry bookkeeping do not balance, and the difference between the two sums is divisible by 9, then it is quite likely that the error is due to a transposition in digits made when transcribing a debit or a credit into the book.

(e) Explain the following number trick: Someone is asked to think of a number; form a new number by reversing the order of the digits; subtract the smaller from the larger number; multiply the difference by any number whatever; scratch out any nonzero digit in the product and announce what is left. The conjurer finds the scratched out digit by calculating the excess for the announced result and then subtracting this excess from 9.

(f) Generalize the theorem of part (a) for an arbitrary base b.

7.10 Double False Position

(a) One of the oldest methods for approximating the real roots of an equation is the rule known as *regula duorum falsorum*, often called the rule of *double false position*. This method seems to have originated in India and was used by the Arabians. In brief the method is this: Let x_1 and x_2 be two numbers lying close to and on each side of a root x of the equation $f(x) = 0$. Then the intersection with the x-axis of the chord joining the points $(x_1, f(x_1))$, $(x_2, f(x_2))$ gives an approximation x_3 to the sought root (see Figure 52). Show that

$$x_3 = [x_2 f(x_1) - x_1 f(x_2)]/[f(x_1) - f(x_2)].$$

The process can now be reapplied with the appropriate pair x_1, x_3 or x_3, x_2.

(b) Compute, by double false position, to three decimal places, the root of $x^3 - 36x + 72 = 0$ which lies between 2 and 3.

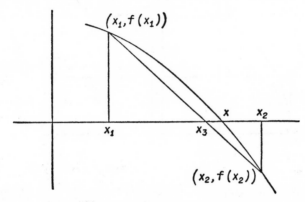

Fig. 52

(c) Compute, by double false position, to three decimal places, the root of $x - \tan x = 0$ which lies between 4.4 and 4.5.

7.11 Khayyam's Solution of Cubics

(a) Given line segments of lengths a, b, n, construct a line segment of length $m = a^3/bn$.

(b) Omar Khayyam was the first to handle every type of cubic that possesses a positive root. Complete the details in the following sketch of Khayyam's geometrical solution of the cubic

$$x^3 + b^2 x + a^3 = cx^2,$$

where a, b, c, x are thought of as lengths of line segments. Khayyam stated this type of cubic rhetorically as "a cube, some sides, and some numbers are equal to some squares."

In Figure 53 construct $AB = a^3/b^2$ (by part (a)) and $BC = c$. Draw a semicircle on AC as diameter and let the perpendicular to AC at B cut it in D. On BD mark off $BE = b$ and through E draw EF parallel to AC. Find G on BC such that $(BG)(ED) = (BE)(AB)$ and complete the rectangle $DBGH$.

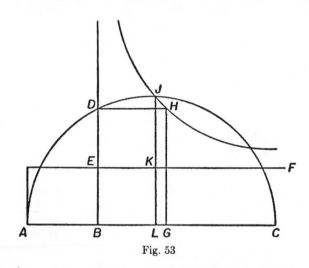

Fig. 53

Through H draw the rectangular hyperbola having EF and ED for asymptotes, and let it cut the semicircle in J. Let the parallel to DE through J cut EF in K and BC in L. Show, successively, that: (1) $(EK)(KJ) = (BG)(ED) = (BE)(AB)$, (2) $(BL)(LJ) = (BE)(AL)$, (3) $(LJ)^2 = (AL)(LC)$, (4) $(BE)^2/(BL)^2 = (LJ)^2/(AL)^2 = LC/AL$, (5) $(BE)^2(AL) = (BL)^2(LC)$, (6) $b^2(BL + a^3/b^2) = (BL)^2(c - BL)$, (7) $(BL)^3 + b^2(BL) + a^3 = c(BL)^2$. Thus BL is a root of the given cubic equation.

7.12 Geometrical Constructions on a Sphere

The Arabians were interested in constructions on a spherical surface. Consider the following problems, to be solved with Euclidean tools and appropriate plane constructions.

(a) Given a material sphere, find its diameter.

(b) On a given material sphere locate the vertices of an inscribed cube.

(c) On a given material sphere locate the vertices of an inscribed regular tetrahedron.

Bibliography

CAJORI, FLORIAN, *A History of Mathematical Notations*, 2 vols. Chicago: The Open Court Publishing Company, 1928–1929.

COOLIDGE, J. L., *A History of Geometrical Methods*. New York: Oxford University Press, 1940.

———, *The Mathematics of Great Amateurs*. Chap. 2, "Omar Khayyam." New York: Oxford University Press, 1949.

DATTA, B., *The Science of the Sulba: A Study in Early Hindu Geometry*. Calcutta: University of Calcutta, 1932.

———, AND A. N. SINGH, *History of Hindu Mathematics*. Bombay: Asia Publishing House, 1962.

HILL, G. F., *The Development of Arabic Numerals in Europe*. New York: Oxford University Press, 1915.

KHAYYAM, OMAR, *Algebra*, ed. and tr. by D. S. Kasir. New York: Teachers College, Columbia University, 1931.

LAMB, HAROLD, *Omar Khayyam, a Life*. New York: Doubleday & Company, Inc., 1936.

LARSEN, H. D., *Arithmetic for Colleges*. New York: The Macmillan Company, 1950.

LOOMIS, E. S., *The Pythagorean Proposition*, 2d ed. Ann Arbor, Mich.: priv. ptd., Edwards Bros., Inc., 1940.

MACFALL, HALDANE, *The Three Students*. New York: Alfred A. Knopf, Inc., 1926.

ORE, OYSTEIN, *Number Theory and Its History*. New York: McGraw-Hill Book Company, Inc., 1948.

SMITH, D. E., AND L. C. KARPINSKI, *The Hindu-Arabic Numerals*. Boston: Ginn & Company, 1911.

STORY, W. E., *Omar Khayyam as a Mathematician* (read at a meeting of the Omar Khayyam Club of America, April 6, 1918). Needham, Mass.: priv. ptd., Rosemary Press, 1919.

WOLFE, H. E., *Introduction to Non-Euclidean Geometry*. New York: Holt, Rinehart and Winston, Inc., 1945.

8

European Mathematics, 500 to 1600

8-1   The Dark Ages

The period starting with the fall of the Roman Empire in the middle of the fifth century and extending into the eleventh century is known as Europe's Dark Ages, for during this period civilization in western Europe reached a very low ebb. Schooling became almost nonexistent, Greek learning all but disappeared, and many of the arts and crafts bequeathed by the ancient world were forgotten. Only the monks of the Catholic monasteries, and a few cultured laymen, preserved a slender thread of Greek and Latin learning. The period was marked by much physical violence and intense religious faith. The old social order gave way and society became feudal and ecclesiastical.

The Romans had never taken to abstract mathematics, but contented themselves with merely practical aspects of the subject associated with commerce and civil engineering. With the fall of the Roman Empire and the subsequent closing down of much of east-west trade and the abandonment of state engineering projects, even these interests waned, and it is no exaggeration to say that very little in mathematics, beyond the development of the Christian calendar, was accomplished in the West during the whole of the half millennium covered by the Dark Ages.

Of the persons charitably credited with playing a role in the history of mathematics during the Dark Ages, we might mention the martyred Roman citizen Boethius, the British ecclesiastical scholars Bede and Alcuin, and the famous French scholar and churchman Gerbert, who became Pope Sylvester II.

The importance of Boethius (*ca.* 475–524) in the story of mathematics rests on the fact that his writings on geometry and arithmetic remained standard texts in the monastic schools for many centuries. These very meager works came to be considered as the height of mathematical achievement, and thus well illustrate the poverty of the subject in Christian Europe during the Dark Ages. For the *Geometry* consists of nothing but the statements of the propositions of Book I and of a few selected propositions of Books III and IV of Euclid's *Elements*, along with some applications to elementary mensuration, and the *Arithmetic* is founded on the tiresome and half mystical, but once highly reputed, work of Nicomachus of four centuries earlier. It is contended by some that part, at least, of the *Geometry* is spurious. With these works, and his writings on philosophy, Boethius became the founder of medieval scholasticism. His high ideals and inflexible integrity led him into political troubles and he suffered a cruel end, for which the Church declared him a martyr.

Bede (*ca.* 673–735), later qualified as Bede *the Venerable*, was born in Northumberland, England, and became one of the greatest of the medieval Church scholars. His numerous writings include some on mathematical subjects, chief of which are his treatises on the calendar and on finger reckoning. Alcuin (735–804), born in Yorkshire, was another English scholar. He was called to France to assist Charlemagne in his ambitious educational project. Alcuin wrote on a number of mathematical topics and is doubtfully credited with a collection of puzzle problems which influenced textbook writers for many centuries (see Problem Study 8.1).

Gerbert (*ca.* 950–1003) was born in Auvergne, France, and early showed unusual abilities. He was one of the first Christians to study in the Moslem schools of Spain and there is evidence that he may have brought back the Hindu-Arabic numerals, without the zero, to Christian Europe. He is said to have constructed abaci, terrestrial and celestial globes, a clock, and perhaps an organ. Such accomplishments corroborated the suspicions of some of his contemporaries that he had traded his soul to the devil. Nevertheless he steadily rose in the Church and was finally elected to the papacy in 999. He was considered a profound scholar and wrote on astrology, arithmetic, and geometry (see Problem Study 8.1 (f)).

8-2 The Period of Transmission

About the time of Gerbert the Greek classics in science and mathematics began to filter into Europe. There followed a period of transmission during which the ancient learning preserved by Moslem culture was passed on to the western Europeans. This took place through Latin translations made by Christian scholars traveling to Moslem centers of learning, through the relations between the Norman kingdom of Sicily and the east, and through western European commercial relations with the Levant and the Arabic world.

The loss of Toledo by the Moors to the Christians in 1085 was followed by an influx of Christian scholars to that city to acquire Moslem learning. Other Moorish centers in Spain were infiltrated and the twelfth century became, in the history of mathematics, a century of translators. One of the earliest Christian scholars to engage in this pursuit was the English monk Adelard of Bath (*ca.* 1120), who studied in Spain and traveled extensively through Greece, Syria, and Egypt. Adelard is credited with Latin translations of Euclid's *Elements* and of al-Khowârizmî's astronomical tables. There are thrilling allusions to the physical risks run by Adelard in his acquisition of Arabic learning; to obtain the jealously guarded knowledge, he disguised himself as a Mohammedan student. Another early translator was the Italian, Plato of Tivoli (*ca.* 1120), who translated the astronomy of al-Battânî, the *Spherics* of Theodosius, and various other works. The most industrious translator of the period was Gherardo of Cremona (1114–1187), who translated into Latin over 90 Arabian works, among which were Ptolemy's *Almagest*, Euclid's *Elements*, and al-Khowârizmî's algebra. We have already, in Section 7-9, mentioned the part played by Gherardo of Cremona in the development of our word *sine*. Other noted translators of the twelfth century were John of Seville and Robert of Chester.

The location and political history of Sicily made that island a natural meeting ground of East and West. Sicily started as a Greek colony, became part of the Roman Empire, linked itself with Constantinople after the fall of Rome, was held by the Arabs for about fifty years in the ninth century, was recaptured by the Greeks, and then taken over by the Normans. During the Norman regime the Greek, Arabian, and Latin tongues were used side by side, and diplomats frequently traveled to Constantinople and Bagdad. Many Greek and Arabian manuscripts in science and mathematics were obtained and translated into Latin. This work was greatly encouraged by the two rulers and patrons of science, Frederick II (1194–1250) and his son Manfred (*ca.* 1231–1266).

Among the first cities to establish mercantile relations with the Arabic world were the Italian commercial centers at Genoa, Pisa, Venice, Milan, and Florence. Italian merchants came in contact with much of eastern civilization, picking up useful arithmetical and algebraical information.

These merchants played an important part in the dissemination of the Hindu-Arabic numeral system.

8-3   Fibonacci and the Thirteenth Century

At the threshold of the thirteenth century appeared Leonardo Fibonacci ("Leonardo, son of Bonaccio"), perhaps the most talented mathematician of the Middle Ages. Also known as Leonardo of Pisa (or Leonardo Pisano), Fibonacci was born about 1175 in the commercial center of Pisa, where his father was connected with the mercantile business. The father's occupation early roused in the boy an interest in arithmetic, and subsequent extended trips to Egypt, Sicily, Greece, and Syria brought him in contact with eastern and Arabic mathematical practices. Thoroughly convinced of the practical superiority of the Hindu-Arabic methods of calculation Fibonacci, in 1202, shortly after his return home, published his famous work called the *Liber abaci*.

The *Liber abaci* is known to us through a second edition which appeared in 1228. The work is devoted to arithmetic and elementary algebra and, though essentially an independent investigation, shows the influence of the algebras of al-Khowârizmî and Abû Kâmil. The book profusely illustrates and strongly advocates the Hindu-Arabic notation and did much to aid the introduction of these numerals into Europe. In the fifteen chapters of the work are explained the reading and writing of the new numerals, methods of calculation with integers and fractions, computation of square and cube roots, and the solution of linear and quadratic equations both by false position and by algebraic processes. Negative and imaginary roots of equations are not recognized and the algebra is rhetorical. Applications are given involving barter, partnership, alligation, and mensurational geometry. The work contains a large collection of problems which served later authors as a storehouse for centuries. We have already, in Section 2-10, mentioned one interesting problem from the collection, which apparently evolved from a much older problem in the Rhind papyrus. Another problem, giving rise to the important *Fibonacci sequence*: $1, 1, 2, 3, 5, \cdots, x, y, x + y, \cdots$, and some other problems from the *Liber abaci*, may be found in Problem Studies 8.2, 8.3, and 8.4.

In 1220 appeared Fibonacci's *Practica geometriae*, a vast collection of material on geometry and trigonometry treated skillfully with Euclidean rigor and some originality, and about 1225 Fibonacci wrote his *Liber quadratorum*, a brilliant and original work on indeterminate analysis, which has marked him as the outstanding mathematician in this field between Diophantus and Fermat. These works were beyond the abilities of most of the contemporary scholars.

Fibonacci's talents came to the attention of the patron of learning, Emperor Frederick II, with the result that Fibonacci was invited to court to partake in a mathematical tournament. Three problems were set by John of Palermo, a member of the emperor's retinue. The first problem was to find a rational number x such that $x^2 + 5$ and $x^2 - 5$ shall each be squares of rational numbers. Fibonacci gave the answer $x = 41/12$, which is correct, since $(41/12)^2 + 5 = (49/12)^2$ and $(41/12)^2 - 5 = (31/12)^2$. The solution appears in the *Liber quadratorum*. The second problem was to find a solution to the cubic equation $x^3 + 2x^2 + 10x = 20$. Fibonacci attempted a proof that no root of the equation can be expressed by means of irrationalities of the form $\sqrt{a + \sqrt{b}}$, or, in other words, that no root can be constructed with straightedge and compasses. He then obtained an approximate answer, which, expressed in decimal notation, is 1.3688081075, and is correct to nine places. The answer appears, without any accompanying discussion, in a work by Fibonacci entitled the *Flos* ("blossom" or "flower") and has excited some wonder. The third problem, also recorded in the *Flos*, is easier and may be found in Problem Study 8.4.

It has been argued that Fibonacci appears greater than he really was because of the lack of equal contemporaries. It is certainly true that the thirteenth century produced very few mathematicians of any stature. Next to Fibonacci, and contemporary with him, was Jordanus Nemorarius, usually identified (but in all likelihood mistakenly) with the German monk Jordanus Saxo who, in 1222, was elected the second general of the rapidly expanding Dominican order. He wrote several works dealing with arithmetic, algebra, geometry, astronomy, and (probably) statics. These prolix works, some of which enjoyed considerable fame at one time, now seem largely trivial. Nemorarius though, was perhaps the first one widely to use letters to represent general numbers, although his practice had little influence on subsequent writers. There is only one instance where Fibonacci did this.

Perhaps mention should also be made of Sacrobosco (John of Holywood, or John of Halifax), Campanus, and Roger Bacon. The first taught mathematics in Paris and wrote a collection of arithmetical rules and a popular compilation of extracts from Ptolemy's *Almagest* and the works of Arabian astronomers. Campanus' chief bid to fame is his Latin translation of Euclid's *Elements*, mentioned in Section 5-3. Roger Bacon, original genius that he was, had little ability in mathematics but was acquainted with many of the Greek works in geometry and astronomy, and, as his eulogies attest, fully appreciated the value of the subject.

It was the early part of the thirteenth century that saw the rise of the universities at Paris, Oxford, Cambridge, Padua, and Naples. Universities later became potent factors in the development of mathematics, many mathematicians being associated with one or more such institutions.

8-4   The Fourteenth Century

The fourteenth century was a mathematically barren one. It was the century of the Black Death, which swept away more than a third of the population of Europe, and in this century the Hundred Years' War, with its political and economical upheavals in northern Europe, got well under way.

The greatest mathematician of the period was Nicole Oresme, who was born in Normandy about 1323. He died in 1382 after a career that carried him from a college professor to a bishop. He wrote five mathematical works and translated some of Aristotle. In one of his tracts appears the first known use of fractional exponents (not, of course, in modern notation), and in another tract he locates points by coordinates, thus foreshadowing modern coordinate geometry. A century later this last tract enjoyed several printings, and it may have influenced Renaissance mathematicians and even Descartes.

Although European mathematics during the Middle Ages was essentially practical, speculative mathematics did not entirely die out. The meditations of scholastic philosophers led to subtle theorizing on motion, infinity, and the continuum, all of which are fundamental concepts in modern mathematics. The centuries of scholastic disputes and quibblings may, to some extent, account for the remarkable transformation from ancient to modern mathematical thinking, and might, as suggested by E. T. Bell, constitute a *submathematical analysis*. From this point of view, Thomas Aquinas, perhaps possessing the acutest mind of the thirteenth century, can well be considered as having played a part in the development of mathematics. Definitely more of the conventional mathematician was Thomas Bradwardine (1290–1349), who died as Archbishop of Canterbury. In addition to speculations on the basic concepts of the continuous and the discrete and on the infinitely large and the infinitely small, Bradwardine wrote four mathematical tracts on arithmetic and geometry.

8-5   The Fifteenth Century

The fifteenth century witnessed the start of the European Renaissance in art and learning. With the collapse of the Byzantine Empire, culminating in the fall of Constantinople to the Turks in 1453, refugees flowed into Italy bringing with them treasures of Greek civilization. Many Greek classics, hitherto known only through the often inadequate Arabic translations, could now be studied from original sources. Also, about the middle of the century, printing was invented and revolutionized the book trade, enabling knowledge to be disseminated at an unprecedented rate. Toward the end of the century, America was discovered and soon the earth was circumnavigated.

Mathematical activity in the fifteenth century was largely centered in the Italian cities and in the central European cities of Nuremberg, Vienna, and Prague, and was concentrated on arithmetic, algebra, and trigonometry. Thus mathematics flourished principally in the growing mercantile cities under the influence of trade, navigation, astronomy, and surveying.

Adhering to chronological order we first mention Nicholas Cusa, who took his name from the city of Cues on the Mosel, where he was born in 1401. The son of a poor fisherman, he rose rapidly in the Church, finally becoming a cardinal. In 1448, he became governor of Rome. He was only incidentally a mathematician but did succeed in writing a few tracts on the subject. He is now remembered along these lines chiefly for his work on calendar reform and his attempts to square the circle and trisect the general angle (see Problem Study 4.14). He died in 1464.

A better mathematician was Georg von Peurbach (1423–1461), who numbered Nicholas Cusa as one of his teachers. After lecturing on mathematics in Italy, he settled in Vienna and made the university there the mathematical center of his generation. He wrote an arithmetic and some works on astronomy, and compiled a table of sines. Most of these works were not published until after his death. He also had started a Latin translation, from the Greek, of Ptolemy's *Almagest*.

The ablest and most influential mathematician of the century was Johann Müller (1436–1476), more generally known, from the Latinized form of his birthplace of Königsberg ("king's mountain"), as Regiomontanus. At a young age he studied under Peurbach in Vienna and was later entrusted with the task of completing the latter's translation of the *Almagest*. He also translated, from the Greek, works of Apollonius, Heron, and Archimedes. His treatise *De triangulis omnimodis*, written about 1464 but posthumously published in 1533, is his greatest publication and was the first systematic European exposition of plane and spherical trigonometry considered independently of astronomy. Regiomontanus traveled considerably in Italy and Germany, finally settling in 1471 at Nuremberg, where he set up an observatory, established a printing press, and wrote some tracts on astronomy. He is said to have constructed a mechanical eagle which flapped its wings and was considered as one of the marvels of the age. In 1475, Regiomontanus was invited to Rome by Pope Sixtus IV to partake in the reformation of the calendar. Shortly after his arrival, at the age of 40, he suddenly died. Some mystery shrouds his death, for, though most accounts claim he probably died of a pestilence, it was rumored that he was poisoned by an enemy.

Regiomontanus' *De triangulis omnimodis* is divided into five books, the first four devoted to plane trigonometry and the fifth to spherical trigonometry. In it he shows much interest in the determination of a triangle satisfying three given conditions. Typical examples considered by him are: (1) Determine a triangle given the difference of two sides, the altitude on the third side, and the difference of the segments into which the altitude divides the third side; (2) Determine a triangle given a side, the altitude on this side, and the ratio of the other two sides; (3) Construct a cyclic quadrilateral given the four sides. Regiomontanus employed rhetorical algebra for solving these problems, finding an unknown part of the figure as a root of a quadratic equation. Although his methods were meant to be considered as general, he always gave specific numerical values to the given parts. The only trigonometric functions employed in *De triangulis omnimodis* are the sine and cosine. Later, however, Regiomontanus computed a table of tangents.

The most brilliant French mathematician of the fifteenth century was Nicolas Chuquet, who was born in Paris but lived and practiced medicine in Lyons. In 1484, he wrote an arithmetic known as *Triparty en la science des nombres*, which was not printed until the nineteenth century. The first of the three parts of this work concerns itself with computation with rational numbers, the second with irrational numbers, and the third with the theory of equations. Chuquet recognized positive and negative integral exponents and syncopated some of his algebra. His work was too advanced, for the time, to exert much influence on his contemporaries. He died around 1500. Some problems from Chuquet may be found in Problem Study 8.7.

In 1494 appeared the first printed edition of the *Summa de arithmetica, geometrica, proportioni et proportionalita*, usually referred to briefly as the *Sūma*, of the Italian friar Luca Pacioli (*ca.* 1445–*ca.* 1509). This work, freely compiled from many sources, purported to be a summary of the arithmetic, algebra, and geometry of the time. It contains little of importance not found in Fibonacci's *Liber abaci* but does employ a superior notation.

The arithmetical portion of the *Sūma* begins with algorisms for the fundamental operations and for extracting square roots. The presentation is rather complete, containing, for example, no less than eight plans for the performance of a multiplication. Mercantile arithmetic is fully dealt with and illustrated by numerous problems; there is an important treatment of double entry bookkeeping. The rule of false position is discussed and applied. In spite of many numerical mistakes, the arithmetical part of the work has become a standard authority on the practices of the time. The algebra in the *Sūma* goes through quadratic equations and contains many problems which lead to such equations. The algebra is syncopated by the use of such abbreviations as p (from *piu*, "more") for plus, m (from *meno*, "less") for minus, *co* (from *cosa*, "thing") for the unknown x, *ce* (from *censo*) for x^2, *cu* (from *cuba*) for x^3, and *cece* (from *censocenso*) for x^4. Equality is sometimes indicated by *ae* (from *aequalis*). Frequently bars appear over the abbreviations, but this was the custom of indicating an omission, as in *Sūma* for *Summa*. The work contains very little of interest in geometry. As with Regiomontanus, algebra is often employed in the solution of geometrical problems.

Pacioli traveled extensively, taught in various places, and wrote a number of other works not all of which were printed. In 1509, he published his *De diuina proportione*, which contains figures of the regular solids thought to have been drawn by Leonardo da Vinci.

The first appearance in print of our present + and − signs is in an arithmetic published in Leipzig in 1489 by Johann Widman (born *ca.* 1460 in Bohemia). Here the signs are not used as symbols of operation but merely to indicate excess and deficiency. Quite likely the plus sign is a contraction of the Latin word *et*, which was frequently used to indicate addition, and it may be that the minus sign is contracted from the abbreviation $\overline{m}$ for minus. Other plausible explanations have been offered. The $+$ and $-$ signs were used as symbols of algebraic operation in 1514 by the Dutch mathematician Vander Hoecke but were probably so used earlier.*

8-6 The Early Arithmetics

With the interest in education which accompanied the Renaissance and with the tremendous increase in commercial activity at the time, hosts of popular textbooks in arithmetic began to appear. Three hundred such books were printed in Europe prior to the seventeenth century. These texts were largely of two types, those written in Latin by classical scholars often attached to the Church schools, and those written in the vernaculars by practical teachers interested in preparing boys for commercial careers. These latter teachers often also served as town surveyors, notaries, and gaugers, and included the influential Rechenmeisters supported by the Hanseatic League, a powerful protective union of commercial towns in the Teutonic countries.

The earliest printed arithmetic is the anonymous and now extremely rare Treviso Arithmetic, published in 1478 in the town of Treviso, located on the trade route linking Venice with the north. It is largely a commercial arithmetic devoted to explaining the writing of numbers, computation with them, and applications to partnership and barter. Like the earlier "algorisms" of the fourteenth century, it also contains some recreational questions.

Far more influential in Italy than the Treviso Arithmetic was the commercial arithmetic written by Piero Borghi. This highly useful work was published in Venice in 1484 and ran through at least seventeen editions, the last appearing in 1557. In 1491 appeared, in Florence, a less important arithmetic by Filippo Calandri, but interesting to us because it contains the first printed example of our modern process of long division and also the first illustrated problems published in Italy. We have already considered Pacioli's *Sūma*, published in 1494, a large portion of which is devoted to arithmetic. Much information regarding the Italian commercial customs of the time may be gleaned from the problems of this book.

*See J. W. L. Glaisher, "On the Early History of the Signs $+$ and $-$ and on the Early German Arithmeticians," *Messenger of Mathematics*, LI (1921–1922), 1–148.

Very influential in Germany was Widman's arithmetic published in 1489 at Leipzig. Another important German arithmetic was that written by Jacob Köbel (1470–1533), a Rechenmeister of Heidelberg. The popularity of this arithmetic, published in 1514, is attested by the fact that it ran through at least 22 editions. But perhaps the most influential of the German commercial arithmetics was that of Adam Riese (ca. 1489–1559), published in 1522. So reputable was this work that even today in Germany the phrase *nach Adam Riese* is used to indicate a correct calculation.

England, too, produced some noted early arithmetics. The first published work in England devoted exclusively to mathematics was an arithmetic written by Cuthbert Tonstall (1474–1559). This book, founded on Pacioli's *Sūma*, was printed in 1522 and was written in Latin. During his eventful life, Tonstall filled a number of ecclesiastical and diplomatic posts. The regard of his contemporaries for his scholarship is indicated by the fact that the first printed edition of Euclid's *Elements* in Greek (1533) was dedicated to him. But the most influential English textbook writer of the sixteenth century was Robert Recorde (ca. 1510–1558). Recorde wrote in English, his works appearing as dialogues between master and student. He wrote at least five books, his first being an arithmetic fancifully entitled *The Grovnd of Artes* and published about 1542. This work enjoyed at least 29 printings. Recorde studied at Oxford and then took a medical degree at Cambridge. He taught mathematics in private classes at both institutions while in residence there and after leaving Cambridge served as physician to Edward VI and Queen Mary. In later life he became "Comptroller of the Mines and Monies" in Ireland. His last years were spent in prison, probably for some misdemeanor connected with his work in Ireland.

8-7 Beginnings of Algebraic Symbolism

Besides his arithmetic, mentioned in the last section, Robert Recorde wrote an astronomy, a geometry, an algebra, a book on medicine, and probably some other works now lost. The book on astronomy, printed in 1551, is called *The Castle of Knowledge* and was one of the first works to introduce the Copernican system to English readers. Recorde's geometry, *The Pathewaie to Knowledge*, was also printed in 1551 and contains an abridgment of Euclid's *Elements*. Of historical interest here is Recorde's algebra, called *The Whetstone of Witte*, published in 1557, for it was in this book that our modern symbol for equality was used for the first time. Recorde justified his adoption of a pair of equal parallel line segments for the symbol of equality "bicause noe 2 thynges can be moare equalle."

Another of our modern algebraic symbols, the familiar radical sign (adopted perhaps because it resembles a small r, for *radix*), was introduced in 1525 by Christoff Rudolff in his book on algebra entitled *Die Coss*. This book was very influential in Germany and an improved edition of the work was brought out by Michael Stifel (1486–1567) in 1553. Stifel has been described as the greatest German algebraist of the sixteenth century. His best-known mathematical work is his *Arithmetica integra*, published in 1544. It is divided into three parts devoted, respectively, to rational numbers, irrational numbers, and algebra. In the first part, Stifel points out the advantages of associating an arithmetical progression with a geometrical one, thus foreshadowing the invention of logarithms nearly a century later. He also gives, in this part, the binomial coefficients up to the seventeenth order. The second part of the book is essentially an algebraic presentation of Euclid's Book X, and the third part deals with equations. Negative roots of an equation are discarded, but the signs $+$, $-$, $\sqrt{\ }$ are used, and often the unknown is represented by a letter.

Stifel was one of the oddest personalities in the history of mathematics. He was originally a monk, was converted by Martin Luther, and became a fanatical reformer. His erratic mind led him to indulge in number mysticism. From an analysis of Biblical writings, he prophesied the end of the world on October 3, 1533 and was forced to take refuge in a prison after ruining the lives of many believing peasants who had abandoned work and property to accompany him to heaven. An extreme example of Stifel's mystical reasoning is his proof, by arithmology, that Pope Leo X was the "beast" mentioned in the *Book of Revelations*.* From LEO DECIMVS he retained the letters LDCIMV. He then added X, for Leo X, and omitted the M, because it stood for mystery. A rearrangement of the letters gave DCLXVI, 666, or the "number of the beast" in the *Book of Revelations*.

Some years later Napier, the inventor of logarithms, showed that 666 stood for the Pope of Rome, and his Jesuit contemporary, Father Bongus, declared that it stood for Martin Luther. During World War I, arithmology

---

*"Let him that hath understanding count the number of the beast: for it is the number of a man; and his number is six hundred three score and six." See W. F. White, *A Scrap-Book of Elementary Mathematics* (Chicago: Open Court Publishing Company, 1927), pp. 180–182.

was used to show that 666 must be interpreted as Kaiser Wilhelm. It has been shown that 666 spells Nero when expressed in the letter symbols of the Aramaic language in which the *Book of Revelations* was originally written.

8-8   Cubic and Quartic Equations

Probably the most spectacular mathematical achievement of the sixteenth century was the discovery, by Italian mathematicians, of the algebraic solution of cubic and quartic equations. The story of this discovery, when told in its most colorful version, rivals any page ever written by Benvenuto Cellini. Briefly told the facts seem to be these. About 1515, Scipione del Ferro (1465–1526), a professor of mathematics at the University of Bologna, solved algebraically the cubic equation $x^3 + mx = n$, probably basing his work on earlier Arabic sources. He did not publish his result but revealed the secret to his pupil Antonio Fior. Now about 1535, Nicolo of Brescia, commonly referred to as Tartaglia (the stammerer) because of a childhood injury which affected his speech, claimed to have discovered an algebraic solution of the cubic equation $x^3 + px^2 = n$. Believing this claim was a bluff, Fior challenged Tartaglia to a public contest of solving cubic equations, whereupon the latter exerted himself and only a few days before the contest found an algebraic solution for cubics lacking a quadratic term. Entering the contest equipped to solve two types of cubic equations, whereas Fior could solve but one type, Tartaglia triumphed completely. Later Girolamo Cardano, an unprincipled genius who taught mathematics and practiced medicine in Milan, upon giving a solemn pledge of secrecy, wheedled the key to the cubic from Tartaglia. In 1545, Cardano published his *Ars magna*, a great Latin treatise on algebra, at Nuremberg, Germany, and in it appeared Tartaglia's solution of the cubic. Tartaglia's vehement protests were met by Lodovico Ferrari, Cardano's most capable pupil, who argued that Cardano had received his information from del Ferro through a third party and accused Tartaglia of plagiarism from the same source. There ensued an acrimonious dispute from which Tartaglia was perhaps lucky to escape alive.

Since the actors in the above drama seem not always to have had the highest regard for truth, one finds a number of variations in the details of the plot.

The solution of the cubic equation $x^3 + mx = n$ given by Cardano in his *Ars magna* is essentially the following. Consider the identity

$$(a - b)^3 + 3ab(a - b) = a^3 - b^3.$$

If we choose a and b so that

$$3ab = m, \qquad a^3 - b^3 = n,$$

then x is given by $a - b$. Solving the last two equations simultaneously for a and b we find

$$a = \sqrt[3]{(n/2) + \sqrt{(n/2)^2 + (m/3)^3}},$$

$$b = \sqrt[3]{-(n/2) + \sqrt{(n/2)^2 + (m/3)^3}},$$

and x is thus determined.

It was not long after the cubic had been solved that an algebraic solution was discovered for the general quartic (or biquadratic) equation. In 1540, the Italian mathematician Zuanne de Tonini da Coi proposed a problem to Cardano which led to a quartic equation (see Problem Study 8.11). Although Cardano was unable to solve the equation, his pupil Ferrari succeeded, and Cardano had the pleasure of publishing this solution also in his *Ars magna*.

Ferrari's method of solving quartics, summarized in modern notation, is as follows. A simple transformation (see Problem Study 8.10 (a)) reduces the complete quartic to one of the form

$$x^4 + px^2 + qx + r = 0.$$

From this we obtain

$$x^4 + 2px^2 + p^2 = px^2 - qx - r + p^2$$

or

$$(x^2 + p)^2 = px^2 - qx + p^2 - r,$$

whence, for arbitrary y,

$$(x^2 + p + y)^2 = px^2 - qx + p^2 - r + 2y(x^2 + p) + y^2$$
$$= (p + 2y)x^2 - qx + (p^2 - r + 2py + y^2).$$

Now let us choose y so that the right member of the above equation is a square. This is the case when*

$$4(p + 2y)(p^2 - r + 2py + y^2) - q^2 = 0.$$

But this is a cubic in y, and may be solved by previous methods. Such a value of y reduces the original problem to nothing but extraction of square roots.

Other algebraic solutions of the general cubic and quartic equations have been given. In the next section we shall consider the methods devised by the sixteenth century French mathematician François Viète. A solution of quartics given by Descartes in 1637 may be found in many of the standard college textbooks on the theory of equations (see Problem Study 10.4 (e)).

Since the solution of the general quartic equation is made to depend on the solution of an associated cubic equation, Euler, about 1750, attempted similarly to reduce the solution of the general quintic equation to that of an associated quartic equation. He failed in this attempt, as did Lagrange about thirty years later. An Italian physician, P. Ruffini (1765–1822), in 1803, 1805, and 1813 supplied a proof of what is now known to be a fact, that the roots of a general fifth, or higher, degree equation cannot be expressed by means of radicals in terms of the coefficients of the equation. This remarkable fact was independently established later, in 1824, by the famous Norwegian mathematician Niels Henrik Abel (1802–1829). Modern developments in the theory of equations are very fascinating, but too advanced to be considered here, and involve such names as Bring, Jerrard, Tschirnhausen, Galois, Jordan, and many others.

Girolamo Cardano is one of the most extraordinary characters in the history of mathematics. He was born in Pavia in 1501 as the illegitimate son of a jurist and developed into a man of passionate contrasts. He commenced his turbulent professional life as a doctor, studying, teaching, and writing mathematics while practicing his profession. He once traveled as far as Scotland and upon his return to Italy successively held important chairs at the Universities of Pavia and Bologna. He was imprisoned for a time for heresy because he published a horoscope of Christ's life. Resigning his chair in Bologna he moved to Rome and became a distinguished astrologer, receiving a pension as astrologer to the papal court. He died in Rome in 1576, by his own hand, one story says, so as to fulfill his earlier astrological pre-

---

*A necessary and sufficient condition for the quadratic $Ax^2 + Bx + C$ to be the square of a linear function is that the discriminant, $B^2 - 4AC$, vanish.

diction of the date of his death. Many stories are told of his wickedness, as when in a fit of rage he cut off the ears of his younger son.

One of the most gifted and versatile men of his time, Cardano wrote a number of works on arithmetic, astronomy, physics, and other subjects. His greatest work is his *Ars magna*, the first great Latin treatise devoted solely to algebra. Here notice is taken of negative roots of an equation and some attention is paid to computations with imaginary numbers. There also occurs a crude method for obtaining an approximate value of a root of an equation of any degree. There is evidence that he was familiar with "Descartes's rule of signs," explained in Problem Study 10.3. As an inveterate gambler, Cardano wrote a gambler's manual in which are considered some interesting questions on probability.

Tartaglia had a hard childhood. He was born about 1499 at Brescia to poor parents. He was present at the taking of Brescia in 1512 by the French, and in the brutalities that accompanied this event suffered a severe saber cut that cleft his jaws and palate and caused an imperfection in his speech. His father died in the massacre, and it was only with great sacrifice that the boy was able to educate himself. It is said that lacking the means to buy paper, he was obliged to use the tombstones in the cemetery as slates. He later earned his livelihood teaching science and mathematics in various Italian cities. He died in Venice in 1557.

Tartaglia was a gifted mathematician. We have already reported his work on the cubic equation. He is also credited with being the first to apply mathematics to the science of artillery fire. He wrote what is generally considered the best Italian arithmetic of the sixteenth century, a two-volume treatise containing full discussion of the numerical operations and commercial customs of the time. He also published editions of Euclid and Archimedes.

In 1572, a few years before Cardano died, Rafael Bombelli published an algebra which contributed noteworthily to the solution of the cubic equation. It is shown in textbooks on the theory of equations that if $(n/2)^2 + (m/3)^3$ is negative, then the cubic equation $x^3 + mx = n$ has three real roots. But in this case, in the Cardano-Tartaglia formula, these roots are expressed by the difference of two cube roots of *complex imaginary numbers*. This seeming anomaly is known as the "irreducible case in cubics" and considerably bothered the early algebraists. Bombelli pointed out the reality of the apparently imaginary roots in the irreducible case. Bombelli also improved current algebraic notation. As an instance, consider his use of a bracket symbol. Thus the compound expression $\sqrt{7 + \sqrt{14}}$ would have been written by Pacioli as $RV\ 7\ \bar{p}\ R\ 14$, where RV, the *radix universalis*, indicates that the square root is to be taken of all that follows; Bombelli would have written this as $R \lfloor 7\,p\,R\,14\,\rfloor$. Bombelli distinguished square and cube roots by writing $R\,q$ and $R\,c$, and indicated $\sqrt{-11}$ by $di\,m\,R\,q\,11$.

8-9 François Viète

The greatest French mathematician of the sixteenth century was François Viète, frequently called by his semi-Latin name of Vieta, a lawyer and member of parliament who devoted most of his leisure time to mathematics. He was born in 1540 at Fontenay and died in 1603 in Paris.

Some entertaining anecdotes are told about Viète. Thus there is the story about the ambassador from the Low Countries who boasted to King Henry IV that France had no mathematician capable of solving a problem proposed in 1593 by his countryman Adrianus Romanus (1561–1615) and which required the solution of a 45th degree equation. Viète was summoned and shown the equation. Recognizing an underlying trigonometric connection he was able, in a few minutes, to give two roots, and then later gave 21 more. The negative roots escaped him. In return Viète challenged Romanus to solve the problem of Apollonius (see Section 6-5), but Romanus was unable to obtain a solution using Euclidean tools. When he was shown the proposer's elegant solution he traveled to Fontenay to meet Viète with the result that a warm friendship developed. Then there is also the story of how Viète successfully deciphered a Spanish code containing several hundred characters and for two years France profited thereby in its war against Spain. So certain was King Philip II that the code was undecipherable that he complained to the Pope that the French were employing magic against his country, "contrary to the practice of the Christian faith." It is said that when absorbed with mathematics Viète would closet himself in his study for days.

Viète wrote a number of works on trigonometry, algebra, and geometry, chief of which are the *Canon mathematicus seu ad triangula* (1579), the *In artem analyticam isagoge* (1591), the *Supplementum geometriae* (1593), *De numerosa potestatum resolutione* (1600), and *De aequationum recognitione et emendatione* (published posthumously in 1615). These works, except the last, were printed and distributed at Viète's own expense.

The first of the above works contains some notable contributions to trigonometry. It is perhaps the first book in western Europe systematically to develop methods for solving plane and spherical triangles with the aid of all six trigonometric functions. Considerable attention is paid to analytical trigonometry (see Problem Study 8.13). Viète obtained expressions for $\cos n\theta$ as a function of $\cos \theta$ for $n = 1, 2, \cdots, 9$, and later suggested a trigonometric solution of the irreducible case in cubics.

Viète's most famous work is his *In artem*, which did much for the development of symbolic algebra. Here Viète introduced the practice of using the vowels to represent unknown quantities and the consonants to represent known ones. Our present custom of using the later letters of the alphabet for unknowns and the early letters for knowns was introduced by Descartes in 1637. Prior to Viète, it was common practice to use different letters or symbols for the various powers of a quantity. Viète used the same letter, properly qualified. Thus our x, x^2, x^3 were written by Viète as A, A *quadratum*, A *cubum*, and by later writers more briefly as A, A q, A c. Viète also qualified the coefficients of a polynomial equation so as to render the equation homogeneous and he used our present $+$ and $-$ signs, but he had no symbol for equality. Thus he would have written

$$5BA^2 - 2CA + A^3 = D$$

as

B 5 in A quad $-$ C plano 2 in A + A cub aequatur D solido.

Note how the coefficients C and D are qualified so as to make each term of the equation three dimensional. Viète used the symbol $=$ between two quantities, not to indicate the equality of the quantities but rather the difference between them.

In *De numerosa*, Viète gives a systematic process, which was in general use until about 1680, for successively approximating to a root of an equation. The method becomes so laborious for equations of high degree that one seventeenth-century mathematician described it as "work unfit for a Christian." Applied to the quadratic equation

$$x^2 + mx = n$$

the method is as follows. Suppose x_1 is a known approximate value of a root of the equation, so that the sought root may be written as $x_1 + x_2$. Substitution in the given equation yields

$$(x_1 + x_2)^2 + m(x_1 + x_2) = n,$$

or

$$x_1{}^2 + 2x_1x_2 + x_2{}^2 + mx_1 + mx_2 = n.$$

Assuming x_2 so small that $x_2{}^2$ may be neglected we obtain

$$x_2 = (n - x_1{}^2 - mx_1)/(2x_1 + m).$$

Now from the improved approximation, $x_1 + x_2$, we calculate in the same way a still better approximation, $x_1 + x_2 + x_3$, and so on. Viète used this method to approximate a root of the sextic equation

$$x^6 + 6000x = 191{,}246{,}976.$$

Viète's posthumously published treatise contains much of interest in the theory of equations. Here we find the familiar transformations for either increasing or multiplying the roots of an equation by a constant. Viète was aware of the expressions for the coefficients of polynomials, up through the fifth degree, as symmetric functions of the roots, and he knew the transformation which rids the general polynomial of its next to the highest degree term. In this treatise is found the following elegant solution of the cubic equation $x^3 + 3ax = 2b$, a form to which any cubic can be reduced. Setting

$$x = a/y - y$$

the given equation becomes

$$y^6 + 2by^3 = a^3,$$

a quadratic in y^3. We thus find y^3, and then y, and then x. Viète's solution of the quartic is similar to Ferrari's. Consider the general depressed quartic

$$x^4 + ax^2 + bx = c,$$

which may be written as

$$x^4 = c - ax^2 - bx.$$

Adding $x^2y^2 + y^4/4$ to both sides yields

$$(x^2 + y^2/2)^2 = (y^2 - a)x^2 - bx + (y^4/4 + c).$$

Now we choose y so that the right member is a perfect square. The condition for this is

$$y^6 - ay^4 + 4cy^2 = 4ac + b^2,$$

a cubic in y^2. Such a y may be found and the problem completed by extracting square roots.

Viète was an outstanding algebraist and so it is no surprise to learn that he applied algebra and trigonometry to his geometry. He contributed to the three famous problems of antiquity by showing that both the trisection and the duplication problems depend upon the solution of cubic equations. In Section 4-8 we have mentioned Viète's calculation of π and his interesting infinite product converging to $2/\pi$, and in Section 6-5 we mentioned his attempted restoration of Apollonius' lost work on *Tangencies*.

In 1594, Viète acquired some unfortunate notoriety by conducting an angry controversy with Clavius on the Gregorian reform of the calendar. Viète's attitude in the matter was wholly unscientific.

8-10   Other Mathematicians of the Sixteenth Century

Our account of the mathematics of the sixteenth century would not be complete without at least a brief mention of some of the other contributors. Of these are the mathematicians Clavius, Cataldi, and Stevin, and the mathematical astronomers Copernicus, Rhaeticus, and Pitiscus.

Christopher Clavius was born in Bamberg, Germany, in 1537 and died in Rome in 1612. He added little of his own to mathematics but probably did more than any other German scholar of the century to promote a knowledge of the subject. He was a gifted teacher and wrote highly esteemed textbooks on arithmetic (1583) and algebra (1608). In 1574 he published an edition of Euclid's *Elements* which is valuable for its extensive scholia. He also wrote on trigonometry and astronomy, and played an important part in the Gregorian reform of the calendar. As a Jesuit, he brought honor to his order.

Pietro Antonio Cataldi was born in Bologna in 1548, taught mathematics and astronomy in Florence, Perugia, and Bologna, and died in the city of his birth in 1626. He wrote a number of mathematical works among which are an arithmetic, a treatise on perfect numbers, an edition of the first six books of the *Elements*, and a brief treatise on algebra. He is credited with taking the first steps in the theory of continued fractions.

The most influential mathematician of the Low Countries in the sixteenth century was Simon Stevin (1548–1620). He became quartermaster general for the Dutch army and directed many public works. He published an arithmetic in which appears one of the earliest expositions of the theory of decimal fractions. Stevin won a high reputation among his contemporaries for his work on statics and hydrostatics.

Astronomy has long contributed to mathematics; in fact at one time the name "mathematician" meant an astronomer. Prominent among the astronomers who stimulated mathematics was Nicolas Copernicus (1473–1543) of Poland. He was educated at the University of Cracow and studied law, medicine, and astronomy at Padua and Bologna. His theory of the universe was completed in 1530 but was not published until the year of his death in 1543. Copernicus' work necessitated the improvement of trigonometry, and Copernicus himself contributed a treatise on the subject.

The leading Teutonic mathematical astronomer of the sixteenth century, and a disciple of Copernicus, was Georg Joachim Rhaeticus (1514–1576). He spent twelve years with hired computers forming two remarkable and still useful trigonometric tables. One was a ten-place table, of all six of the trigonometric functions, for every $10''$ of arc; the other was a fifteen-place table for sines for every $10''$ of arc, along with first, second, and third differences. Rhaeticus was the first to define the trigonometric functions as ratios of the sides of a right triangle. It was because of the importunities of Rhaeticus that Copernicus' great work was dramatically published just before the author died.

Rhaeticus' table of sines was edited and perfected in 1593 by Bartholomäus Pitiscus (1561–1613), a German clergyman with a preference for mathematics. His very satisfactory treatise on trigonometry was the first work on the subject to bear this title.

In summarizing the mathematical achievements of the sixteenth century, we can say that symbolic algebra was well started, computation with the Hindu-Arabic numerals became standardized, decimal fractions were developed, the cubic and quartic equations were solved and the theory of equations generally advanced, negative numbers were becoming accepted, trigonometry was perfected and systematized, and some excellent tables were computed. The stage was set for the remarkable strides of the next century.

It is interesting to note here that the first work on mathematics printed in the New World appeared in 1556 at Mexico City; it was a small commercial compendium by Juan Diez.

Problem Studies

8.1 Problems from the Dark Ages

Alcuin of York (*ca.* 775) was perhaps the compiler of the Latin collection entitled *Problems for the Quickening of the Mind.* Solve the following five problems from this collection.

(a) If 100 bushels of corn be distributed among 100 people in such a manner that each man receives 3 bushels, each woman 2, and each child 1/2 of a bushel, how many men, women, and children were there?

(b) Thirty flasks, ten full, ten half-empty, and ten entirely empty are to be divided among three sons so that flasks and contents should be shared equally. How may this be done?

(c) A dog chasing a rabbit, which has a start of 150 feet, jumps 9 feet every time the rabbit jumps 7. In how many leaps does the dog overtake the rabbit?

(d) A wolf, a goat, and a cabbage must be moved across a river in a boat holding only one besides the ferryman. How must he carry them across so that the goat shall not eat the cabbage, nor the wolf the goat?

(e) A dying man wills that if his wife, being with child, gives birth to a son, the son shall inherit 3/4 and the widow 1/4 of the property; but if a daughter is born, she shall inherit 7/12 and the widow 5/12 of the property. How is the property to be divided if both a son and a daughter are born? (This problem is of Roman origin. The solution given in Alcuin's collection is not acceptable.)

(f) In his *Geometry*, Gerbert solved the problem, considered very difficult at the time, of determining the legs of a right triangle whose hypotenuse and area are given. Solve this problem.

(g) Gerbert expressed the area of an equilateral triangle of side a as $(a/2)(a - a/7)$. Show that this is equivalent to taking $\sqrt{3} = 1.714$.

8.2 The Fibonacci Sequence

(a) Show that the following problem, found in the *Liber abaci*, gives rise to the *Fibonacci sequence*: 1, 1, 2, 3, 5, 8, $\cdots$, x, y, $x + y$, $\cdots$.

How many pairs of rabbits can be produced from a single pair in a year if every month each pair begets a new pair which from the second month on becomes productive?

If u_n represents the nth term of the Fibonacci sequence show that

(b) $u_{n+1}u_{n-1} = u_n^2 + (-1)^n$, $n \geq 2$,

(c) $u_n = [(1 + \sqrt{5})^n - (1 - \sqrt{5})^n]/2^n\sqrt{5}$,

(d) $\lim\limits_{n \to \infty} (u_n/u_{n+1}) = (\sqrt{5} - 1)/2$,

(e) u_n and u_{n+1} are relatively prime.

There is an extensive literature concerning the Fibonacci sequence. For some of the more eccentric applications to dissection puzzles, art, phyllotaxis, and the logarithmic spiral, see, for example, E. P. Northrop, *Riddles in Mathematics*.

8.3 Problems from the *Liber abaci*

Solve the following problems found in the *Liber abaci* (1202). The first one was posed to Fibonacci by a magister in Constantinople; the second was designed to illustrate the rule of three; the third is an example of an inheritance problem which reappeared in later works by Chuquet and Euler.

(a) If A gets from B 7 denarii, then A's sum is fivefold B's; if B gets from A 5 denarii, then B's sum is sevenfold A's. How much has each?

(b) A certain king sent 30 men into his orchard to plant trees. If they could set out 1000 trees in 9 days, in how many days would 36 men set out 4400 trees?

(c) A man left to his oldest son one bezant and a seventh of what was left; then, from the remainder, to his next son he left two bezants and a seventh of what was left; then, from the new remainder, to his third son he left three bezants and a seventh of what was left. He continued this way, giving each son one bezant more than the previous son and a seventh of what remained. By this division it developed that the last son received all that was left and all the sons shared equally. How many sons were there and how large was the man's estate?

8.4 Further Problems of Fibonacci

(a) Show that the squares of the numbers $a^2 - 2ab - b^2$, $a^2 + b^2$, $a^2 + 2ab - b^2$ are in arithmetic progression. If $a = 5$ and $b = 4$ the common difference is 720, and the first and third squares are $41^2 - 720 = 31^2$ and $41^2 + 720 = 49^2$. Dividing by 12^2 we obtain Fibonacci's solution to the first of the tournament problems, namely, find a rational number x such that $x^2 + 5$ and $x^2 - 5$ are each squares of rational numbers (see Section 8-3). The problem is insolvable if the 5 is replaced by 1, 2, 3, or 4. Fibonacci showed that if x and h are integers such that $x^2 + h$ and $x^2 - h$ are perfect squares, then h must be divisible by 24. As examples we have $5^2 + 24 = 7^2$, $5^2 - 24 = 1^2$ and $10^2 + 96 = 14^2$, $10^2 - 96 = 2^2$.

(b) Find a solution to the following problem, which is the third of the tournament problems solved by Fibonacci: Three men possess a pile of money, their shares being 1/2, 1/3, 1/6. Each man takes some money from the pile until nothing is left. The first man then returns 1/2 of what he took, the second 1/3, and the third 1/6. When the total so returned is divided equally among the men it is found that each then possesses what he is entitled to. How much money was in the original pile, and how much did each man take from the pile?

(c) Solve the following problem given by Fibonacci in the *Liber abaci*. This problem reappeared in a remarkable number of variations. It contains the essence of the idea of an annuity.

A man entered an orchard through seven gates, and there took a certain number of apples. When he left the orchard he gave the first guard half the apples that he had and one apple more. To the second guard he gave half his remaining apples and one apple more. He did the same to each of the remaining five guards, and left the orchard with one apple. How many apples did he gather in the orchard?

8.5   Star-Polygons

A *regular star-polygon* is the figure formed by connecting with straight lines every ath point, starting with some given one, of the n points which divide a circumference into n equal parts, where a and n are relatively prime and $n > 2$. Such a star-polygon is represented by the symbol $\{n/a\}$, and is sometimes called a regular *n-gram*. When $a = 1$ we have a regular polygon. Star-polygons made their appearance in the ancient Pythagorean school, where the $\{5/2\}$ star-polygon, or pentagram, was used as a badge of recognition. Star-polygons also occur in the geometry of Boethius and the translations of Euclid from the Arabic by Adelard and Campanus. Bradwardine developed some of their geometric properties. They were considered also by Regiomontanus, Charles de Bouelles (1470–1533), and Johann Kepler (1571–1630).

(a) Construct, with the aid of a protractor, the star-polygons {5/2}, {7/2}, {7/3}, {8/3}, {9/2}, {9/4}, {10/3}.

(b) Let $\phi(n)$, called the *Euler ϕ function*, denote the number of numbers less than n and prime to it. Show that there are $[\phi(n)]/2$ regular n-grams.

(c) Show that if n is prime there are $(n-1)/2$ regular n-grams.

(d) Show that the sum of the angles at the "points" of the regular $\{n/a\}$ star-polygon is given by $(n-2a)180°$. (This result was given by Bradwardine.)

8.6   Problems from Regiomontanus

Solve the following three problems found in Regiomontanus' *De triangulis omnimodis* (1464):

(a) Determine a triangle given the difference of two sides, the altitude on the third side, and the difference of the segments into which the altitude divides the third side. (The numerical values given by Regiomontanus are 3, 10, and 12.)

(b) Determine a triangle given a side, the altitude on this side, and the ratio of the other two sides. (The numerical values given by Regiomontanus are 20, 5, and 3/5.)

(c) Construct a cyclic quadrilateral given the four sides.

8.7   Problems from Chuquet

Solve the following problems adapted from Chuquet's *Triparty en la science des nombres* (1484):

(a) A merchant visited three fairs. At the first he doubled his money and spent $30, at the second he tripled his money and spent $54, at the third he quadrupled his money and spent $72, and then had $48 left. How much money had he at the start?

(b) A carpenter agrees to work under the conditions that he is to be paid $5.50 every day he works, while he must pay $6.60 every day he does not work. At the end of 30 days he finds he has paid out as much as he has received. How many days did he work?

(c) Two wine merchants enter Paris, one of them with 64 casks of wine, the other with 20. Since they have not enough money to pay the customs duties, the first pays 5 casks of wine and 40 francs, and the second pays 2 casks of wine and receives 40 francs in change. What is the price of each cask of wine and the duty on it?

(d) Chuquet gave the *regle des nombres moyens*, which says that if a, b, c, d are positive numbers then $(a+b)/(c+d)$ lies between a/c and b/d. Prove this.

8.8   Problems from Pacioli

Solve the following two problems found in Pacioli's *Sūma* (1494). The second problem is an elaboration of the popular "frog in the well problem" and has had many variants.

(a) The radius of the inscribed circle of a triangle is 4 and the segments into which one side is divided by the point of contact are 6 and 8. Determine the other two sides.

(b) A mouse is at the top of a poplar tree that is 60 feet high and a cat is on the ground at its foot. The mouse descends 1/2 of a foot each day and at night it turns back 1/6 of a foot. The cat climbs one foot a day and slips back 1/4 of a foot each night. The tree grows 1/4 of a foot between the cat and the mouse each day and shrinks 1/8 of a foot every night. How long will it take the cat to reach the mouse?

8.9   Early Commercial Problems

Solve the following problems found in early European arithmetics.

(a) This problem, from Buteo's arithmetic of 1559, is based upon difficulties of the early Roman navigators.

Two ships 20,000 stadia apart weighed anchor to sail straight toward each other. It happened that the first one set sail at daybreak with the north wind blowing. Toward evening, when it had gone 1200 stadia, the north wind fell and the southwest wind rose. At this time, the other ship set sail and and sailed 1400 stadia during the night. The first ship, however, was driven back 700 stadia by the contrary wind, but with the morning north wind it was driven ahead in the usual manner of outward sailing while the other went back 600 stadia. Thus alternately, night and day, the ships were carried along by a favorable wind and then driven back by an unfavorable one. I ask how many stadia the ships sailed in all and when they met?

(b) Here is a problem given by Tartaglia to illustrate the important matter of exchange.

If 100 lire of Modon money amounts to 115 lire in Venice, and if 180 lire in Venice comes to 150 in Corfu, and if 240 lire Corfu money is worth as much as 360 lire in Negroponte, what is the value in Modon coinage of 666 lire Negroponte money?

(c) The early arithmetics gave many problems involving custom duties. Following is a problem of this sort adapted from Clavius' arithmetic of 1583.

A merchant bought 50,000 pounds of pepper in Portugal for 10,000 scudi, paying a tax of 500 scudi. He carried it to Italy at a cost of 300 scudi and there paid another duty of 200 scudi. The transportation from the coast to

Florence cost 100 scudi and he was obliged to pay an impost of 100 scudi to that city. Lastly the government demanded a tax from each merchant of 1000 scudi. Now he is perplexed to know what price to charge per pound so that, after all these expenses, he may make a profit of one-tenth of a scudi a pound.

(d) In a practical manual for merchants written by the Florentine Ghaligai in 1521 occurs the following problem concerning profit and loss.

A man bought a number of bales of wool in London, each bale weighing 200 pounds, English measure, and each bale cost him 24 fl. He sent the wool to Florence and paid carriage duties, and other expenses, amounting to 10 fl. a bale. He wishes to sell the wool in Florence at such a price as to make 20 per cent on his investment. How much should he charge a hundredweight if 100 London pounds are equivalent to 133 Florentine pounds?

(e) Interest problems were very common. Here is one from Fibonacci's *Liber abaci* of 1202.

A certain man puts one denarius at interest at such a rate that in five years he has two denarii, and in five years thereafter the money doubles. I ask how many denarii he would gain from this one denarius in 100 years.

(f) The following problem is from Humphrey Baker's *The Well Spring of Sciences* (1568) and concerns itself with partnership.

Two marchauntes haue companied together, the first hath layde in the first of Januarie, 640 li. The seconde can lay in nothing vntill the firste of April. I demaund how much he shall lay in, to the end that he may take halfe the gaynes. (Assume that the partnership is to last for one year from the date of the first man's investment.)

(g) Here is essentially an annuity problem from Tartaglia's *General trattato* of 1556. It should be borne in mind that this problem was proposed before the invention of logarithms.

A merchant gave a university 2814 ducats on the understanding that he was to be paid back 618 ducats a year for nine years, at the end of which the 2814 ducats should be considered as paid. What compound interest was he getting on his money?

8.10 Cubic Equations

(a) Show that the transformation $x = z - a_1/na_0$ converts the n-ic equation

$$a_0x^n + a_1x^{n-1} + a_2x^{n-2} + \cdots + a_n = 0$$

into an equation in z which lacks the $(n-1)$st degree term.

(b) By part (a) the transformation $x = z - b/3a$ converts the cubic equation $ax^3 + bx^2 + cx + d = 0$ into one of the form $z^3 + 3Hz + G = 0$. Find H and G in terms of a, b, c, d.

(c) Derive the Cardano-Tartaglia formula,

$$x = \sqrt[3]{(n/2) + \sqrt{(n/2)^2 + (m/3)^3}} - \sqrt[3]{-(n/2) + \sqrt{(n/2)^2 + (m/3)^3}},$$

for solving the cubic equation $x^3 + mx = n$ (see Section 8-8).

(d) Solve $x^3 + 63x = 316$, for one root, by both the Cardano-Tartaglia formula and Viète's method.

(e) As an example of the irreducible case in cubics solve $x^3 - 63x = 162$ by the Cardano-Tartaglia formula. Then show that $(-3 + 2\sqrt{-3})^3 = 81 + 30\sqrt{-3}$ and $(-3 - 2\sqrt{-3})^3 = 81 - 30\sqrt{-3}$, whence the root given by the formula is -6 in disguise.

8.11 Quartic Equations

(a) Cardano solved the particular quartic $13x^2 = x^4 + 2x^3 + 2x + 1$ by adding $3x^2$ to both sides. Do this and solve the equation for all four roots.

(b) Da Coi in 1540 proposed the following problem to Cardano: "Divide 10 into three parts such that they shall be in continued proportion and that the product of the first two shall be 6." If the three parts be denoted by a, b, c, we have

$$a + b + c = 10, \qquad ac = b^2, \qquad ab = 6.$$

Show that when a and c are eliminated we obtain the quartic equation

$$b^4 + 6b^2 + 36 = 60b.$$

It was in trying to solve this quartic that Cardano's pupil Ferrari discovered his general method.

(c) Obtain, by both Ferrari's and Viète's methods, the cubic equations associated with the quartic of part (b).

8.12 Sixteenth-Century Notation

(a) Write, in Bombelli's notation, the expression

$$\sqrt{\left[\sqrt[3]{(\sqrt{68} + 2)} - \sqrt[3]{(\sqrt{68} - 2)} \right]}.$$

(b) Write, in modern notation, the following expression which occurs in Bombelli's work:

$$R\,c\lfloor\,4\,p\,di\,m\,R\,q\,11\,\rfloor\,p\,R\,c\lfloor\,4\,m\,di\,m\,R\,q\,11\,\rfloor.$$

(c) Write, in Viète's notation,

$$A^3 - 3BA^2 + 4CA = 2D.$$

8.13   Problems from Viète

(a) Establish the following identities given by Viète in his *Canon mathematicus seu ad triangula* (1579):

$$\sin \alpha = \sin (60° + \alpha) - \sin (60° - \alpha),$$

$$\csc \alpha + \cot \alpha = \cot \alpha/2,$$

$$\csc \alpha - \cot \alpha = \tan \alpha/2.$$

(b) Express $\cos 5\theta$ as a function of $\cos \theta$.

(c) Starting with $x_1 = 200$ approximate, by Viète's method, a root of $x^2 + 7x = 60{,}750$.

(d) Find the x_2, of Viète's method of successive approximations, for the cubic equation $x^3 + px^2 + qx = r$ (see Section 8-9).

8.14   Problems from Clavius

Solve the following recreational problems found in Clavius' algebra of 1608.

(a) In order to encourage his son in the study of arithmetic, a father agrees to pay his boy 8 cents for every problem correctly solved and to fine him 5 cents for each incorrect solution. At the end of 26 problems neither owes anything to the other. How many problems did the boy solve correctly?

(b) If I were to give 7 cents to each of the beggars at my door I would have 24 cents left. I lack 32 cents of being able to give them 9 cents apiece. How many beggars are there, and how much money have I?

(c) A servant is promised $100 and a cloak as his wages for a year. After 7 months he leaves this service and receives the cloak and $20 as his due. How much is the cloak worth?

Bibliography

CAJORI, FLORIAN, *A History of Mathematical Notations*, 2 vols. Chicago: The Open Court Publishing Company, 1928–1929.

CARDAN, JEROME, *The Book of My Life*, tr. from the Latin by Jean Stoner. New York: Dover Publications, Inc., 1963.

CROSBY, H. L., JR., *Thomas of Bradwardine His Tractatus de Proportionibus: Its Significance for the Development of Mathematical Physics.* Madison, Wis.: University of Wisconsin Press, 1955.

CUNNINGTON, SUSAN, *The Story of Arithmetic, a Short History of Its Origin and Development.* London: Swan Sonnenschein & Company, Ltd., 1904.

DAVID, F. N., *Games, Gods and Gambling.* New York: Hafner Publishing Company, 1962.

DAY, M. S., *Scheubel as an Algebraist, Being a Study of Algebra in the Middle of the Sixteenth Century, Together with a Translation of and a Commentary upon an Unpublished Manuscript of Scheubel's Now in the Library of Columbia University.* New York: Teachers College, Columbia University, 1926.

HAMBIDGE, JAY, *Dynamic Symmetry in Composition.* New Haven: Yale University Press, 1923.

———, *Practical Applications of Dynamic Symmetry*, ed. by Mary C. Hambidge. New Haven: Yale University Press, 1932.

INFELD, LEOPOLD, *Whom the Gods Love, the Story of Evariste Galois.* New York: McGraw-Hill Book Company, Inc., 1948.

KRAITCHIK, MAURICE, *Mathematical Recreations.* New York: W. W. Norton & Company, Inc., 1942.

MORLEY, HENRY, *Jerome Cardan, The Life of Girolamo Cardano of Milan, Physician*, 2 vols. London: Chapman & Hall, Ltd., 1854.

NICOMACHUS OF GERASA, *Introduction to Arithmetic*, tr. by M. L. D'Ooge, with *Studies in Greek Arithmetic*, by F. E. Robbins and L. C. Karpinski. Ann Arbor, Mich.: University of Michigan Press, 1938.

NORTHROP, E. P., *Riddles in Mathematics.* Princeton, N. J.: D. Van Nostrand Company, Inc., 1944.

ORE, OYSTEIN, *Number Theory and Its History.* New York: McGraw-Hill Book Company, Inc., 1948.

———, *Cardano, The Gambling Scholar.* Princeton, N. J.: Princeton University Press, 1953.

———, *Niels Henrik Abel, Mathematician Extraordinary.* Minneapolis: University of Minnesota Press, 1957.

ORESME, NICOLE, *An Abstract of Nicholas Orême's Treatise on the Breadths of Forms*, tr. by C. G. Wallis. Annapolis, Md.: St. John's Book Store, 1941.

SMITH, D. E., *Rara Arithmetica*. Boston: Ginn & Company, 1908.

——, *A Source Book in Mathematics*. New York: McGraw-Hill Book Company, Inc., 1929.

SULLIVAN, J. W. N., *The History of Mathematics in Europe, from the Fall of Greek Science to the Rise of the Conception of Mathematical Rigour*. New York: Oxford University Press, 1925.

WATERS, W. G., *Jerome Cardan, a Biographical Study*. London: Lawrence & Bullen, 1898.

YATES, R. C., *The Trisection Problem*. Ann Arbor, Mich.: Edwards Bros., Inc., 1947.

2

SEVENTEENTH CENTURY AND LATER EUROPEAN MATHEMATICS

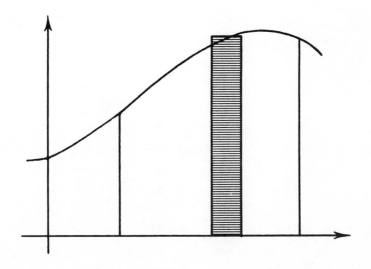

9

The Dawn of Modern Mathematics

9-1 The Seventeenth Century

The seventeenth century is outstandingly conspicuous in the history of mathematics. Early in the century Napier revealed his invention of logarithms, Harriot and Oughtred contributed to the notation and codification of algebra, Galileo founded the science of dynamics, and Kepler announced his laws of planetary motion. Later in the century Desargues and Pascal opened a new field of pure geometry, Descartes launched modern analytic geometry, Fermat laid the foundations of modern number theory, and Huygens made distinguished contributions to the theory of probability and other fields. Then, toward the end of the century, after a host of seventeenth-century mathematicians had prepared the way, the epoch-making creation of the calculus was made by Newton and Leibniz. We thus see that during the seventeenth century many new and vast fields were opened up for mathematical investigation.

The great impetus given to mathematics in the seventeenth century was shared by all intellectual pursuits, and was largely due, no doubt, to the political, economic, and social advances of the time. The century witnessed strong gains in the struggle for human rights, saw machines well advanced from the amusing toys of Heron's day to objects of increasing economic importance, and observed a growing spirit of intellectual internationalism and scientific skepticism. The more favorable political atmosphere of northern Europe, and the general conquering of the cold and darkness of the long winter months by adequate advances in heating and lighting, probably largely account for the northward shift of mathematical activity in the seventeenth century from Italy to France and England.

241

It is only fair to note here two facts that will tend to make our treatment of the history of the mathematics in the second part of this book a somewhat unbalanced presentation. The first of these is that mathematical activity began to grow at so great a rate that henceforth many names must be omitted that might have been considered in a less productive period. The second fact is that, with the unfolding of the seventeenth century, there occurred an increasing amount of mathematical research that cannot be considered as elementary, and which therefore is beyond the matter to be discussed in this book. Our discussion of seventeenth-century mathematics, along with some developments of more recent times, will occupy the concluding chapters of the book. In the present chapter, and the following one, we shall consider those developments that can be appreciated without a knowledge of the calculus. The most important parts of the omitted portions will be considered in Chapter XI, which will contain a sketch of the development of the ideas of the calculus from their beginnings in Greek antiquity up to the remarkable contributions made by Newton and Leibniz and their immediate precursors in the second half of the seventeenth century.

9-2   Logarithms

Many of the fields in which numerical calculations are important, such as astronomy, navigation, trade, engineering, and war, made ever increasing demands that these computations be performed more quickly and accurately. These increasing demands were met successively by four remarkable inventions: the Hindu-Arabic notation, decimal fractions, logarithms, and the modern computing machines. It is now time to consider the third of these great laborsaving devices, the invention of logarithms by John Napier in the early seventeenth century. The fourth invention will be considered below in Section 9-10.

John Napier (1550–1617) lived most of his life at the imposing family estate of Merchiston Castle, near Edinburgh, Scotland, and expended much of his energies in the political and religious controversies of his day. He was violently anti-Catholic and championed the causes of John Knox and James I. In 1593, he published a bitter and widely read attack on the Church of Rome entitled *A Plaine Discouery of the whole Reuelation of Saint Iohn*, in which he endeavored to prove that the Pope was Antichrist and that the

Creator proposed to end the world in the years between 1688 and 1700. The book ran through 21 editions, at least ten of them during the author's lifetime, and Napier sincerely believed that his reputation with posterity would rest upon this book. He also wrote prophetically of various infernal war engines and of "devises of sayling under water," accompanying the writings with plans and diagrams. Some of his war chariots are remarkably like a modern tank, and one of them was to contain a great chopping mouth that would destroy anything in its path. It is no wonder that his remarkable ingenuity and imagination led some to believe he was mentally unbalanced and others to regard him as a dealer in the black art. Many stories, probably unfounded, are told in support of these views. Thus there was the time he announced that his coal black rooster would identify for him which of his servants was stealing from him. The servants were sent one by one into a darkened room with instructions to pat the rooster on the back. Unknown to the servants, Napier had coated the bird's back with lampblack, and the guilty servant, fearing to touch the rooster, returned with clean hands.

As relaxation from his political and religious polemics, Napier amused himself with the study of mathematics and science, with the result that four products of his genius are now recorded in the history of mathematics. These are (1) the invention of logarithms; (2) a clever mnemonic, known as the *rule of circular parts*, for reproducing the formulas used in solving right spherical triangles; (3) at least two trigonometric formulas of a group of four known as *Napier's analogies*, useful in the solution of oblique spherical triangles; and, (4) the invention of a device, called *Napier's rods*, or *Napier's bones*, used for mechanically multiplying, dividing, and taking square roots of numbers. We turn now to the first, and most remarkable, of these four contributions; for a discussion of the other three, see Problem Studies 9.2 and 9.3.

As we know today, the power of logarithms as a computing device lies in the fact that by them multiplication and division are reduced to the simpler operations of addition and subtraction. A forerunner of this idea is apparent in the formula

$$\sin A \sin B = \{\cos (A - B) - \cos (A + B)\}/2,$$

well known in Napier's time, and it is quite probable that Napier's line of thought started with this formula, since otherwise it is difficult to account for his initial restriction of logarithms to those of the sines of angles. Napier labored at least twenty years upon his theory, and, whatever the genesis of his idea, his final definition of a logarithm is as follows. Consider a line segment AB and an infinite ray DE, as shown in Figure 54. Let points C and F start moving simultaneously from A and D, respectively, along these

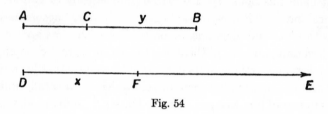

Fig. 54

lines, with the same initial rate. Suppose C moves with a velocity always numerically equal to the distance CB, and that F moves with a uniform velocity. Then Napier defined DF to be the logarithm of CB. That is, setting $DF = x$ and $CB = y$,

$$x = \text{Nap log } y.$$

In order to avoid the nuisance of fractions, Napier took the length of AB as 10^7, for the best tables of sines available to him extended to seven places. From Napier's definition, and through the use of knowledge not available to Napier, it develops that*

$$\text{Nap log } y = 10^7 \log_{1/e} (y/10^7),$$

so that the frequently made statement that Napierian logarithms are natural logarithms is actually without basis. One observes that the Napierian logarithm decreases as the number increases, contrary to what happens with natural logarithms.

*The result is easily shown with the aid of a little calculus. Thus we have $AC = 10^7 - y$, whence

$$\text{velocity of } C = -dy/dt = y.$$

That is, $dy/y = -dt$, or integrating, $\ln y = -t + C$. Evaluating the constant of integration by substituting $t = 0$, we find that $C = \ln 10^7$, whence

$$\ln y = -t + \ln 10^7.$$

Now

$$\text{velocity of } F = dx/dt = 10^7,$$

so that $x = 10^7 t$. Therefore

$$\text{Nap log } y = x = 10^7 t = 10^7 (\ln 10^7 - \ln y)$$
$$= 10^7 \ln (10^7/y) = 10^7 \log_{1/e} (y/10^7).$$

It further develops that, over a succession of equal periods of time, *y decreases* in *geometric* progression while *x increases* in *arithmetic* progression. Thus we have the fundamental principle of a system of logarithms, the association of a geometric and an arithmetic progression. It now follows, for example, that if $a/b = c/d$, then

$$\text{Nap log } a - \text{Nap log } b = \text{Nap log } c - \text{Nap log } d,$$

which is one of the many results established by Napier.

Napier published his discussion of logarithms in 1614 in a brochure entitled *Mirifici logarithmorum canonis descriptio* (A Description of the Wonderful Law of Logarithms). The work contains a table giving the logarithms of the sines of angles for successive minutes of arc. The *Descriptio* roused immediate and widespread interest, and in the year following its publication Henry Briggs (1561–1631), professor of geometry at Gresham College in London, and later professor at Oxford, traveled to Edinburgh to pay his respects to the great inventor of logarithms. It was upon this visit that both Napier and Briggs agreed that the tables would be more useful if they were altered so that the logarithm of 1 would be 0 and the logarithm of 10 would be an appropriate power of 10. Thus were born the so-called *Briggsian*, or *common*, logarithms of today. Logarithms of this sort, which are essentially logarithms to the base 10, owe their superior utility in numerical computations to the fact that our number system also is based on 10. For a number system having some other base b it would, of course, be most convenient for computational purposes to have tables of logarithms also to the base b.

Briggs devoted all his energies toward the construction of a table upon the new plan, and in 1624 published his *Arithmetica logarithmica*, containing a 14-place table of common logarithms of the numbers from 1 to 20,000 and from 90,000 to 100,000. The gap from 20,000 to 90,000 was later filled in, with help, by Adriaen Vlacq (1600–1666), a Dutch bookseller and publisher. In 1620, Edmund Gunter (1581–1626), one of Briggs's colleagues, published a seven-place table of the common logarithms of the sines and tangents of angles for intervals of a minute of arc. It was Gunter who invented the words *cosine* and *cotangent*; he is known to engineers for his "Gunter's chain." Briggs and Vlacq published four fundamental tables of logarithms, the results of which have only recently been superseded when, between 1924 and 1949, extensive 20-place tables were calculated in England in partial celebration of the tercentenary of the discovery of logarithms.

The word *logarithm* means "ratio number," and was adopted by Napier after first using the expression *artificial number*. Briggs introduced the word *mantissa*, which is a late Latin term of Etruscan origin, originally meaning an "addition" or "makeweight," and which in the sixteenth century came to mean "appendix." The term *characteristic* was also suggested by Briggs and was used by Vlacq. It is curious that it was customary in early tables of common logarithms to print the characteristic as well as the mantissa, and that it was not until the eighteenth century that the present custom of printing only the mantissas was established.

Napier's wonderful invention was enthusiastically adopted throughout Europe. In astronomy, in particular, the time was overripe for such a discovery; as Laplace asserted, the invention of logarithms "by shortening the labors doubled the life of the astronomer." Bonaventura Cavalieri, about whom we shall have more to say in Chapter XI, did much to bring logarithms into vogue in Italy. A similar service was rendered by Johann Kepler in Germany and Edmund Wingate in France. Kepler will be considered more fully below in Section 9-6; Wingate, who spent many years in France, became the most prominent seventeenth-century British textbook writer on elementary arithmetic.

Napier's only rival for priority of invention of logarithms was the Swiss instrument maker Jobst Bürgi (1552–1632). Bürgi conceived and constructed a table of logarithms independently of Napier, publishing his results in 1620, six years after Napier had announced his discovery to the world. Though both men had conceived the idea of logarithms long before publishing, it is generally believed that Napier had the idea first. Whereas Napier's approach was geometrical, Bürgi's was algebraic. Nowadays a logarithm is universally regarded as an exponent. Thus if $n = b^x$, we say x is the logarithm of n to the base b. From this definition the laws of logarithms follow immediately from the laws of exponents. One of the paradoxes in the history of mathematics is the fact that logarithms were discovered before exponents were in use.

9-3 The Savilian and Lucasian Professorships

Since so many distinguished British mathematicians have held either a Savilian professorship at Oxford or a Lucasian professorship at Cambridge, a brief reference to these professorships is desirable.

Sir Henry Savile was one time warden of Merton College at Oxford, later provost of Eton, and a lecturer on Euclid at Oxford. In 1619, he founded two professorial chairs at Oxford, one in geometry and one in astronomy. Henry Briggs was the first occupant of the Savilian chair of geometry at Oxford. The earliest professorship of mathematics established in Great Britain was a chair in geometry founded by Sir Thomas Gresham in 1596 at Gresham College in London. Briggs also had the honor of being the first to occupy this chair. John Wallis, Edmund Halley, and Sir Christopher Wren are other seventeenth-century incumbents of Savilian professorships.

Henry Lucas, who represented Cambridge in parliament in 1639–1640, willed the university resources for the founding in 1663 of the professorship that bears his name. Isaac Barrow was elected the first occupant of this chair in 1664, and six years later was succeeded by Isaac Newton.

9-4   Harriot and Oughtred

Thomas Harriot (1560–1621) was another mathematician who lived the longer part of his life in the sixteenth century but whose outstanding publication appeared in the seventeenth century. He is of special interest to Americans because in 1585 he was sent by Sir Walter Raleigh to the New World to survey and map what was then called Virginia but is now North Carolina. As a mathematician, Harriot is usually considered the founder of the English school of algebraists. His great work in this field, the *Artis analyticae praxis*, was not published until ten years after his death and deals largely with the theory of equations. This work did much toward setting the present standards for a textbook on the subject. It includes a treatment of equations of the first, second, third, and fourth degrees, the formation of equations having given roots, the relations between the roots and the coefficients of an equation, the familiar transformations of an equation into another having roots bearing some specific relation to the roots of the original equation, and the numerical solution of equations. Much of this material is found, of course, in the works of Viète, but Harriot's is a more complete and better systematized treatment. Harriot followed Viète's plan of using vowels for unknowns and consonants for constants, but he adopted the lower case rather than the upper case letters. He improved on Viète's notation for powers by representing a^2 by aa, a^3 by aaa, and so forth. He was also the first to use the signs $>$ and $<$ for "is greater than" and "is less than," respectively, but these symbols were not immediately accepted by other writers.

Harriot has been erroneously credited with several other mathematical innovations and discoveries, such as a well-formed analytic geometry (before Descartes' publication of 1637), the statement that any polynomial of degree n has n roots, and "Descartes' rule of signs." Some of these errors of authorship seem due to insertions, made by later writers, among some of Harriot's preserved manuscripts. Thus there are eight volumes of Harriot's manuscripts in the British Museum, but the part dealing with analytic geometry has been shown by D. E. Smith to be an interpolation by a later hand.

Harriot was also prominent as an astronomer, having discovered sunspots and having observed the satellites of Jupiter independently of Galileo and at about the same time.

In the same year, 1631, that Harriot's posthumous work on algebra appeared, there also appeared the first edition of William Oughtred's popular *Clavis mathematicae*, a work on arithmetic and algebra that did much toward spreading mathematical knowledge in England. William Oughtred (1574–1660) was one of the most influential of the seventeenth-century English writers on mathematics. Although by profession an Episcopal minister, he gave free private lessons to pupils interested in mathematics. Among such pupils were John Wallis, Christopher Wren, and Seth Ward, later famous, respectively, as a mathematician, an architect, and an astronomer. It is said that Oughtred died in a transport of joy when he heard the news of the restoration of Charles II. To this Augustus De Morgan once remarked, "It should be added, by way of excuse, that he was eighty-six years old."

In his writings, Oughtred placed emphasis on mathematical symbols, giving over 150 of them. Of these only three have come down to present times: the cross ($\times$) for multiplication, the four dots (: :) used in a proportion, and our frequently used symbol for difference between ($\sim$). The cross as a symbol for multiplication, however, was not readily adopted because, as Leibniz objected, it too closely resembles x. Although Harriot on occasion used the dot ($\cdot$) for multiplication, this symbol was not prominently used until Leibniz adopted it. Leibniz also used the cap symbol ($\cap$) for multiplication, a symbol which is used today to indicate multiplication in the theory of sets. The Anglo-American symbol for division ($\div$) is also of seventeenth-century origin, having first appeared in print in 1659 in an algebra by the Swiss Johann Heinrich Rahn (1622–1676). The symbol became known in England some years later when this work was translated. This symbol for division has long been used in continental Europe to indicate subtraction. Our familiar signs, in geometry, for similar (∼), and for congruent (≅), are due to Leibniz.

Besides the *Clavis mathematicae*, Oughtred published *The Circles of Proportion* (1632) and *Trigonometrie* (1657). The second work is of some historical importance because of its early attempt to introduce abbreviations for the names of the trigonometric functions. The first work describes a circular slide rule. Oughtred, however, was not the first to describe in print a slide rule of the circular type, and an argument of priority of invention

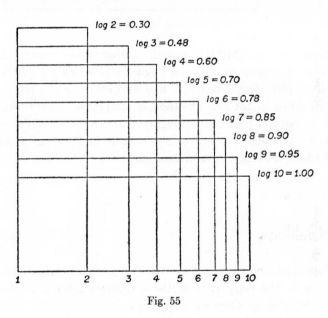

Fig. 55

rests between him and Richard Delamain, one of his pupils. But Oughtred does seem unquestionably to have invented, about 1622, the straight logarithmic slide rule. In 1620, Gunter constructed a logarithmic scale, or a line of numbers on which the distances are proportional to the logarithms of the numbers indicated (see Figure 55), and mechanically performed multiplications and divisions by adding and subtracting segments of this scale with the aid of a pair of dividers. The idea of carrying out these additions and subtractions by having two like logarithmic scales, one sliding along the other as shown in Figure 56, is due to Oughtred. Although Oughtred invented such a simple slide rule as early as 1622, he did not describe it in print until 1632. A runner for the slide rule was suggested by Isaac Newton in 1675, but was not actually constructed until nearly a century later. Several slide rules for special purposes, such as for commercial transactions, for measuring timber, and so forth, were devised in the seventeenth century. The log log scale was invented in 1815, and it was in 1850 that the French army officer Amédée Mannheim (1831–1906) standardized the modern slide rule.

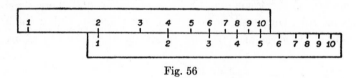

Fig. 56

It is believed that Oughtred was the author of the remarkable anonymous 16-page Appendix to the 1618 English edition by Edward Wright of Napier's *Descriptio*. Here appears the first use of the cross for multiplication, the first invention of the radix method of calculating logarithms, and the first table of natural logarithms. Oughtred also wrote a work on *gauging* (the science of computing the capacities of casks and barrels) and he translated and edited a French work on mathematical recreations.

9-5   Galileo

There were two outstanding astronomers who contributed notably to mathematics in the early part of the seventeenth century: the Italian, Galileo Galilei, and the German, Johann Kepler.

Galileo, the son of an impoverished Florentine nobleman, was born in Pisa in 1564. After a start as a medical student, he secured parental permission to devote himself to science and mathematics, fields in which he possessed strong natural talent. It was while still a medical student at the University of Pisa that Galileo made the historically famous observation that the great pendulous lamp of the cathedral there oscillated to and fro with a period independent of the size of the arc of oscillation.* He later showed the period of a pendulum to be also independent of the weight of the pendulum's bob. When 25, Galileo was appointed professor of mathematics

*This is only approximately true, the approximation being very close in the case of small amplitudes of oscillation.

at Pisa, and while holding this appointment is said to have performed experiments from the leaning tower there showing that, contrary to the teachings of Aristotle, heavy bodies do not fall faster than light ones. He arrived at the law that the distance a body falls is proportional to the square of the time of falling, in accordance with the familiar formula $s = \frac{1}{2}gt^2$. Because of local controversies, Galileo resigned his chair in 1591 and the following year accepted a professorship of mathematics at Padua, where there was an atmosphere more friendly to scientific pursuits. Here, for nearly 18 years, Galileo continued his experiments and his teaching, and won a widespread fame. It was while at Padua that Galileo heard of the discovery, in about 1607, of the telescope by the lens grinder Johann Lippersheim of Holland, and he set about making some instruments of his own, producing one that had a magnifying power of more than 30 diameters. With his telescope he observed sunspots, the mountains on the moon, the phases of Venus, Saturn's rings, and the four bright satellites of Jupiter. The observation of the sunspots contradicted the Aristotelian view that the sun was without blemish, and that of the satellites of Jupiter corroborated the Copernican theory of the solar system. These discoveries roused the opposition of the Church and finally, in 1633, three years after his publication of a book which supported the Copernican theory, Galileo was summoned to appear before the Inquisition, and there forced to recant his scientific findings. Not many years later he became blind. He died in 1642.

To Galileo we owe the modern spirit of science as a harmony between experiment and theory. He founded the mechanics of freely falling bodies and laid the foundation of dynamics in general, a foundation upon which Isaac Newton was able later to build the science. He was the first to realize the parabolic nature of the path of a projectile in a vacuum and speculated on laws involving momentum. He invented the first modern type microscope and the onetime very popular sector compasses (see Problem Study 9.6). Very interesting, historically, are statements made by Galileo showing that he grasped the idea of equivalence of infinite classes (see Problem Study 9.7), a fundamental point in Cantor's nineteenth-century theory of sets, which has been so influential in the development of modern analysis. These statements, and the bulk of Galileo's ideas in dynamics, can be found in his *Discorsi e dimostrazioni matematiche intorno a due nuove scienze*, published in Leyden in 1638. It would seem that Galileo was jealous of his famous contemporary, Johann Kepler, for although Kepler had announced all three of his important laws of planetary motion by 1619, these laws were completely ignored by Galileo.

9-6 Kepler

Johann Kepler was born near Stuttgart in 1571 and educated at the University of Tübingen with the original intention of becoming a Lutheran minister. His deep interest in astronomy led him to change his plans, and in 1594, when in his early twenties, he accepted a lectureship at the University of Grätz in Austria. In 1599, he became assistant to the famous but quarrelsome Danish-Swedish astronomer Tycho Brahe, who had moved to Prague as court astronomer to Kaiser Rudolph II. Shortly after, in 1601, Brahe suddenly died and Kepler inherited both his master's position and his vast and very accurate collection of astronomical data on the motion of the planets. With extraordinary pertinacity and after much computational labor, trial and error, and false solutions, Kepler was able finally to formulate in 1609 his first two laws, and then ten years later in 1619, his third law, of planetary motion.

These laws of planetary motion are landmarks in the history of astronomy and mathematics, for in the effort to justify them Isaac Newton was led to create modern celestial mechanics. The three laws are

I. *The planets move about the sun in elliptical orbits with the sun at one focus.*

II. *The radius vector joining a planet to the sun sweeps over equal areas in equal intervals of time.*

III. *The square of the time of one complete revolution of a planet about its orbit is proportional to the cube of the orbit's semimajor axis.*

The empirical discovery of these laws from Brahe's mass of data constitutes one of the most remarkable inductions ever made in science. It is very interesting that 1800 years after the Greeks had developed the properties of the conics there should occur such an illuminating practical application of them.

Kepler was one of the precursors of the calculus. In order to compute the areas involved in his second law of planetary motion, he had to resort to a crude form of the integral calculus. He also, in his *Stereometria doliorum vinorum* (Solid Geometry of Wine Barrels, 1615), applied crude integration procedures to the finding of the volumes of 93 solids obtained by rotating segments of conic sections about an axis in their plane. Among these solids were the torus and two solids which he called *the apple* and *the lemon*, the last two solids being those obtained by revolving a major and a minor arc, respectively, of a circle about the arc's chord as an axis. Kepler became interested in this matter upon observing some of the poor methods in use by the wine gaugers of his time. It is quite possible that Cavalieri was influenced by this work of Kepler when he later carried the refinement of the infinitesimal calculus a stage further with his *method of indivisibles*. We shall return to a discussion of all this in Chapter XI.

Notable contributions were made by Kepler to the subject of polyhedra. He seems to have been the first to recognize an *antiprism* (obtained from a prism by rotating the top base in its own plane so as to make its vertices correspond to the sides of the lower base, and then joining in zigzag fashion the vertices of the two bases). He also discovered the cuboctahedron, rhombic dodecahedron, and rhombic triakontahedron.* The second of these polyhedra occurs in nature as a garnet crystal. Of the four possible regular star-polyhedra two were discovered by Kepler and the other two in 1809 by Louis Poinsot (1777–1859), a pioneer worker in geometrical mechanics. The Kepler-Poinsot star-polyhedra are space analogues of the regular star-polygons in the plane (see Problem Study 8.5). Kepler also interested himself in the problem of filling the plane with regular polygons (not necessarily all alike) and space with regular polyhedra (see Problem Study 9.9).

Kepler solved the problem of determining the type of conic determined by a given vertex, the axis through this vertex, and an arbitrary tangent with its point of contact, and he introduced the word *focus* into the geometry of conics. He approximated the perimeter of an ellipse of semiaxes a and b by the use of the formula $\pi(a + b)$. He also laid down a so-called *principle of continuity* which essentially postulates the existence at infinity in a plane of certain ideal points and an ideal line, having many of the properties of ordinary points and lines. Thus he explained that a line can be considered as closed at infinity, that two parallel lines should be regarded as intersecting at infinity, and that a parabola may be regarded as a limiting case of either an ellipse or a hyperbola in which one of the foci has retreated to infinity. This concept was much extended in 1822 by the French geometer Poncelet when he made an effort to find in geometry a "real" justification for imaginaries that occur elsewhere in mathematics.

Kepler's work is often a blend of mystical and highly fanciful speculation combined with a truly deep grasp of scientific truths. It is sad that his life was made almost unendurable by a multiplication of worldly misfortunes. His favorite child died of smallpox, his wife went mad and died, his mother

---

*Construction patterns for these solids can be found in Miles C. Hartley, *Patterns of Polyhedrons*, rev. ed. (Ann Arbor, Mich.; priv. ptd., Edwards Bros., Inc., 1957).

was charged with witchcraft, he himself was accused of heterodoxy, and his stipend was always in arrears. One report says that his second marriage was even less fortunate than his first although he took the precaution to analyze carefully the merits and demerits of eleven girls before choosing the wrong one. He was forced to augment his income by casting horoscopes, and he died in 1630 while on a journey to obtain some of his long overdue salary.

9-7 Desargues

In 1639, nine years after Kepler's death, there appeared in Paris a remarkably original but little heeded treatise on the conic sections.* It was written by Gérard Desargues, an engineer, architect, and onetime French army officer, who was born in Lyons in 1593 and who died in the same city about 1662. The work was so generally neglected by other mathematicians that it was soon forgotten and all copies of the publication disappeared. Two centuries later, when the French geometer Michel Chasles (1793–1880) wrote his still standard history of geometry, there was no means of estimating the value of Desargues' work. Six years later, however, in 1845, Chasles happened upon a manuscript copy of the treatise, made by Desargues' pupil, Philippe de la Hire (1640–1718), and since that time the work has been regarded as one of the classics in the early development of synthetic projective geometry.

There are several reasons that can be advanced to account for the initial neglect of Desargues' little volume. It was overshadowed by the more supple analytic geometry introduced by Descartes two years earlier. Geometers were generally expending their energies either developing this new powerful tool or trying to apply infinitesimals to geometry. Also, Desargues adopted an unfortunate and eccentric style of writing. He introduced some 70 new terms, many of a recondite botanical origin, of which only one, *involution,* has survived, and, curiously enough, this one was preserved because it was the one piece of Desargues' technical jargon that was singled out for the sharpest criticism and ridicule by his reviewer.

Desargues wrote other books besides the one on conic sections, one of

Brouillon projet d'une atteinte aux événemens des rencontres d'un cone avec un plan. (Proposed Draft of an Attempt to Deal with the Events of the Meeting of a Cone with a Plane.)

them being a treatise on how to teach children to sing well. But it is the
little book on conic sections that marks him as the most original contributor
to synthetic geometry in the seventeenth century. Starting with Kepler's
doctrine of continuity, the work develops many of the fundamental theorems
on involution, harmonic ranges, homology, poles and polars, and perspective
—topics familiar to those who have taken one of our present-day courses in

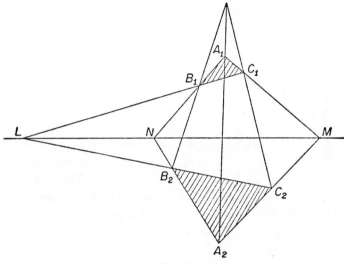

Fig. 57

projective geometry.* Interesting is a remark to the effect that the notion
of poles and polars may be extended to spheres and to certain other surfaces
of the second degree. It is likely that Desargues was aware of only a few of
the surfaces of second degree, many of these surfaces probably remaining
unknown until their complete enumeration by Euler in 1748. Elsewhere
we find Desargues' fundamental two-triangle theorem: *If two triangles, in
the same plane or not, are so situated that lines joining pairs of corresponding
vertices are concurrent, then the points of intersection of pairs of corresponding
sides are collinear, and conversely* (see Figure 57).

Desargues, when he was in his thirties and living in Paris, made a
considerable impression on his contemporaries through a series of gratuitous
lectures. His work was appreciated by Descartes, and Blaise Pascal once
credited Desargues as being the source of much of his inspiration. La Hire,

*That some of these concepts were known to the ancient Greeks has been pointed
out in Chapter 6.

with considerable labor, tried to show that all the theorems of Apollonius' *Conic Sections* can be derived from the circle by Desargues' method of central projection. In spite of all this, however, the new geometry took little hold in the seventeenth century, and the subject lay practically dormant until the early part of the nineteenth century, when enormous interest in the subject developed and great advances were made by such men as Gergonne, Poncelet, Brianchon, Dupin, Chasles, and Steiner.

9-8   Some Later Developments in Projective Geometry

The term *pole*, in the sense used in projective geometry, was first introduced in 1810 by the French mathematician Servois, and the corresponding term *polar* by Gergonne three years later. The idea of poles and polars was later elaborated by Gergonne and Poncelet into a regular method, out of which grew the elegant *principle of duality* of projective geometry (see Problem Study 9.12). Projective geometry was freed of any metrical basis by Staudt in 1847.

About 1872 Felix Klein announced his fruitful *Erlanger Programm* for the codification of geometries. Somewhat oversimply stated, this claims a geometry to be the investigation of those properties of figures which remain unchanged when the figures are subjected to a group of transformations. For plane Euclidean geometry the group of transformations is the totality of all rotations and translations in the plane. Since areas and line lengths of figures remain invariant when the figures are subjected to such transformations, areas and line lengths are properly regarded as subject matter in Euclidean geometry. Also, a circle remains a circle, and the midpoint of a line segment remains the mid-point of the line segment, under such transformations, and so circles and mid-points of line segments form part of the content of Euclidean geometry. Now plane projective geometry studies those properties of figures which remain invariant when the figures are subjected to the group of projective transformations, in which are the transformations of central projection, where a plane of figures may be projected from a point outside the plane onto another plane not through the chosen center of projection, as in Figure 58. In this way, for example, a circle may be projected into an ellipse. Areas and line lengths no longer remain unchanged under such transformations, and therefore do not constitute subject matter of projective geometry. Circles and mid-points also are not concepts of projective geometry. But it can be shown that collinear points transform into collinear points, and concurrent lines into concurrent lines, whence collinearity of points and concurrency of lines do constitute subject matter of projective geometry. In Section 6-9, it was pointed out

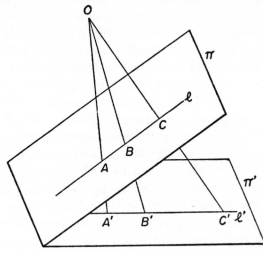

Fig. 58

that Pappus was aware of the invariance of a cross ratio under projection. Klein's synthesis of geometry is still useful in the domains where it applies, and many geometries can be considered neatly in this way, but shortly after the development of general relativity in 1916, geometries came to light which cannot be fitted into the codification and a new point of view upon the matter was developed based upon the idea of abstract space with a superimposed structure that may or may not be definable in terms of some transformation group. We cannot go into this any further here.

9-9   Pascal

One of the few contemporaries of Desargues who showed a real appreciation of his work was Blaise Pascal, a mathematical genius of high order. Pascal was born in the French province of Auvergne in 1623 and very early showed phenomenal ability in mathematics. At the age of 12 he discovered, quite unassisted, many of the theorems of elementary geometry; at 14 he participated in the weekly gatherings of a group of French mathematicians from which the French Academy ultimately arose in 1666; when 16 he arrived at some new and deep theorems in the projective geometry of these curves; two to three years later he invented the first computing machine and began to apply his unusual talents to mechanics and physics; a few years later, in 1648, he wrote a comprehensive manuscript on conic sections.

This astonishing and precocious activity came suddenly to an end in 1650, when, suffering from frail health, Pascal decided to abandon his researches in mathematics and science and to devote himself to religious contemplation. Three years later, however, he returned briefly to mathematics. At this time he wrote his *Traité du triangle arithmétique*, conducted several experiments on fluid pressure, and in correspondence with Fermat assisted in laying the foundations of the mathematical theory of probability. But late in 1654 he received what he regarded as a strong intimation that these renewed activities were not pleasing to God. The divine hint occurred when his runaway horses dashed over the parapet of the bridge at Neuilly and he himself was saved only by the miraculous breaking of the traces. Fortified with a reference to the accident written on a small piece of parchment henceforth carried next to his heart, he dutifully went back to his religious meditations.

Only once again, in 1658, did Pascal return to mathematics. While suffering with toothache some geometrical ideas occurred to him, and his teeth suddenly ceased to ache. Regarding this as a sign of divine will, he obediently applied himself assiduously for eight days toward developing his ideas, producing in this time a fairly full account of the geometry of the cycloid curve and solving some problems that subsequently, when issued as challenge problems, baffled other mathematicians. His famous *Provincial Letters* and his *Pensées*, which are read today as models of early French literature, were written toward the close of his brief life. He died in Paris in 1662 at the age of 39. We might add here that his father, Étienne Pascal (1588–1640), was also an able mathematician; it is for the father that the *limaçon of Pascal* is named (see Problem Study 4.7 (c)).

Pascal has been described as the greatest "might-have-been" in the history of mathematics. With such unusual talents and such deep geometrical intuition he should have produced, under more favorable conditions, a great deal more. But his health was such that most of his life was spent racked with physical pain, and from early manhood he also suffered the mental torments of a religious neurotic.

Pascal's manuscript on conic sections was founded on the work of Desargues and is now lost, but it was seen by Descartes and Leibniz. Descartes could not believe that the work was written by the boy and felt that it must have been written by the father instead. Here occurred Pascal's famous "mystic hexagram" theorem of projective geometry: *If a hexagon be inscribed in a conic, then the points of intersection of the three pairs of opposite sides are collinear, and conversely* (see Figure 59). He probably established the theorem, in Desargues' fashion, by first proving it

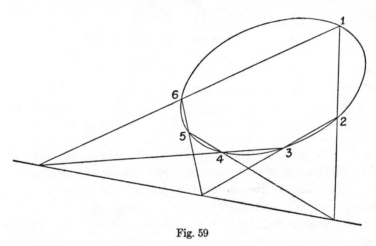

Fig. 59

true for a circle and then passing by projection to any conic section. Although the theorem is one of the richest in the whole of projective geometry (see Problem Study 9.11), we probably should take lightly the often told tale that Pascal himself deduced over 400 corollaries from it. The manuscript was never published, and probably never completed, but in 1640 Pascal did print a one-page broadside, entitled *Essay pour les coniques*, announcing some of his findings. Only two copies of this famous leaflet are known to be still in existence, one at Hanover among the papers of Leibniz, and the other in the Bibliotheque Nationale at Paris. Pascal's "mystic hexagram" theorem is involved in the third lemma of the leaflet.

Pascal's *Traité du triangle arithmétique* was written in 1653 but was not printed until 1665. He constructed his "arithmetical triangle" as indicated in the adjoined figure. Any element (in the second or a following row) is obtained as the sum of all those elements of the preceding row lying just above or to the left of the desired element. Thus, in the fourth row,

$$35 = 15 + 10 + 6 + 3 + 1.$$

The triangle, which may be of any order, is obtained by drawing a diagonal as indicated in the figure. The student of college algebra will recognize that the numbers along such a diagonal are the successive coefficients in a binomial expansion. For example, the numbers along the fifth diagonal,

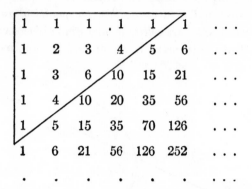

namely 1, 4, 6, 4, 1, are the successive coefficients in the expansion of $(a + b)^4$. The finding of binomial coefficients was one of the uses to which Pascal put his triangle. He also used it, particularly in his discussions on probability, for finding the number of combinations of n things taken r at a time (see Problem Study 9.13 (g)), which he correctly stated to be

$$\frac{n!}{r!\,(n-r)!},$$

where $n!$ is our present-day notation* for the product

$$n(n-1)(n-2)\cdots(3)(2)(1).$$

There are many relations involving the numbers of the arithmetic triangle, several of which were developed by Pascal (see Problem Study 9.13). Pascal was not the originator of the arithmetic triangle, for such an array had appeared in a number of prior works, the oldest known reference being a work of 1303 by the Chinese algebraist Chu Shï-kié. It is because of Pascal's development of many of the triangle's properties and because of the applications which he made of these properties that the array has become known

*The symbol $n!$, called *factorial n*, was introduced in 1808 by Christian Kramp (1760–1826) of Strasbourg, who chose this symbol so as to circumvent printing difficulties incurred by a previously used symbol.

as *Pascal's triangle.* In Pascal's treatise on the triangle appears one of the earliest acceptable statements of the method of mathematical induction.

Although the Greek philosophers of antiquity discussed necessity and contingency at length, it is perhaps correct to say that there was no mathematical treatment of probability until the latter part of the fifteenth century and the early part of the sixteenth century, when some of the Italian mathematicians attempted to evaluate the chances in certain gambling games, like that of dice. Cardano, as was noted in Section 8-8, wrote a brief gambler's guidebook in which some of the aspects of mathematical probability are involved. But it is generally agreed that the one problem to which can be credited the origin of the science of probability is the so-called *problem of the points.* This problem requires the determination of the division of the stakes of an interrupted game of chance between two supposedly equally skilled players, knowing the scores of the players at the time of interruption and the number of points needed to win the game. Pacioli, in his *Sūma* of 1494, was one of the first writers to introduce the problem of the points into a work on mathematics. The problem was also discussed by Cardano and Tartaglia. But a real advance was not made until the problem was proposed, in 1654, to Pascal, by the Chevalier de Méré, an able and experienced gambler whose theoretical reasoning on the problem did not agree with his observations. Pascal became interested in the problem and communicated it to Fermat. There ensued a remarkable correspondence between the two men,* in which the problem was correctly but differently solved by each. Pascal solved the general case, obtaining many results through a use of the arithmetical triangle. Thus it was that in their correspondence Pascal and Fermat laid the foundations of the science of probability.

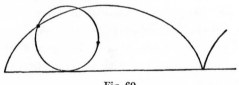

Fig. 60

Pascal's last mathematical work was that on the cycloid, the curve traced by a point on the circumference of a circle as the circle rolls along a straight line (see Figure 60). This curve, which is very rich in mathematical and physical properties, played an important role in the early development

*The correspondence appears in D. E. Smith, *A Source Book in Mathematics* (New York: McGraw-Hill Book Company, Inc., 1929).

of the methods of the calculus. Galileo was one of the first to call attention to the curve and once recommended that it be used for the arches of bridges. Shortly after, the area under one arch of the curve was found, and methods of drawing tangents to the curve were discovered. These discoveries led mathematicians to consider questions concerned with the surface and volume of revolution obtained by rotating a cycloidal arch about various lines. Such problems, as well as others concerned with the centroids of the figures formed, were solved by Pascal, and some of the results were issued by him as challenge problems to other mathematicians. Pascal's solutions were effected by the pre-calculus method of indivisibles and were equivalent to the evaluation of a number of definite integrals encountered in present-day calculus classes. The cycloid has so many attractive properties, and has engendered so many quarrels, that it has been called "the Helen of geometry."

Pascal sometimes wrote under the nom de plume of Lovis de Montalte, or its anagram, Amos Dettonville.

9-10　Calculating Machines

The invention of the first calculating machine is attributed to Pascal, who, in 1642, devised an adding machine to assist his father in the auditing of the government accounts at Rouen. The instrument was able to handle numbers not exceeding six digits. It contained a sequence of engaging dials, each marked from 0 to 9, so designed that when one dial of the sequence turned from 9 to 0 the preceding dial of the sequence automatically turned one unit. Thus the "carrying" process of addition was mechanically accomplished. Pascal manufactured over 50 machines, some of which are still preserved in the Conservatoire des Arts et Métiers at Paris. It is interesting that Pascal has also been credited with the invention of the one-wheeled wheelbarrow as we know it today.

Later in the century, Leibniz (1671) in Germany and Sir Samuel Morland (1673) in England invented machines that multiplied. Similar attempts were made by a number of others, but most of these machines proved to be slow and impractical. In 1820, Thomas de Colmar, although not familiar with Leibniz's work, transformed a Leibniz type of machine into one which could perform subtractions and divisions. This machine proved to be the prototype of almost all commercial machines built before

1875, and of many developed since that time. In 1875, the American Frank Stephen Baldwin was granted a patent for the first practical calculating machine that could perform the four fundamental operations of arithmetic without any resetting of the machine. In 1878, Willgodt Theophile Odhner, a Swede, was granted a United States patent on a machine very similar in design to that of Baldwin. Today there are several makes of electrically operated desk calculators, such as the Friden, Marchant, and Monroe, which have essentially the same basic construction as the Baldwin machine.

About 1812, the English mathematician Charles Babbage (1792–1871) began to consider the construction of a machine to aid in the calculation of mathematical tables. He resigned the Lucasian professorship at Cambridge in order to devote all his energies to the construction of his machine. In 1823, after investing and losing his own personal fortune in the venture, he secured financial aid from the British government and set to work to make a *difference engine* capable of employing 26 significant figures and of computing and printing successive differences out to the sixth order. But Babbage's work did not progress satisfactorily, and ten years later the governmental aid was withdrawn. Babbage thereupon abandoned his difference engine and commenced work on a more ambitious machine which he called his *analytic engine*, which was intended to execute completely automatically a whole series of arithmetical operations assigned to it at the start by the operator. This machine, also, was never completed, largely because the necessary precision tools were as yet not made.

The Babbage project, however, provided the inspiration for the remarkable giant mechanical brains that have come into existence in recent years. The first direct descendent of the Babbage analytic engine is the great *IBM Automatic Sequence Controlled Calculator* (the ASCC) completed at Harvard University in 1944 as a joint enterprise by the University and the International Business Machines Corporation under contract for the Navy Department. The machine is 51 feet long, 8 feet high, with two panels six feet long, and weighs about five tons. An improved second model of the ASCC was made for use, beginning in 1948, at the Naval Proving Ground, Dahlgren, Virginia. Another descendent of Babbage's efforts is the *Electronic Numerical Integrator and Calculator* (the ENIAC), a multipurpose electronic computer completed in 1945 at the University of Pennsylvania under contract with the Ballistic Research Laboratory of the Army's Aberdeen Proving Ground. This machine occupies a 30 by 50 foot room and weighs about 30 tons. These amazing high-speed computing machines, along with similar projects, like the *Selective Sequence Electronic Calculator* (SSEC) of the

International Business Machines Corporation, the *Electronic Discrete Variable Calculator* (EDVAC) of the University of Pennsylvania, the MANIAC of the Institute for Advanced Study at Princeton, the *Universal Automatic Computor* (UNIVAC) of the Bureau of Standards, and the various *differential analyzers,* presaged machines of even more fantastic accomplishment. Every half-dozen or so years, a new generation of machines seems to eclipse in speed, reliability, and memory those of the preceding generation. The following table of comparisons of calculations of π performed on electronic computers shows the rapid increase in computational speed that has taken place.

| | | | *DECIMAL* | |
AUTHOR	*MACHINE*	*DATE*	*PLACES*	*TIME*
Reitwiesner	ENIAC	1949	2037	70 hours
Nicholson and Jeenel	NORC	1954	3089	13 minutes
Felton	Pegasus	1958	10000	33 hours
Genuys	IBM 704	1958	10000	100 minutes
Genuys	IBM 704	1959	16167	4.3 hours
Shanks and Wrench	IBM 7090	1961	100265	8.7 hours

It has been predicted by Shanks and Wrench that the generation of electronic computers succeeding the IBM 7090 will possess speeds 100 times as fast, reliability 100 times as great, and memories 10 times as large as possessed by the IBM 7090, and that with these machines the computation of π to 1,000,000 decimal places will not be difficult.

Because of the immense progress in computing machinery, numerical analysis has received a tremendous stimulus in recent times and has become a subject of ever-growing importance.

Problem Studies

9.1 Logarithms

(a) Using the familiar laws of exponents, establish the following useful properties of logarithms:

(1) $\log_a mn = \log_a m + \log_a n,$

(2) $\log_a (m/n) = \log_a m - \log_a n$,

(3) $\log_a (m^r) = r \log_a m$,

(4) $\log_a \sqrt[s]{m} = (\log_a m)/s$.

(b) Show that

(1) $\log_b N = \log_a N/\log_a b$ (with this formula we may compute logarithms to a base b when we have available a table of logarithms to some base a),

(2) $\log_N b = 1/\log_b N$,

(3) $\log_N b = \log_{1/N} (1/b)$.

(c) By extracting the square root of 10, then the square root of the result thus obtained, and so on, the following table can be constructed:

$$10^{1/2} = 3.16228 \qquad 10^{1/256} = 1.00904$$

$$10^{1/4} = 1.77828 \qquad 10^{1/512} = 1.00451$$

$$10^{1/8} = 1.33352 \qquad 10^{1/1024} = 1.00225$$

$$10^{1/16} = 1.15478 \qquad 10^{1/2048} = 1.00112$$

$$10^{1/32} = 1.07461 \qquad 10^{1/4096} = 1.00056$$

$$10^{1/64} = 1.03663 \qquad 10^{1/8192} = 1.00028$$

$$10^{1/128} = 1.01815 \qquad \cdot \quad \cdot \quad \cdot \quad \cdot \quad \cdot \quad \cdot$$

With this table, we may compute the common logarithm of any number between 1 and 10, and hence, by adjusting the characteristic, of any positive number whatever. Thus, let N be any number between 1 and 10. Divide N by the largest number in the table which does not exceed N. Suppose the divisor is $10^{1/p_1}$ and that the quotient is N_1. Then $N = 10^{1/p_1}N_1$. Treat N_1 in the same fashion, and continue the process, obtaining

$$N = 10^{1/p_1}10^{1/p_2} \cdots 10^{1/p_n}N_n.$$

Let us stop when N_n differs from unity only in the sixth decimal place. Then, to five places,

$$N = 10^{1/p_1}10^{1/p_2} \cdots 10^{1/p_n}$$

and

$$\log N = 1/p_1 + 1/p_2 + \cdots + 1/p_n.$$

Compute, in this manner, log 4.26.

9.2   Napier and Spherical Trigonometry

(a) There are ten formulas which are useful for solving right spherical triangles. There is no need to memorize these formulas for it is easy to reproduce them by means of two rules devised by Napier. In Figure 61 is

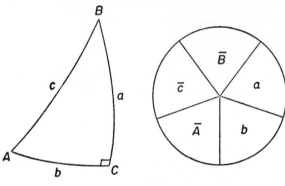

Fig. 61

pictured a right spherical triangle, lettered in conventional manner. To the right of the triangle appears a circle divided into five parts, containing the same letters as the triangle, except C, arranged in the same order. The bars on c, B, A mean *the complement of* (thus $\bar{B}$ means $90° - B$). The angular quantities a, b, $\bar{c}$, $\bar{A}$, $\bar{B}$ are called the *circular parts*. In the circle there are two circular parts contiguous to any given part, and two parts not contiguous to it. Let us call the given part the *middle part*, the two contiguous parts the *adjacent parts*, and the two noncontiguous parts the *opposite parts*. Napier's rules may be stated as follows:

I. The sine of any middle part is equal to the product of the cosines of the two opposite parts.

II. The sine of any middle part is equal to the product of the tangents of the two adjacent parts.

By applying each of these rules to each of the circular parts obtain the ten formulas used for solving right spherical triangles.

(b) The formula connecting the sides a, b, c of a right spherical triangle is called the *Pythagorean relation* for the triangle. Find the Pythagorean relation for a right spherical triangle.

(c) The following formulas are known as *Napier's analogies* (the word analogy being used in its archaic sense of "proportion"):

$$\frac{\sin \frac{1}{2}(A - B)}{\sin \frac{1}{2}(A + B)} = \frac{\tan \frac{1}{2}(a - b)}{\tan \frac{1}{2}c},$$

$$\frac{\cos \frac{1}{2}(A - B)}{\cos \frac{1}{2}(A + B)} = \frac{\tan \frac{1}{2}(a + b)}{\tan \frac{1}{2}c},$$

$$\frac{\sin \frac{1}{2}(a - b)}{\sin \frac{1}{2}(a + b)} = \frac{\tan \frac{1}{2}(A - B)}{\cot \frac{1}{2}C},$$

$$\frac{\cos \frac{1}{2}(a - b)}{\cos \frac{1}{2}(a + b)} = \frac{\tan \frac{1}{2}(A + B)}{\cot \frac{1}{2}C}.$$

These formulas, which are analogous to the law of tangents in plane trigonometry, may be used to solve oblique spherical triangles for which the given parts are two sides and the included angle or two angles and the included side.

(1) Find A, C, b for a spherical triangle in which $a = 125° 38'$, $c = 73° 24'$, $B = 102° 16'$.

(2) Find A, B, c for a spherical triangle in which $a = 93° 8'$, $b = 46° 4'$, $C = 71° 6'$.

9.3 Napier's Rods

The difficulty that was so widely experienced in the multiplication of large numbers led to mechanical ways of carrying out the process. Very celebrated in its time was Napier's invention, known as *Napier's rods*, or *Napier's bones*, and described by the inventor in his work *Rabdologiae*, published in 1617. In principle the invention is the same as the Arabian lattice, or grating, method which we described in Section 7-2, only in the invention the process is carried out with the aid of rectangular strips of bone, metal, wood, or cardboard, prepared beforehand. For each of the ten digits one should have some strips, like the one shown to the left below for 6, bearing the various multiples of that digit. To illustrate the use of these strips in multiplication, let us select the example chosen by Napier in the *Rabdologiae*, the multiplication of 1615 by 365. Put strips headed 1, 6, 1, 5 side by side as

```
        8075
        9690
        4845
        ─────────
        589475 Ans.
```

$3\,(1615) = 4845$

$5\,(1615) = 8075$

$6\,(1615) = 9690$

shown to the right in the figure. The results of multiplying 1615 by the 5, the 6, and the 3 of 365 are then easily read off as 8075, 9690, and 4845, some simple diagonal additions of two digits being necessary to obtain these results. The final product is then gotten by an addition, as illustrated in the figure.

Make a set of Napier's rods and perform some multiplications.

9.4 The Slide Rule

(a) Construct, with the aid of tables, a logarithmic scale, to be designated as the D scale, about 10 inches long. Use the scale, along with a pair of dividers, to perform some multiplications and divisions.

(b) Construct two logarithmic scales, to be called C and D scales, of the same size. By sliding C along D perform some multiplications and divisions. (Refer to the laws of logarithms (Problem Study 9.1) for a suggestion.)

(c) Construct a logarithmic scale half as long as the above D scale and designate by A two of these short scales placed end to end. Show how the A and D scales may be used for extracting square roots.

(d) How would one design a scale to be used with the D scale for extracting cube roots?

(e) Construct a scale just like the C and D scales only running in the reverse direction, and call it the CI (C *inverted*) scale. Show how the CI and D scales may be used for performing multiplications. What is the advantage of the CI and D scales over the C and D scales for this purpose?

9.5 Freely Falling Bodies

Assuming that all bodies fall with the same constant acceleration g, Galileo showed that the distance d a body falls is proportional to the square of the time t of falling. Establish the following stages of Galileo's argument.

(a) If v is the velocity at the end of time t, then $v = gt$.

(b) If v and t refer to one falling body and V and T to a second falling body, then $v/V = t/T$, whence the right triangle having legs of numerical lengths v and t is similar to the right triangle having legs of numerical lengths V and T.

(c) Since the increase in velocity is uniform, the average velocity of fall is $\frac{1}{2}v$, whence $d = \frac{1}{2}vt$ = area of right triangle with legs v and t.

(d) $d/D = t^2/T^2$. Show also that $d = \frac{1}{2}gt^2$.

Galileo illustrated the truth of this final law by observing the times of descent of bodies rolling down inclined planes.

9.6 Sector Compasses

About 1597, Galileo perfected the *sector compasses*, an instrument which enjoyed considerable popularity for more than two centuries. The instrument consists of two arms fastened together at one end by a pivot joint, as shown in Figure 62. On each arm there is a simple scale radiating from the pivot and having the zero of the scale at the pivot. In addition to these simple scales other scales often appeared, some of which will be described below. Many problems can readily be solved using the simple scales of the compasses, the only theory required being that of similar triangles.

(a) Show how the sector compasses may be used to divide a given line segment into five equal parts.

(b) Show how the sector compasses may be used to change the scale of a drawing.

(c) Show how the sector compasses may be used to find the fourth proportional x to three given quantities a, b, c (that is, to find x where $a : b = c : x$), and thus applied to problems in foreign exchange.

(d) Galileo illustrated the use of his sector compasses by finding the amount of money that should have been invested 5 years ago at 6 per cent, compounded annually, to amount to 150 scudi today. Try to solve this problem with the sector compasses.

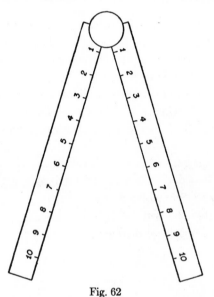

Fig. 62

Among the additional scales frequently found on the arms of sector compasses was one (the *line of areas*) marked according to the squares of the numbers involved, and used for finding squares and square roots of numbers. Another scale (the *line of volumes*) was marked according to the cubes of the numbers involved. Another gave the chords of arcs of specified numbers of degrees for a circle of unit radius, and served engineers as a protractor. Still another (called the *line of metals*) contained the medieval symbols for gold, silver, iron, copper, and so forth, spaced according to the densities of these metals, and was used to solve such problems as finding the diameter of an iron sphere having its weight equal to that of a given copper sphere.

The sector compasses are neither as accurate nor as easy to manipulate as the slide rule.

9.7 Some Simple Paradoxes from Galileo's *Discorsi*

Explain the following two geometrical paradoxes considered by Galileo in his *Discorsi* of 1638.

(a) Suppose the large circle of Figure 63 has made one revolution in rolling along the straight line from A to B, so that AB is equal to the cir-

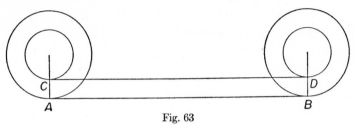

Fig. 63

cumference of the large circle. Then the small circle, fixed to the large one, has also made one revolution, so that CD is equal to the circumference of the small circle. It follows that *the two circles have equal circumferences!*

(b) Let $ABCD$ be a square and HE any line parallel to BC, cutting the diagonal BD in G, as shown in Figure 64. Let circle $B(C)$ cut HE in F, and

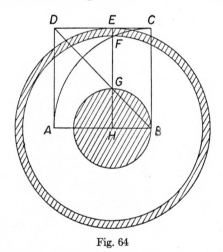

Fig. 64

draw the three circles $H(G)$, $H(F)$, $H(E)$. First show that the area of circle $H(G)$ is equal to the area of the ring between circles $H(F)$ and $H(E)$. Then let H approach B so that in the limit circle $H(G)$ becomes the point B and the ring becomes the circumference $B(C)$. We now conclude that *the single point B is equal to the whole circumference $B(C)$!*

(c) Explain the remark in the *Discorsi* that "neither is the number of squares less than the totality of all numbers, nor the latter greater than the former."

9.8 Kepler's Laws

(a) Where is a planet in its orbit when its speed is greatest?

(b) Check, approximately, Kepler's third law using the following modern figures. (*A.U.* is an abbreviation for *astronomical unit*, the length of the semimajor axis of the earth's orbit.)

Planet	Time in years	Semimajor axis
Mercury	0.241	0.387 *A.U.*
Venus	0.615	0.723 *A.U.*
Earth	1.000	1.000 *A.U.*
Mars	1.881	1.524 *A.U.*
Jupiter	11.862	5.202 *A.U.*
Saturn	29.457	9.539 *A.U.*

(c) What would be the period of a planet having a semimajor axis of 100 *A.U.*?

(d) What would be the semimajor axis of a planet having a period of 125 years?

(e) Two hypothetical planets are moving about the sun in elliptical orbits having equal semimajor axes. The semiminor axis of one, however, is half that of the other. How do the periods of the two planets compare?

(f) The moon revolves about the earth in 27.3 days in an elliptical orbit whose semimajor axis is 60 times the earth's radius. What would be the period of a hypothetical satellite revolving very close to the earth's surface?

9.9 Mosaics

A very interesting problem of mosaics is to fill the plane with congruent regular polygons. Let n be the number of sides of each polygon. Then the interior angle at each vertex of such a polygon is $(n - 2)180°/n$. Prove this statement.

(a) If we do not permit a vertex of one polygon to lie on a side of another, show that the number of polygons at each vertex is given by

$2 + 4/(n - 2)$, and hence that we must have $n = 3$, 4, or 6. Construct illustrative mosaics.

(b) If we insist that a vertex of one polygon lie on a side of another, show that the number of polygons clustered at such a vertex is given by $1 + 2/(n - 2)$, whence we must have $n = 3$ or 4. Construct illustrative mosaics.

(c) Construct mosaics containing (1) two sizes of equilateral triangles, the larger having a side twice that of the smaller, (2) two sizes of squares, the larger having a side twice that of the smaller, (3) congruent equilateral triangles and congruent regular dodecagons, (4) congruent equilateral triangles and congruent regular hexagons, (5) congruent squares and congruent regular octagons.

(d) Suppose we have a mosaic composed of regular polygons of three different kinds at each vertex. If the three kinds of polygons have p, q, r sides, respectively, show that

$$1/p + 1/q + 1/r = 1/2.$$

One integral solution of this equation is $p = 4$, $q = 6$, $r = 12$. Construct a mosaic of the type under consideration and composed of congruent squares, congruent regular hexagons, and congruent regular dodecagons.

9.10 Proving Theorems by Projection

(a) If l is a given line in a given plane π, and O is a given center of projection (not on π), show how to find a plane π' such that the projection of l onto π' will be the line at infinity on π'. (The operation of selecting a suitable center of projection O and plane of projection π' so that a given line on a given plane shall project into the line at infinity on π' is called the operation of "projecting a given line to infinity.")

(b) Show that under the projection of part (a), the line at infinity in π will project into the intersection of π' with the plane through O parallel to π.

(c) Let UP, UQ, UR be three concurrent coplanar lines, cut by two lines OX and OY in P_1, Q_1, R_1 and P_2, Q_2, R_2 respectively (see Figure 65). Prove that the intersections of Q_1R_2 and Q_2R_1, R_1P_2 and R_2P_1, P_1Q_2 and P_2Q_1 are collinear.

(d) Prove that if $A_1B_1C_1$ and $A_2B_2C_2$ are two coplanar triangles such that B_1C_1 and B_2C_2 meet in L, C_1A_1 and C_2A_2 meet in M, A_1B_1 and A_2B_2 meet in N, where L, M, N are collinear, then A_1A_2, B_1B_2, C_1C_2 are concurrent. (This is the converse part of the statement of Desargues's two-triangle theorem as given in Section 9-7.)

(e) Show that by parallel projection (a projection where the center of projection is at infinity) an ellipse may always be projected into a circle.

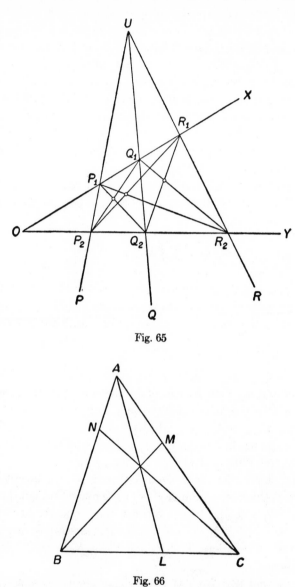

Fig. 65

Fig. 66

(f) In 1678, the Italian Giovanni Ceva (*ca.* 1647–1736) published a work containing the following theorem (see Figure 66) now known by his name: *The three lines which join three points L, M, N on the sides BC, CA, AB of a triangle ABC to the opposite vertices are concurrent if and only if*

$$(AN/NB)(BL/LC)(CM/MA) = +1.$$

This is a companion theorem to Menelaus' theorem, stated in Section 6-6. Using Ceva's theorem prove that the lines joining the vertices of a triangle to the opposite points of contact of the inscribed circle are concurrent. Then, by means of part (e), prove that the lines joining the vertices of a triangle to the opposite points of contact of an inscribed ellipse are concurrent.

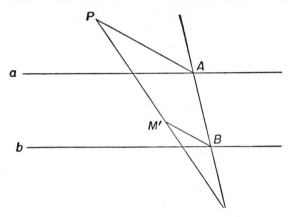

(g) La Hire invented the following interesting mapping of the plane onto itself (see Figure 67): Draw any two parallel lines a and b, and select a point P in their plane. Through any second point M of the plane draw a line cutting a in A and b in B. Then the map M' of M will be taken as the intersection with MP of the parallel to AP through B. (1) Show that M' is independent of the particular line MAB through M used in determining it. (2) Generalize La Hire's mapping to the situation where a and b need not be parallel.

9.11 Pascal's Theorem

The consequences of Pascal's "mystic hexagram" theorem are very numerous and attractive, and an almost unbelievable amount of research has been expended on the configuration. There are 60 possible ways of forming a hexagon from 6 points on a conic and, by Pascal's theorem, to each hexagon corresponds a *Pascal line*. These 60 Pascal lines pass three by three through 20 points, called *Steiner points*, which in turn lie four by four on 15 lines, called

Plücker lines. The Pascal lines also concur three by three in another set of points, called *Kirkman points*, of which there are 60. Corresponding to each Steiner point, there are three Kirkman points such that all four lie upon a line, called a *Cayley line.* There are 20 of these Cayley lines, and they pass four by four through 15 points, called *Salmon points.* There are many further extensions and properties of the configuration, and the number of different proofs that have been supplied for the "mystic hexagram" theorem itself is now legion. In this Problem Study we shall consider a few of the many corollaries of the "mystic hexagram" theorem which can be obtained by making some of the six points coincide with one another. For simplicity we shall number the points 1, 2, 3, 4, 5, 6. Then Pascal's theorem says that the intersections of the pairs of lines 12, 45; 23, 56; 34, 61 are collinear if and only if the six points lie on a conic.

(a) If a pentagon 12345 is inscribed in a conic, show that the pairs of lines 12, 45; 23, 51; 34 and the tangent at 1, intersect in three collinear points.

(b) Given five points, draw at any one of them the tangent to the conic determined by the five points.

(c) Given four points of a conic and the tangent at any one of them, construct further points on the conic.

(d) Show that the pairs of opposite sides of a quadrangle inscribed in a conic, together with the pairs of tangents at opposite vertices, intersect in four collinear points.

(e) Show that if a triangle is inscribed in a conic, then the tangents at the vertices intersect the opposite sides in three collinear points.

(f) Given three points on a conic and the tangents at two of them, construct the tangent at the third.

9.12　Principle of Duality

In plane projective geometry, there is a remarkable symmetry between points and lines, when the ideal elements at infinity are utilized, such that if in a true proposition about "points" and "lines" we should interchange the roles played by these words, and perhaps smooth out the language, we obtain another true proposition about "lines" and "points." As a simple example, consider the following two propositions related in this way:

Any two distinct points determine one and only one line on which they both lie.

Any two distinct lines determine one and only one point through which they both pass.

This symmetry, which results in the pairing of the propositions of plane projective geometry, is a far-reaching principle known as the *principle of duality*. Once the principle of duality is established, then the proof of one proposition of a dual pair carries with it the proof of the other. Let us dualize Pascal's theorem. We first restate Pascal's theorem in a form which is perhaps more easily dualized.

The six vertices of a hexagon lie on a conic if and only if the points of intersection of the three pairs of opposite sides lie on a line.

Dualizing this we obtain:

The six sides of a hexagon are tangent to a conic if and only if the lines joining the three pairs of opposite vertices intersect in a point.

This theorem was first published by C. J. Brianchon (1785–1864), when a student at the École Polytechnique in Paris, in 1806, nearly 200 years after Pascal had stated his theorem.

(a) Dualize 9.11 (a).

(b) Given 5 lines, find on any one of them the point of contact of the conic touching the 5 lines.

(c) Given 4 tangents to a conic and the point of contact of any one of them, construct further tangents to the conic.

(d) Dualize 9.11 (d).

(e) Dualize 9.11 (e).

(f) Given 3 tangents to a conic and the points of contact of 2 of them, construct the point of contact of the third.

(g) Dualize Desargues' two-triangle theorem.

There are several ways in which the principle of duality may be proved. It is possible to give a set of postulates for projective geometry which are themselves arranged in dual pairs. It follows that any theorem derived from such a set of postulates may be dualized by simply dualizing the steps in the proof. The principle may also be proved analytically once the idea of "coordinates" of a line and "equation" of a point are established (see Section 10-3). Finally, the student who is familiar with the elementary notion of poles and polars with respect to some base conic will realize that, under the correspondence between poles and polars so set up, to each figure consisting of lines and points is associated a dual figure consisting of points and lines. It was in this last way that the principle of duality was first established.

9.13 Pascal's Triangle

Establish the following relations, all of which were developed by Pascal, involving the numbers of the arithmetic triangle.

(a) Any element (not in the first row or the first column) of the arithmetic triangle is equal to the sum of the element just above it and the element just to the left of it.

(b) Any given element of the arithmetic triangle, decreased by 1, is equal to the sum of all the elements above the row and to the left of the column containing the given element.

(c) The mth element in the nth row is $(m + n - 2)!/(m - 1)!(n - 1)!$, where, by definition, $0! = 1$.

(d) The element in the mth row and nth column is equal to the element in the nth row and mth column.

(e) The sum of the elements along any diagonal is twice the sum of the elements along the preceding diagonal.

(f) The sum of the elements along the nth diagonal is 2^{n-1}.

(g) Let us be given a group of n objects. Any set of r of these objects, considered without regard to order, is called a *combination of the* n *objects taken* r *at a time*, or, more briefly, as an *r-combination* of the n objects. We shall use the symbol $C(n,r)$ to denote the number of such combinations. Thus, the 2-combinations of the four letters a, b, c, d are

$$ab, \ ac, \ ad, \ bc, \ bd, \ cd,$$

whence $C(4,2) = 6$. It is shown in textbooks on college algebra that

$$C(n,r) = \frac{n!}{r!\,(n-r)!} \cdot$$

Show that $C(n,r)$ appears at the intersection of the nth diagonal and the $(r + 1)st$ column of the arithmetic triangle.

Bibliography

BARLOW, C. W. C., AND G. H. BRYAN, *Elementary Mathematical Astronomy*. London: University Tutorial Press, 1923.

BELL, E. T., *Men of Mathematics*. New York: Simon and Schuster, Inc., 1937.

BERKELEY, E. C., *Giant Brains; or, Machines That Think*. New York: John Wiley & Sons, Inc., 1949.

CAJORI, FLORIAN, *A History of the Logarithmic Slide Rule and Allied Instruments*. New York: McGraw-Hill Book Company, Inc., 1909.

————, *William Oughtred, a Great Seventeenth-Century Teacher of Mathematics.* Chicago: The Open Court Publishing Company, 1916.

COOLIDGE, J. L., *A History of Geometrical Methods.* New York: Oxford University Press, 1940.

————, *The Mathematics of Great Amateurs.* New York: Oxford University Press, 1949.

COXETER, H. S. M., *Regular Polytopes.* New York: Pitman Publishing Corporation, 1949.

DAVID, F. N., *Games, Gods and Gambling.* New York: Hafner Publishing Company, 1962.

DREYER, J. L. E., *Tycho Brahe, a Picture of Scientific Life and Work in the Seventeenth Century.* London: A. & C. Black, Ltd., 1890.

FAHIE, J. J., *Galileo, His Life and Work.* London: John Murray, 1903.

GADE, J. A., *The Life and Times of Tycho Brahe.* Princeton, N. J.: Princeton University Press, 1947.

GALILEI, GALILEO, *Dialogues Concerning Two New Sciences,* tr. by Henry Crew and Alfonso de Salvio, introd. by Antonio Favaro. New York: The Macmillan Company, 1914.

HACKER, S. G., *Arithmetical View Points.* Pullman, Wash.: Mimeographed at Washington State College, 1948.

HARTLEY, MILES C., *Patterns of Polyhedrons.* Ann Arbor, Mich.: priv. ptd., Edwards Bros., Inc., 1957.

HOOPER, ALFRED, *Makers of Mathematics.* New York: Random House, Inc., 1948.

IVINS, W. M., JR., *Art and Geometry.* Cambridge, Mass.: Harvard University Press, 1946.

KNOTT, C. G., *Napier Tercentenary Memorial Volume.* London: Longmans, Green & Co., Ltd., 1915.

KRAITCHIK, MAURICE, *Mathematical Recreations.* New York: W. W. Norton & Company, Inc., 1942.

MORRISON, PHILIP, AND EMILY MORRISON, *Charles Babbage and His Calculating Engines* (selected writings of Charles Babbage and others). New York: Dover Publications, Inc., 1961.

MUIR, JANE, *Of Men and Numbers, the Story of the Great Mathematicians.* New York: Dodd, Mead & Company, Inc., 1961.

NORTHROP, E. P., *Riddles in Mathematics.* Princeton, N. J.: D. Van Nostrand Company, Inc., 1944.

SMITH, D. E., *A Source Book in Mathematics.* New York: McGraw-Hill Book Company, Inc., 1929.

SULLIVAN, J. W. N., *The History of Mathematics in Europe, from the Fall of Greek Science to the Rise of the Conception of Mathematical Rigour.* New York: Oxford University Press, 1925.

TODHUNTER, ISAAC, *A History of the Mathematical Theory of Probability, from the Time of Pascal to that of Laplace.* New York: Chelsea Publishing Company, 1949.

TURNBULL, H. W., *The Great Mathematicians.* New York: New York University Press, 1961.

10

Analytic Geometry and Other Pre-Calculus Developments

10-1 Analytic Geometry

While Desargues and Pascal were opening the new field of projective geometry, Descartes and Fermat were conceiving ideas of modern analytic geometry. There is a fundamental distinction between the two studies, for the former is a *branch* of geometry whereas the latter is a *method* of geometry. There are few academic experiences that can be more thrilling to the student of elementary college mathematics than his introduction to this new and powerful method of attacking geometrical problems. The essence of the idea, it will be recalled, is the establishment of a correspondence between plane curves and algebraic equations in two variables, so that for each curve there is a definite equation $f(x,y) = 0$, and for each such equation there is a definite plane curve. Likewise, a correspondence is established between the algebraic and analytic properties of the equation $f(x,y) = 0$ and the geometric properties of the associated curve. Geometry is cleverly reduced to algebra and analysis.

There are differences of opinion as to who invented analytic geometry, even as to what age should be credited with the invention, and the matter certainly cannot be settled without an agreement as to just what constitutes analytic geometry. We have seen that the ancient Greeks indulged in a good deal of geometric algebra, and it is well known that the idea of coordinates was used in the ancient world by the Egyptians and the Romans in surveying and by the Greeks in map making. Particularly strong in the favor of the Greeks is the fact that Apollonius derived the bulk of his geometry of the conic sections from the geometrical equivalents of certain Cartesian equations of these curves, an idea which seems to have originated with Menaechmus. We also noted, in Section 8-4, that in the fourteenth century Nicole Oresme anticipated another aspect of analytic geometry when he represented certain laws by graphing the dependent variable (*latitudo*) against the independent one (*longitudo*), as the latter variable was permitted to take on small increments. Still, all this seems a far cry from what we today think of as analytic geometry, and it may be correct to regard the decisive contributions made in the seventeenth century by Descartes and Fermat as the essential origin of at least the modern spirit of the subject.

10-2   Descartes

René Descartes was born near Tours in 1596. When eight, he was sent to the Jesuit school at La Flèche. It was there that he developed (at first because of delicate health) his lifelong habit of lying in bed till late in the morning. These meditative hours of morning rest were later regarded by Descartes as his most productive periods. In 1612, Descartes left school and shortly after went to Paris, where, with Mersenne and Mydorge (see Section 10-7), he devoted some time to the study of mathematics. In 1617, he commenced several years of soldiering by first joining the army of Prince Maurice of Orange. Upon quitting military life he spent four or five years traveling through Germany, Denmark, Holland, Switzerland, and Italy. After resettling for a couple of years in Paris, where he continued his mathematical studies and his philosophical contemplations, and where for a while he took up the construction of optical instruments, he decided to move to Holland, then at the height of its power. There he lived for 20 years, devoting his time to philosophy, mathematics, and science. In 1649, he reluctantly went to Sweden at the invitation of Queen Christina. A few months later he contracted inflammation of the lungs and died in Stockholm early in 1650.

It was during his stay of 20 years in Holland that Descartes accomplished his writing. He spent the first four years writing *Le monde*, a physical account of the universe, but this was prudently abandoned and left incomplete when Descartes heard of Galileo's condemnation by the Church. He turned to the writing of a philosophical treatise on universal science under the title of *Discours de la méthode pour bien conduire sa raison et chercher la vérité dans les sciences* (A Discourse on the Method of Rightly Conducting the Reason and Seeking Truth in the Sciences); this was accompanied by three appendices entitled *La dioptrique, Les météores,* and *La géométrie.* The *Discours,* with the appendices, was published in 1637, and it is in the last of the three appendices that Descartes' contributions to analytic geometry appear. In 1641, Descartes published a work called *Meditationes,* devoted to a lengthy explanation of the philosophic views sketched in the *Discours,* and in 1644, he issued his *Principia philosophiae,* which contains some inaccurate laws of nature and an inconsistent cosmological theory of vortices.

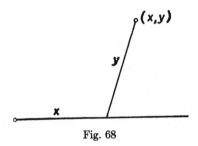

Fig. 68

La géométrie, the famous third appendix of the *Discours,* occupies about 100 pages of the complete work, and is itself divided into three parts. The first part contains an explanation of some of the principles of algebraic geometry and shows a real advance over the Greeks. To the Greeks a variable corresponded to the length of some line segment, the product of two variables to the area of some rectangle, and the product of three variables to the volume of some rectangular parallelepiped. Beyond this the Greeks could not go. To Descartes, on the other hand, x^2 did not suggest an area, but rather the fourth term in the proportion $1 : x = x : x^2$, and as such is representable by an appropriate line length which can easily be constructed when x is known. Using a unit segment we can, in this way, represent any power of a variable, or the product of any number of variables, by a line length, and actually construct the line length with Euclidean tools when the values of the variables are assigned. With this arithmetization of geometry Descartes, in the first part of *La géométrie,* marks off x on a given axis and then a length y at a fixed angle to this axis, and endeavors to construct points whose x's and y's satisfy a given relation (see Figure 68). For example, if we have the relation $y = x^2$, then for each value of x we are able to construct the corresponding y as the fourth term of the above proportion. Descartes shows special interest in obtaining such relations for curves which are defined kinematically. As an application of his method he discusses the problem: If $p_1, \cdots, p_m, p_{m+1}, \cdots, p_{m+n}$ are the lengths of $m + n$ line segments drawn from a point P to $m + n$ given lines, making given angles with these lines, and if

$$p_1 p_2 \cdots p_m = k p_{m+1} p_{m+2} \cdots p_{m+n},$$

where k is a constant, find the locus of P. The ancient Greeks solved this problem for the cases where m and n do not exceed 2 (see Section 6-9), but

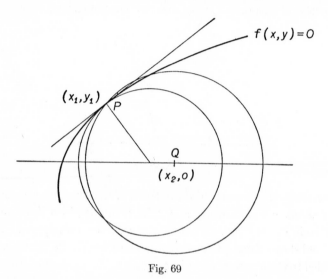

Fig. 69

the general problem had remained a baffling one. Descartes easily shows that higher cases of the problem lead to loci of degrees greater than 2, and in certain cases he is able actually to construct with Euclidean tools points of the loci (see Problem Study 10.2 (a)). That Descartes' analytic geometry can cope with the general problem is a fine tribute to the power of the new method. It is said that it was Descartes' attempt to solve this problem that inspired his invention of analytic geometry.

The second part of *La géométrie* deals among other things with a now obsolete classification of curves and with an interesting method of constructing tangents to curves. The method of drawing tangents is as follows (see Figure 69). Let the equation of the given curve be $f(x,y) = 0$ and let (x_1, y_1) be the coordinates of the point P of the curve at which we wish to construct a tangent. Let Q, having coordinates $(x_2,0)$, be a point on the x-axis. Then the equation of the circle with Q as center and passing through P is

$$(x - x_2)^2 + y^2 = (x_1 - x_2)^2 + y_1{}^2.$$

If we eliminate y between this equation and the equation $f(x,y) = 0$ we obtain an equation in x leading to the abscissas of the points where the circle cuts the given curve. We now determine x_2 so that this equation in x will have a *pair* of roots equal to x_1. This condition fixes Q as the intersection of the x-axis and the normal to the curve at P, since the circle is now tangent to the given curve at P. Once this circle is drawn, we may easily construct the required tangent. As an example of the method consider the construction of the tangent to the parabola $y^2 = 4x$ at the point $(1,2)$. Here we have

$$(x - x_2)^2 + y^2 = (1 - x_2)^2 + 4.$$

The elimination of y gives

$$(x - x_2)^2 + 4x = (1 - x_2)^2 + 4,$$

or

$$x^2 + 2x(2 - x_2) + (2x_2 - 5) = 0.$$

The condition that this quadratic equation have two equal roots is that its discriminant vanish, that is, that

$$(2 - x_2)^2 - (2x_2 - 5) = 0,$$

or

$$x_2 = 3.$$

The circle with center $(3,0)$ and passing through the point $(1,2)$ of the curve may now be drawn, and the required tangent finally constructed. This method of constructing tangents is applied by Descartes to a number of different curves, including one of the quartic ovals named after him.* Here we have a general process which tells us exactly what to do to solve our problem, but it must be confessed that in the more complicated cases the required algebra may be quite forbidding. This is a well-recognized fault

---

*A *Cartesian oval* is the locus of a point whose distances, r_1 and r_2, from two fixed points satisfy the relation $r_1 + mr_2 = a$, where m and a are constants. The central conics will be recognized as special cases.

with elementary analytic geometry—we often know what to do but lack the technical ability to do it. There are, of course, much better methods than the above for finding tangents to curves.

The third part of *La géométrie* concerns itself with the solution of equations of degree greater than two. Use is made of what we now call "Descartes's rule of signs," a rule for determining limits to the number of positive and the number of negative roots possessed by a polynomial (see Problem Study 10.3). In *La géométrie*, Descartes fixed the custom of employing the first letters of the alphabet to denote known quantities, and the last letters to denote unknown ones. He also introduced our present system of indices (such as a^3, a^4, and so forth), which is a great improvement over Viète's way of designating powers, and he realized that a letter might represent any quantity, positive or negative. Here we also find the first use of the method of indeterminate coefficients. Thus, in the example of the last paragraph, we used the vanishing of the discriminant to determine the value of x_2 so that the quadratic equation

$$x^2 + 2x(2 - x_2) + (2x_2 - 5) = 0$$

should have both roots equal to 1. As an illustration of the method of indeterminate coefficients we might accomplish this by saying that we want

$$x^2 + 2(2 - x_2)x + (2x_2 - 5) \equiv (x - 1)^2 \equiv x^2 - 2x + 1,$$

whence we must have, by equating coefficients of like powers of x,

$$2(2 - x_2) = -2 \quad \text{and} \quad 2x_2 - 5 = 1.$$

Either of these leads to $x_2 = 3$.

La géométrie is not in any sense a systematic development of the method of analytics, and the reader must pretty much construct the method for himself from certain isolated statements. There are 32 figures in the text, but in none do we find the coordinate axes explicitly set forth. The work was written with intentional obscurity and as a result was too difficult to be widely read. In 1649 a Latin translation appeared with explanatory notes by F. de Beaune, edited with commentary by Frans van Schooten the Younger. This, and the revised 1659–1661 edition, had a wide circulation. A hundred years or more later the subject achieved the familiar form found in our present-day college textbooks. The words *coordinates, abscissa,* and *ordinate* as now technically used in analytic geometry were contributed by Leibniz in 1692.

There are several legends describing the initial flash that led Descartes to the contemplation of analytic geometry. According to one story, it occurred to him in a dream. Another story, perhaps on a par with the story of Isaac Newton and the falling apple, is that the first idea came to him when watching a fly crawling about the ceiling near a corner of his room. He noticed that the path of the fly could be described if one knew a relation connecting the fly's distances from the two adjacent walls. Even though this latter story may be apocryphal, it has good pedagogical value.

Of the other two appendices to the *Discours,* one was devoted to optics and the other to an explanation of numerous meteorological, or atmospheric, phenomena, including the rainbow.

Among other mathematical items credited to Descartes is the first announcement of the relation $v - e + f = 2$, connecting the numbers of vertices v, edges e, and faces f of a convex polyhedron (see Problem Study 3.12). He was the first to discuss the so-called folium of Descartes, a nodal cubic curve found in many of our calculus texts, but he did not completely picture the curve. In correspondence he considered parabolas of higher order ($y^n = px, n > 2$) and gave a remarkably neat construction of a tangent to the cycloid.

10-3   Remarks on Some Later Developments in Analytic Geometry

There are plane coordinate systems other than the rectangular and oblique Cartesian systems. As a matter of fact, one can invent coordinate systems rather easily. All one needs is an appropriate frame of reference along with some accompanying rules telling us how to locate a point in the plane by means of an ordered set of numbers referred to the frame of reference. Thus, for the rectangular Cartesian system, the frame of reference consists of two perpendicular axes, each carrying a scale, and we are all familiar with the rules telling us how to locate a point with respect to this frame by the ordered pair of real numbers representing the signed distances of the point from the two axes. The Cartesian systems are much the commonest systems in use, and have been developed enormously. Much terminology, like our classification of curves into linear, quadratic, cubic, and so forth, stems from our use of this system. Some curves, however, such as many spirals, have intractable equations when referred to a Cartesian frame, whereas they enjoy relatively simple equations when referred to some other skillfully designed coordinate system. Particularly useful in the case of spirals is the polar coordinate system, where, it will be recalled, the frame of reference is an infinite ray and where a point is located by a pair of real numbers, one of which represents a distance and the other an angle. The idea of polar coordinates seems to have been introduced in 1691 by Jakob Bernoulli (1654–1705).* Further coordinate systems were little investigated until toward the close of the eighteenth century, when geometers were led to break away from the Cartesian systems in situations where the peculiar necessities of a problem indicated some other algebraic apparatus as more suitable. After all, coordinates were made for geometry and not geometry for coordinates.

An interesting development in coordinate systems was inaugurated by Julius Plücker (1801–1868) in 1829, when he noted that our fundamental element need not be the point, but can be any geometric entity. Thus, if we choose the straight line as our fundamental element, we might locate any straight line not passing through the origin of a given rectangular Cartesian frame of reference by recording, say, the x and y intercepts of the given line. Plücker actually chose the negative reciprocals of these intercepts as the location numbers of the line and considerably exploited the analytic geometry of these so-called *line coordinates*. A point now, instead of having coordinates, possesses a linear equation, namely the equation satisfied by the coordinates of all the lines passing through the point (see Problem Study 10.6). The double interpretation of a pair of coordinates as either point coordinates or line coordinates and of a linear equation as either the equation of a line or the equation of a point, furnishes the basis of Plücker's analytical proof of the principle of duality of projective geometry. A curve may be regarded either as the locus of its points or as the envelope of its tangents (see Figure 70). If, instead of points or straight lines, we should choose circles as fundamental elements, then we would require an ordered triple of numbers to determine one of our elements completely. For example, on a rectangular Cartesian frame of reference, we might take the two Cartesian coordinates of the circle's center along with the circle's radius. Ideas such as these led to considerable

*See, however, C. B. Boyer, "Newton as an Originator of Polar Coordinates," *American Mathematical Monthly*, February, 1949, pp. 73–78.

generalization and the development of a dimension theory. *The dimensionality of a geometry was considered as the number of coordinates needed to locate a fundamental element of the geometry.* According to this concept, the plane is two dimensional in points, and also in lines, but is three dimensional in circles. It can be shown that the plane is five dimensional if the totality of all conic sections in the plane should be chosen as the manifold of fundamental elements. Dimension theory has, of course, developed far beyond this elementary concept, and is today a subject of considerable extent and depth.

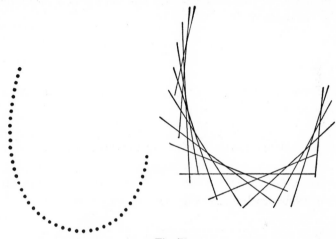

Fig. 70

Although Descartes had mentioned solid analytic geometry, he did not elaborate it. Others, like the younger Frans van Schooten, La Hire, and Johann Bernoulli, suggested our familiar solid analytic geometry, but it was not until 1700 that the subject was first systematically developed, by Antoine Parent (1666–1716) in a paper presented to the French Academy. A. C. Clairaut (1713–1765), in 1731, was the first to write analytically on nonplanar curves in space. Euler later advanced the whole subject well beyond elementary stages. These initial workers chose the point as fundamental element. Although space is three dimensional in points it may be shown that it is four dimensional in lines and also in spheres. It is three dimensional, however, in planes (see Problem Study 10.7).

The first nebulous notions of a hyperspace which is n dimensional $(n > 3)$ in points are lost in the dimness of the past and were confused by metaphysical considerations. A bold plunge into the study of such spaces was not made until the middle of the nineteenth century, by Arthur Cayley

(1821–1895) in 1843, Hermann Grassmann (1809–1877) in 1844, and Bernhard Riemann (1826–1866) in 1854. Today a *real* space (or manifold) of n dimensions is defined as the set of all ordered n-tuples $(x_1,x_2,\cdots,x_n)$ of n real numbers $x_1, x_2, \cdots, x_n$, each of which is permitted to range over a prescribed class of real numbers. Any particular such n-tuple is called a *point* of the space. Relations among these points are defined by formulas analogous to the formulas holding for the corresponding relations among points in, say, Cartesian point spaces of two and three dimensions. Thus, since the distance between the two points (x_1,x_2) and (y_1,y_2) in a two dimensional rectangular Cartesian system is given by

$$\sqrt{(x_1 - y_1)^2 + (x_2 - y_2)^2},$$

and that between the two points (x_1,x_2,x_3) and (y_1,y_2,y_3) in a three dimensional rectangular Cartesian system is given by

$$\sqrt{(x_1 - y_1)^2 + (x_2 - y_2)^2 + (x_3 - y_3)^2},$$

we *define* the *distance* between the two points $(x_1,\cdots,x_n)$ and $(y_1,\cdots,y_n)$ in an n dimensional rectangular Cartesian system to be

$$\sqrt{(x_1 - y_1)^2 + \cdots + (x_n - y_n)^2}.$$

An n dimensional geometry of this sort is thus really a purely analytical study which employs geometric terminology. In other words, some of the most important developments in modern mathematics have their origin in the early studies of analytic geometry.

10-4   Fermat

At the same time that Descartes was formulating the basis of modern analytic geometry, the subject was also occupying the attention of another French mathematical genius, Pierre de Fermat. Fermat's claim to priority rests on a letter written to Roberval in September, 1636, in which it is stated that the ideas of the writer were even then seven years old. The details of the work appear in the posthumously published paper *Isogoge ad locus planos et solidos*. Here we find the equation of a general straight line and of a circle, and a discussion of hyperbolas, ellipses, and parabolas. In a work on tangents and quadratures, completed before 1637, Fermat defined many new curves analytically. Where Descartes suggested a few new curves, generated by mechanical motion, Fermat proposed many new ones, defined by algebraic equations. The curves $x^m y^n = a$, $y^n = ax^m$, and $r^n = a\theta$ are still known as *hyperbolas*, *parabolas*, and *spirals of Fermat*. Fermat also proposed, among others, the cubic curve later known as the *witch of Agnesi*, named after Maria Gaetana Agnesi (1718–1799), a versatile woman distinguished as a mathematician, linguist, philosopher, and somnambulist. Thus, where to a large extent Descartes began with a locus and then found its equation, Fermat started with the equation and then studied the locus. These are the two inverse aspects of the fundamental principle of analytic geometry. Fermat's work is written in Viète's notation and thus has an archaic look when compared with Descartes' more modern symbolism. Fermat's use of infinitesimals in geometry, and particularly his application of them to questions of maxima and minima, will be considered in the next chapter.

Fermat was born near Toulouse in 1601 (?) and died in Castres in 1665. He was the son of a leather merchant and received his early education at home. At the age of 30 he obtained the post of councillor for the local parliament at Toulouse and there discharged his duties with modesty and punctiliousness. Working as a humble and retiring lawyer, he devoted the bulk of his leisure time to the study of mathematics. Although he published very little during his lifetime, he was in scientific correspondence with many leading mathematicians of his day and in this way considerably influenced his contemporaries. He enriched so many branches of mathematics with so many important contributions that he has been called the greatest French mathematician of the seventeenth century.

Of Fermat's varied contributions to mathematics, the most outstanding is the founding of the modern theory of numbers. In this field, Fermat possessed extraordinary intuition and ability. It was probably the Latin translation of Diophantus' *Arithmetica*, made by Bachet de Méziriac in 1621, that first directed Fermat's attention to number theory. Many of Fermat's contributions to the field occur as marginal statements made in his copy of Bachet's work. In 1670, five years after his death, these notes were incorporated in a new, but unfortunately carelessly printed, edition of the *Arithmetica*, brought out by his son Clément-Samuel. Many of the unproved theorems announced by Fermat have later been shown to be correct. The following examples illustrate the tenor of Fermat's investigations.

(1) *If p is a prime and a is prime to p, then $a^{p-1} - 1$ is divisible by p.* For example, if $p = 5$ and $a = 2$, then $a^{p-1} - 1 = 15 = (5)(3)$. This theorem, known as the *little Fermat theorem*, was given by Fermat without proof in a letter to Frénicle de Bessy, dated October 18, 1640. The first published proof of it was given by Euler in 1736 (see Problem Study 10.8).

(2) *Every odd prime can be expressed as the difference of two squares in one and only one way.* Fermat gave a simple proof of this. If p is an odd prime, then one easily verifies that

$$p = [(p + 1)/2]^2 - [(p - 1)/2]^2.$$

On the other hand, if $p = x^2 - y^2$, then $p = (x + y)(x - y)$. But since p is prime its only factors are p and 1. Hence $x + y = p$ and $x - y = 1$, or $x = (p + 1)/2$ and $y = (p - 1)/2$.

(3) *A prime of the form $4n + 1$ can be represented as the sum of two squares.* For example, $5 = 4 + 1, 13 = 9 + 4$, $17 = 16 + 1$, $29 = 25 + 4$. This theorem was first stated by Fermat in a letter to Mersenne, dated December 25, 1640. The first published proof was given by Euler in 1754, who, moreover, succeeded in showing that the representation is unique.

(4) *A prime of the form $4n + 1$ is only once the hypotenuse of an integral sided right triangle; its square is twice; its cube is three times; and so forth.* For example, consider $5 = 4(1) + 1$. Now $5^2 = 3^2 + 4^2$; $25^2 = 15^2 + 20^2$ $= 7^2 + 24^2$; $125^2 = 75^2 + 100^2 = 35^2 + 120^2 = 44^2 + 117^2$.

(5) *Every nonnegative integer can be represented as the sum of four or fewer squares.* This difficult theorem was established by Lagrange in 1770.

(6) *The area of an integral sided right triangle cannot be a square number.* This also was established later by Lagrange.

(7) *There is only one solution in integers of $x^2 + 2 = y^3$, and only two of $x^2 + 4 = y^3$.* This problem was issued as a challenge problem to English mathematicians. The solutions are $x = 5$, $y = 3$, for the first equation, and $x = 2$, $y = 2$ and $x = 11$, $y = 5$, for the second equation.

(8) *There do not exist positive integers x, y, z such that $x^4 + y^4 = z^2$.*

(9) *There do not exist positive integers x, y, z, n such that $x^n + y^n = z^n$, when $n > 2$.* This famous conjecture is known as *Fermat's last theorem*. It was stated by Fermat in the margin of his copy of Bachet's translation of

Diophantus, at the side of Problem 8 of Book II: "To divide a given square number into two squares." Fermat's marginal note reads, "To divide a cube into two cubes, a fourth power, or in general any power whatever into two powers of the same denomination above the second is impossible, and I have assuredly found an admirable proof of this, but the margin is too narrow to contain it." Whether Fermat really possessed a sound demonstration of this problem will probably forever remain an enigma. Many of the most prominent mathematicians since his time have tried their skill on the problem, but the general conjecture still remains open. There is a proof given elsewhere by Fermat for the case $n = 4$, and Euler supplied a proof (later perfected by others) for $n = 3$. About 1825, independent proofs for the case $n = 5$ were given by Legendre and Dirichlet, and in 1839, Lamé proved the theorem for $n = 7$. Very significant advances in the study of the problem were made by the German mathematician E. Kummer (1810–1893). In 1843, Kummer submitted a purported proof to Dirichlet, who pointed out an error in the reasoning. Kummer then returned to the problem with renewed vigor, and a few years later, after developing an important allied subject in higher algebra called the *theory of ideals*, derived very general conditions for the insolubility of the Fermat relation. Almost all important subsequent progress on the problem has been based on Kummer's investigations. It is now known that Fermat's last theorem is certainly true for all $n < 4003$,* and for many other special values of n. In 1908, the German mathematician P. Wolfskehl bequeathed 100,000 marks to the Academy of Science at Göttingen as a prize for the first complete proof of the theorem. The result was a deluge of alleged proofs by glory- and money-seeking laymen, and ever since then the problem has haunted amateurs somewhat as does the trisection of an arbitrary angle and the squaring of the circle. Fermat's last theorem has the peculiar distinction of being the mathematical problem for which the greatest number of incorrect proofs have been published.

(10) We have already mentioned, in Section 6-4, Fermat's conjecture that $f(n) = 2^{2^n} + 1$ is prime for all nonnegative integral n. The conjecture proved to be incorrect when Euler showed that $f(5)$ is a composite number.

In 1897, a paper was found in the library at Leyden, among the manuscripts of Christiaan Huygens, in which Fermat describes a general method by which he may have made many of his discoveries. The method is known as Fermat's *method of infinite descent* and is particularly useful in establishing negative results. In brief the method is this. To prove that a certain rela-

*This was shown in 1955, with the aid of the SWAC, by J. L. Selfridge, C. A. Nicol, and H. S. Vandiver.

tion connnecting positive integers is impossible, assume, on the contrary, that the relation can be satisfied by some particular set of positive integers. From this assumption, show that the same relation then holds for another set of smaller positive integers. Then, by a reapplication, the relation must hold for another set of still smaller positive integers, and so on ad infinitum. Since the positive integers cannot be decreased in magnitude indefinitely, it follows that the assumption at the start is untenable, and therefore the original relation is impossible. To make the method clear, let us apply it by proving anew that $\sqrt{2}$ is irrational. Suppose $\sqrt{2} = a/b$, where a and b are positive integers. Now

$$\sqrt{2} + 1 = 1/(\sqrt{2} - 1),$$

whence

$$\frac{a}{b} + 1 = \frac{1}{\dfrac{a}{b} - 1} = \frac{b}{a - b},$$

and

$$\sqrt{2} = \frac{a}{b} = \frac{b}{a - b} - 1 = \frac{2b - a}{a - b} = \frac{a_1}{b_1}, \quad \text{say.}$$

But, since $1 < \sqrt{2} < 2$, after replacing $\sqrt{2}$ by a/b and then multiplying through by b, we have $b < a < 2b$. Now, since $a < 2b$, it follows that $0 < 2b - a = a_1$. And since $b < a$, it follows that $a_1 = 2b - a < a$. Thus a_1 is a positive integer less than a. By a reapplication of our procedure we find $\sqrt{2} = a_2/b_2$, where a_2 is a positive integer less than a_1. The process may be repeated indefinitely. Since the positive integers cannot be decreased in magnitude indefinitely it follows that $\sqrt{2}$ cannot be rational.

We have already mentioned, in Section 9-9, the Pascal-Fermat correspondence which laid the foundations of the science of probability. It will be recalled that it was the so-called *problem of the points* that started the matter: "Determine the division of the stakes of an interrupted game of chance between two supposedly equally skilled players, knowing the scores of the players at the time of interruption and the number of points needed to win the game." Fermat discussed the case where one player A needs 2 points to win, and the other player B needs 3 points. Here is Fermat's solution for this particular case. Since it is clear that four more trials will decide the game, let a indicate a trial where A wins and b a trial where B wins, and consider the 16 combinations of the two letters a and b taken 4 at a time:

$a\ a\ a\ a$	$a\ a\ a\ b$	$a\ b\ b\ a$	$b\ b\ a\ b$
$b\ a\ a\ a$	$b\ b\ a\ a$	$a\ b\ a\ b$	$b\ a\ b\ b$
$a\ b\ a\ a$	$b\ a\ b\ a$	$a\ a\ b\ b$	$a\ b\ b\ b$
$a\ a\ b\ a$	$b\ a\ a\ b$	$b\ b\ b\ a$	$b\ b\ b\ b$

The cases where a appears 2 or more times are favorable to A; there are 11 of them. The cases where b appears 3 or more times are favorable to B; there are 5 of them. Therefore the stakes should be divided in the ratio 11 : 5.

10-5   Huygens

The great Dutch genius, Christiaan Huygens, lived an uneventful but remarkably productive life. He was born at The Hague in 1629 and studied at Leyden under Frans van Schooten the Younger. In 1651, when he was 22, he published a paper pointing out fallacies committed by Saint-Vincent in his work on the quadrature of the circle. This was followed by a number of tracts dealing with the quadrature of the conics and with Snell's trigonometric improvement of the classical method of computing π (see Section 4-8). In 1654, he and his brother devised a new and better way of grinding and polishing lenses, and as a consequence Huygens was able to settle a number of questions in observational astronomy, such as the nature of Saturn's appendages. Huygens' work in astronomy led him, a couple of years later, to invent the pendulum clock, so that he might have more exact means of measuring time.

In 1657, Huygens wrote the first formal treatise on probability, based on the Pascal-Fermat correspondence. This was the best account of the subject until the posthumous appearance, in 1713, of Jakob Bernoulli's *Ars conjectandi*, which contained a reprint of the earlier treatise by Huygens. Many interesting and not easy problems were solved by Huygens, and he introduced the important concept of "mathematical expectation."[*] He showed, among other things, that if p is the probability of a person winning

---

[*]If p denotes the probability that a person will win a certain sum s, then sp is called his *mathematical expectation*.

a sum a, and q that of winning a sum b, then he may expect to win the sum $ap + bq$.

In 1665, Huygens moved to Paris in order to benefit from a pension offered to him by Louis XIV. While there, in 1668, he communicated to the Royal Society of London a paper in which he demonstrates experimentally that the combined momentum of two bodies in a given direction is the same before and after a collision.

In 1673 appeared, in Paris, Huygens' greatest publication, his *Horologium oscillatorium*. This work is in five parts, or chapters. The first part concerns itself with the pendulum clock that the author had invented in 1656. The second part is devoted to a discussion of bodies falling freely in a vacuum, sliding on a smooth inclined plane, or sliding along a smooth curve. Here is shown the isochronous property of an inverted cycloid—that a heavy particle will reach the bottom of an inverted cycloidal arch in the same length of time no matter from what point on the arch it begins its descent. In the third part occurs a treatment of evolutes and involutes. The *evolute* of a plane curve is the envelope of the normals to the curve, and any curve having a given curve for its evolute is called an *involute* of that given curve. As applications of his general theory, Huygens finds the evolute of a parabola and of a cycloid. In the former case he obtains a semicubical parabola, and in the latter another cycloid of the same size. In the fourth part of the *Horologium* is found a treatment of the compound pendulum with a proof that the center of oscillation and the point of suspension are interchangeable. The last part of the work concerns itself with the theory of clocks. Here we find a description of the cycloidal pendulum (see Problem Study 10.9), in which the period of oscillation is the same no matter how great or how small the amplitude of the oscillation, something which is only approximately true of the period of oscillation of a simple pendulum. This last part closes with 13 theorems related to centrifugal force in circular motion, proving, among other things, the now familiar fact that for uniform circular motion the magnitude of the centrifugal force is directly proportional to the square of the linear speed and inversely proportional to the radius of the circle. In 1675, under Huygens' directions, was made the first watch regulated by a balance spring; it was presented to Louis XIV.

Huygens returned to Holland in 1681, constructed some lenses of very large focal lengths, and invented the achromatic eyepiece for telescopes. In 1689, he visited England and made the acquaintance of Isaac Newton, whose work he greatly admired. Shortly after his return to Holland in the following year, he published a treatise expounding the wave theory of light, and was able on the basis of this theory to deduce geometrically the laws of reflection and refraction, and to explain the phenomenon of double refraction. Newton, on the other hand, supported the emission theory of light, and his greater eminence caused contemporary scientists to favor that theory to the wave theory.

Huygens also wrote a number of minor tracts. He rectified the cissoid of Diocles, investigated the geometry of the catenary (the curve assumed by a perfectly flexible inextensible chain of uniform linear density hanging from two supports not in the same vertical line), wrote on the logarithmic curve, gave in modern form, for polynomials, Fermat's rule for maxima and minima, and made numerous applications of mathematics to physics.

Like many of the demonstrations given by Newton, Huygens' proofs are almost entirely accomplished, with great care to rigor, by the methods of Greek geometry. To read his works one would little realize that he was acquainted with the powerful new methods of analytic geometry and the calculus. Huygens died in the city of his birth in 1695.

10-6   Some Seventeenth-Century Italian Mathematicians

There are some lesser mathematicians of the seventeenth century whose works should be mentioned, if only briefly. We do this in the present and the following three sections, treating the men by geographical areas.

We have noted that Galileo valued the cycloid for the graceful form it would give to arches in architecture. He also, in 1599, attempted to ascertain the area under one arch of the curve by balancing a cycloidal template against circular templates of the size of the generating circle. He incorrectly concluded that the area under an arch was very nearly, but not exactly, three times the area of the circle. The first published mathematical demonstration that the area is *exactly* three times that of the generating circle was furnished, in 1644, by his pupil, Evangelista Torricelli (1608–1647), by using early infinitesimal methods. Fermat proposed to Torricelli the problem of determining a point in the plane of a triangle such that the sum of its distances from the three vertices should be a minimum. Torricelli's solution was published in 1659 by his pupil Viviani. This point, now known as the *isogonic* *center* of the triangle, was the first notable point of the triangle discovered in times more recent than that of Greek mathematics. An elegantly simple analysis of the problem was later furnished by Jacob Steiner.* In 1640 Torricelli found the length of an arc of the logarithmic spiral. This curve had also been rectified two years earlier by Descartes, and was the first curve after the circle to be rectified. Torricelli is, of course, much better known for his contributions to physics, where he developed the theory of the barometer and worked on such questions as the value of the acceleration due to gravity, the theory of projectiles, and the motion of fluids.

Another of Galileo's disciples who interested himself in both physics and geometry, was Vincenzo Viviani (1622–1703), a man highly honored during his lifetime. Among his geometrical accomplishments was that of determining the tangent to the cycloid, but several had solved this problem previously. In 1692, he proposed the following problem, which attracted wide attention: A hemispherical dome has four equal windows of such size that the rest of the surface can be exactly squared; show how this is possible. Correct solutions were furnished by a number of eminent contemporary mathematicians. Viviani solved the trisection problem by using an equilateral hyperbola.

Mention should perhaps be made of the Cassini family, several members of which contributed notably to astronomy and made skillful applications of mathematics to this field. The *Cassinian curve,* which is the locus of a point the product of whose distances from two fixed foci is a constant, was studied by Giovanni Domenico Cassini (1625–1712) in 1680 in connection with work on the relative motions of the earth and sun. In a family of confocal Cassinian curves is found the figure-eight-shaped lemniscate of Bernoulli, a fact not noted until the end of the eighteenth century. The Cassinian curves can be found as the intersections of a torus by planes parallel to the axis of the torus.

10-7   Some Seventeenth-Century French Mathematicians

An early noteworthy European Diophantist was the Frenchman Bachet de Méziriac (1581–1638). His charming and classic *Problèmes plaisants et délectables,* which appeared in 1612 and again, enlarged, in 1624, contains

---

*See, for example, R. A. Johnson, *Modern Geometry* (Boston: Houghton Mifflin Company, 1929), pp. 218–225, and Richard Courant and H. E. Robbins, *What Is Mathematics?* (New York: Oxford University Press, 1941), pp. 354–361.

many arithmetical tricks and questions which have reappeared in practically all subsequent collections of mathematical puzzles and recreations. In 1621, he published an edition of the Greek text of Diophantus' *Arithmetica*, along with a Latin translation and notes. It was in a copy of this work that Fermat made his famous marginal notes.

Another number theorist, and a voluminous writer in many fields, was the Minimite friar Marin Mersenne (1588–1648). He maintained a constant correspondence with the greatest mathematicians of his day and served admirably, in those prejournal times, as a clearinghouse for mathematical ideas. He edited the works of many of the Greek mathematicians and wrote on a variety of subjects. He is especially known in connection with the so-called Mersenne primes, or prime numbers of the form $2^p - 1$, which he discussed in a couple of places in his work *Cogitata physico-mathematica* of 1644. The connection between Mersenne primes and perfect numbers was pointed out in Section 3-3. The Mersenne prime for $p = 4253$ is the first known prime to possess more than 1000 digits in its decimal expansion.

Claude Mydorge (1585–1647), a friend of Descartes, was a geometer and a physicist. He published some works on optics and a synthetic treatment of the conic sections in which he simplified many of Apollonius' prolix proofs. He left an interesting manuscript containing the statements and solutions of over a thousand geometrical problems and edited the popular *Récréations mathématiques* of Leurechon.

Another geometer and physicist, and another whose extensive correspondence served as a medium for the intercommunication of mathematical ideas, was Gilles Persone de Roberval (1602–1675). He became well known for his method of drawing tangents and his discoveries in the field of higher plane curves. He endeavored to consider a curve as generated by a point whose motion is compounded from two known motions. Then the resultant of the velocity vectors of the two known motions gives the tangent line to the curve. For example, in the case of a parabola, we may consider the two motions as away from the focus and away from the directrix. Since the distances of the moving point from the focus and the directrix are always equal to each other, the velocity vectors of the two motions must also be equal to each other. It follows that the tangent at a point of the parabola bisects the angle between the focal radius to the point and the perpendicular through the point to the directrix (see Figure 71). This idea of tangents was also held by Torricelli, and an argument of priority ensued. Roberval also claimed to be the inventer of Cavalieri's pre-calculus method of indivisibles and to have squared the cycloid before Torricelli. These matters of priority are difficult to settle, for Roberval was consistently tardy in disclosing his discoveries.

This tardiness has been explained by the fact that for 40 years, starting in 1634, Roberval held a professorial chair at the Collège Royale. This chair automatically became vacant every three years, to be filled by open competition in mathematical contests in which the questions were set by the outgoing incumbent. In any event, Roberval successfully employed the method of indivisibles to the finding of a number of areas, volumes, and centroids.

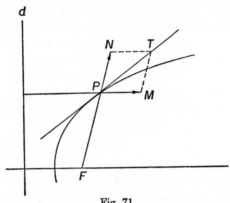

Fig. 71

We have already, in Section 9-7, said something of the work of Phillipe de la Hire (1640–1718). He has been described as a man of scattered genius, having been a painter, an architect, an astronomer, and a mathematician. In addition to his work on conic sections described earlier, he wrote on graphical methods, various types of higher plane curves, and magic squares. He constructed maps of the earth by *globular projection*, where the center of projection is not at a pole of the sphere, as in Ptolemy's stereographic projection (see Problem Study 6.10), but on the radius produced through a pole to a distance of $r \sin 45°$ outside the sphere.

10-8   Some Seventeenth-Century British Mathematicians

Great Britain had its share of lesser mathematicians in the seventeenth century. We have already mentioned William, Viscount Brouncker (1620–1684) elsewhere. He was one of the founders, and the first president, of the Royal Society of London, and maintained relations with Wallis, Fermat, and other leading mathematicians. He wrote on the rectification of the parabola and the cycloid, and had no qualms in using infinite series to express quantities which he could not determine otherwise. Thus he proved that the area bounded by the rectangular hyperbola $xy = 1$, the x-axis, and the ordinates $x = 1$ and $x = 2$, is equal to

$$1/(1)(2) + 1/(3)(4) + 1/(5)(6) + \cdots$$

and to

$$1 - 1/2 + 1/3 - 1/4 + \cdots.$$

Brouncker was the first British writer to investigate and use properties of continued fractions. We have given, in Section 4-8, his interesting continued fraction development of $4/\pi$.

The Scotch mathematician, James Gregory (1638–1675), has also been mentioned elsewhere (Section 4-8). He became successively, in 1668 and 1674, professor of mathematics at St. Andrews and at Edinburgh. He was equally interested in physics and published a work on optics in which is described the reflecting telescope now known by his name. In mathematics, he expanded in infinite series arc tan x, tan x, and arc sec x (1667) and was one of the first to distinguish between convergent and divergent series. He gave an ingenious but unsatisfactory proof that the Euclidean quadrature of the circle is impossible. The series

$$\text{arc tan } x = x - x^3/3 + x^5/5 - x^7/7 + \cdots,$$

which has played so great a part in calculations of π, is known by his name. He died at an early age, shortly after going blind from the eyestrain induced by his astronomical observations. It is interesting that his nephew, David Gregory (1661–1708), also served as professor of mathematics at Edinburgh, from 1684 to 1691, after which he was appointed Savilian professor of astronomy at Oxford. He too was interested in optics, writing on that subject as well as on geometry and the Newtonian theory.

It has been said that but for London's Great Fire of 1666, Sir Christopher Wren (1632–1723) would have been known as a mathematician instead of as an architect. He was Savilian professor of astronomy at Oxford from 1661 to 1673, and, for a time, president of the Royal Society. He wrote on the laws of collision of bodies, on subjects connected with optics, the resistance of fluids, and other topics in mathematical physics and celestial mechanics. He is credited with the discovery, in 1669, of the two systems of rulings on a hyperboloid of one sheet. He was the first (1658) to show that an arch of the cycloid is equal in length to eight times the radius of the generating circle. But after the great fire Wren took such a prominent part in rebuilding St. Paul's cathedral and some 50 or more other churches and public buildings that his fame as an architect overshadowed his reputation as a mathematician.

Mention should perhaps also be made of Robert Hooke (1635–1703) and Edmund Halley (1656–1742), although these men achieved fame in allied fields rather than in mathematics itself. For almost 40 years, Hooke served as professor of geometry at Gresham College. He is known to every student of elementary physics by his law relating the stress and strain of a stretched elastic string. He invented the conical pendulum and attempted to find the law of force (later shown by Newton to be the inverse square law) under which the planets revolve about the sun. He and Huygens both designed watches regulated by a balance spring. The other man, Halley, succeeded Wallis as Savilian professor of geometry, and later became astronomer royal. He conjecturally restored the lost Book VIII of Apollonius' *Conic Sections* and edited various works of the ancient Greeks, translating some of these from the Arabic even though he did not know a single word of the language. He also compiled a set of mortality tables of the sort now basic in the life insurance business. His major original contributions, however, were chiefly in astronomy, and of excellent quality. He was as kind and generous in his dealings with other scholars as Hooke was jealous and irritable. Much of his work was done in the eighteenth century.

10-9   Some Seventeenth-Century Mathematicians of Germany and the Low Countries

The auspicious progress made in mathematics by Germany during the sixteenth century did not continue in the seventeenth century. The Thirty Years' War (1618–1648) and the subsequent unrest in the Teutonic countries made the century unsuited there to intellectual progress. Kepler and Leibniz stand out as the only first-class German mathematicians of the period, and the only minor German mathematician whom we shall mention here is Ehrenfried Walther von Tschirnhausen (1651–1708). Tschirnhausen devoted much time to mathematics and physics, leaving his impress on the study of curves and the theory of equations. In 1682, he introduced and studied *catacaustic curves*, such a curve being the envelope of light rays, emitted from a point source, after reflection from a given curve. The special sinusoidal spiral, $a = r \cos^3 (\theta/3)$, is known as *Tschirnhausen's cubic*. The general sinusoidal spiral, $r^n = a \cos n\theta$, where n is rational, was studied by Colin Maclaurin in 1718 (see Problem Study 10.10). In the theory of equations, Tschirnhausen is particularly known for a transformation which converts an nth degree polynomial equation in x into an nth degree polynomial equation in y in which the coefficients of y^{n-1} and y^{n-2} are both zero. Later, in 1834, G. B. Jerrard found a Tschirnhausen transformation that converts an nth degree polynomial equation in x into an nth degree polynomial equation in y in which the coefficients of y^{n-1}, y^{n-2}, y^{n-3} are all zero. This transformation, as applied to a quintic polynomial, had been given earlier, in 1786, by E. S. Bring and is of importance in the transcendental solution of the quintic equation by means of elliptic functions.

In spite of strenuous times the geographic region now known as the Low Countries produced a number of lesser mathematicians in the seventeenth century. Willebrord Snell (1580 or 1581–1626) has already been mentioned in connection with his work on the mensuration of the circle. He was an infant prodigy and it is said that by the age of 12 he had acquainted himself with the standard mathematical works of his time. The name *loxodrome*, for a path on a sphere which makes a constant angle with the meridians, is due to Snell, and he was an early investigator of the properties of polar spherical triangles. These latter were first discussed by Viète.

Albert Girard (1595–1632), who seems to have lived chiefly in Holland, also interested himself in spherical geometry and trigonometry. In 1626, he published a treatise on trigonometry which contains the earliest use of our abbreviations *sin, tan, sec* for sine, tangent, and secant. He gave the expression for the area of a spherical triangle in terms of its spherical excess. Girard was also an algebraist of considerable power. He edited the works of Simon Stevin.

Grégoire de Saint-Vincent (1584–1667) was a prominent circle squarer of the seventeenth century. He applied pre-calculus methods to various quadrature problems.

Frans van Schooten the Younger (1615–1660 or 1661), a professor of mathematics who edited two Latin editions of Descartes' *La géométrie*, taught mathematics to Huygens, Hudde, and Sluze. He wrote on perspective and edited Viète's works. His father, Frans van Schooten the Elder, and his half brother, Petrus van Schooten, also were professors of mathematics.

Johann Hudde (1633–1704) was burgomaster of Amsterdam. He wrote on maxima and minima and the theory of equations. In the latter subject he gave an ingenious rule for finding multiple roots of a polynomial which is equivalent to our present method, where we find the roots of the highest common factor of the polynomial and its derivative.

René Francois Walter de Sluze (1622–1685), a canon in the Church, wrote numerous tracts in mathematics. He discussed spirals, points of inflection, and the finding of geometric means. The family of curves, $y^n = k(a - x)^p x^m$, where the exponents are positive integers, are called, after him, *pearls of Sluze*.

We conclude with Nicolaus Mercator (*ca.* 1620–1687), who was born in Holstein, then a part of Denmark, but who spent most of his life in England. He edited Euclid's *Elements*, wrote on trigonometry, astronomy, the computation of logarithms, and cosmography. The series

$$\ln (1 + x) = x - x^2/2 + x^3/3 - x^4/4 + \cdots,$$

which was independently discovered by Saint-Vincent, is sometimes referred to as *Mercator's series*. It converges for $-1 < x \leqq 1$, and can be used very satisfactorily for computing logarithms (see Problem Study 10.13). The familiar map of a sphere known as *Mercator's projection*, in which loxodromes appear as straight lines, is not due to Nicolaus Mercator, but to Gerhardus Mercator (1512–1594).

10-10   Academies, Societies, and Periodicals

The great increase that occurred in scientific and mathematical activity at a time when no periodicals existed led to the formation of a number of discussion circles with regular times of meeting. Some of these groups finally crystallized into academies, the first of which was established in Naples in 1560, followed by the Accademia dei lincei in Rome in 1603. Then, following the northward swing of mathematical activity in the seventeenth century, the Royal Society was founded in London in 1662 and the French Academy in Paris in 1666. These academies constituted centers where scholarly papers could be presented and discussed.

But the need for periodicals for the prompt dissemination of new scientific and mathematical findings was increasingly felt, until today the extent of such literature has become enormous. One count claims that prior to 1700 there were only 17 periodicals containing mathematical articles, the first of these having appeared in 1665. In the eighteenth century, 210 such periodicals appeared, and in the nineteenth century, the number of new journals of this sort reached 950. Many of these, however, often contained little relating to pure mathematics. Perhaps the oldest of the current journals devoted chiefly or entirely to advanced mathematics is the French *Journal de l'École Polytechnique*, launched in 1794. A number of more elementary mathematics journals were started earlier, but many of these aimed to entertain the subscriber with puzzles and problems, rather than to advance mathematical knowledge. Some of our current high-grade mathematical periodicals were started during the first half of the nineteenth century. Foremost among these are the German journal entitled *Journal für die reine und angewandte Mathematik*, first published in 1826 by A. L. Crelle, and the French journal entitled *Journal de mathématiques pures et appliquées*, which appeared in 1836 under the editorship of J. Liouville. These two journals are frequently called *Crelle's Journal* and *Liouville's Journal*, after the names of their founders. In England the *Cambridge Mathematical Journal* was founded in 1839, became the *Cambridge and Dublin Mathematical Journal* from 1846 to 1854, and in 1855 took the title of *Quarterly Journal of Pure and Applied Mathematics*. The *American Journal of Mathematics* was established in 1878 under the editorship of J. J. Sylvester. The earliest permanent periodicals devoted to the interests of teachers of mathematics, rather than to mathematical research, are the *Archiv der Mathematik und Physik*, founded in 1841, and the *Nouvelles annales de mathématiques*, founded a year later.

In the second half of the nineteenth century, there was a powerful development which increased the number of high-quality mathematics journals. This was the formation of a number of large mathematical societies having regular periodicals as their official organs. The earliest of these societies was the London Mathematical Society, organized in 1865, and which immediately began to publish its *Proceedings*. This society has become the national mathematical society of England. Seven years later, the Société Mathématique de France was established in Paris, and its official journal is known as its *Bulletin*. In Italy, in 1884, the mathematical society Circolo Matematico di Palermo was organized, and three years later it began to publish its *Rendiconti*. About this time the Edinburgh Mathematical Society was founded in Scotland, and has since maintained its *Proceedings*. The American Mathematical Society was organized, under a different name, in 1888, and began to issue its *Bulletin*, then later, in 1900, its *Transactions*, and more recently, in 1950, its *Proceedings*. Germany was the last of the leading mathematical countries to organize a national mathematical society, but in 1890 the Deutsche Mathematiker-Vereinigung was organized, which, in 1892, began the publication of its *Jahresbericht*. This last journal carried a number of extensive reports on modern developments in different fields of mathematics, such a report sometimes running into many hundreds of pages. These reports may be regarded as forerunners of the later large encyclopedias of mathematics. The excellent mathematics journals of the Soviet Union, though of later origin, are not to be ignored.

Today almost every country has its mathematical society, and many have additional associations devoted to various levels of mathematical instruction. These societies and associations have become potent factors in the organization and development of research activity in mathematics and in the improvement of methods of teaching the subject.

Very valuable to researchers is the journal *Mathematical Reviews*, sponsored by a number of mathematical organizations located both in the United States and abroad. This journal appeared in 1940 and contains abstracts and reviews of the current mathematical literature of the world.

Problem Studies

10.1 Geometric Algebra

(a) Given a unit segment and a segment of length x, construct with straightedge and compasses segments of lengths x^2, x^3, x^4, $\cdots$.

(b) Given a unit segment and segments of lengths x, y, z construct segments of lengths xy and xyz.

(c) Given a unit segment show that, if $f(x)$ and $g(x)$ are polynomials in x having coefficients represented by given line segments, we may construct a segment of length $y = f(x)/g(x)$ corresponding to any line segment chosen for x.

(d) Let us be given a quadratic equation $x^2 - gx + h = 0$, $g > 0$,

$h > 0$. On a line segment of length g as diameter draw a semicircle C, and then draw a line parallel to the diameter of C at a distance $\sqrt{h}$ from it, cutting C in a point P. From P drop a perpendicular upon the diameter of C, dividing the diameter into two parts r and s. Show that r and s represent the roots of the given quadratic equation. Solve $x^2 - 7x + 12 = 0$ by this method.

(e) Let us be given a quadratic equation $x^2 + gx - h = 0$, $g > 0$, $h > 0$. On a segment of length g as diameter draw a circle C, and then draw a tangent to C and mark off on it from the point of contact a length equal to $\sqrt{h}$. From the other extremity of this tangent segment, draw the secant passing through the center of C. Denoting the whole secant by r and its external segment by s show that $-r$ and s represent the roots of the given quadratic equation. Solve $x^2 + 4x - 21 = 0$ by this method.

10.2 Descartes' La géométrie

(a) Given five lines $L_1, \cdots, L_5$, arranged as in Figure 72. Let p_i denote the distance of a point P from line L_i. Taking L_5 and L_4 as x and y axes, find the equation of the locus of a point P moving such that

$$p_1 p_2 p_3 = a p_4 p_5.$$

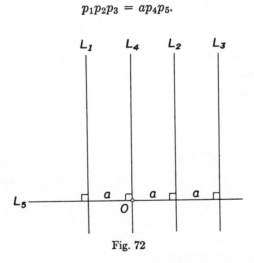

Fig. 72

(The locus is a cubic which Newton called a *Cartesian parabola* and which has also sometimes been called a *trident*; it appears frequently in *La géométrie*.)

(b) Show that we may with Euclidean tools construct as many points as we wish on the locus in part (a).

(c) Given any four lines L_1, L_2, L_3, L_4, let p_i denote the distance of a point P from line L_i. Show that the locus of P such that $p_1p_2 = kp_3p_4$ is a conic.

(d) Carry through Descartes' method of drawing a tangent at a general point (x_1, y_1) of the parabola $y^2 = 2mx$, and show that it leads to the fact that the subnormal (the projection upon the axis of the parabola of the segment of the normal lying between the curve and the axis) is of constant length, equal to half the latus rectum of the parabola.

10.3   Descartes' Rule of Signs

If c_1, c_2, $\cdots$, c_m are any m nonzero real numbers, and if two consecutive terms of this sequence have opposite signs, we say that these two terms present a *variation* of sign. With this concept we may state "Descartes' rule of signs," a proof of which may be found in any textbook on the theory of equations, as follows: *Let $f(x) = 0$ be a polynomial equation with real coefficients and arranged in descending powers of x. The number of positive roots of the equation is either equal to the number of variations of signs presented by the coefficients of $f(x)$, or less than this number of variations by a positive even number. The number of negative roots is either equal to the number of variations of sign presented by the coefficients of $f(-x)$, or less than this number of variations by a positive even number. A root of multiplicity m is counted as m roots.* Investigate the nature of the roots of the following equations by means of Descartes' rule of signs:

(a) $x^9 + 3x^8 - 5x^3 + 4x + 6 = 0$,

(b) $2x^7 - 3x^4 - x^3 - 5 = 0$,

(c) $3x^4 + 10x^2 + 5x - 4 = 0$.

(d) Show that $x^n - 1 = 0$ has exactly two real roots if n is even, and only one real root if n is odd.

(e) Show that $x^5 + x^2 + 1 = 0$ has four imaginary roots.

(f) Prove that if p and q are real, and $q \neq 0$, the equation $x^3 + px + q = 0$ has two imaginary roots when p is positive.

(g) Prove that if the roots of a polynomial equation are all positive, the signs of the coefficients are alternately positive and negative.

10.4   Problems from Descartes

(a) Draw the graph of the folium of Descartes,

$$x^3 + y^3 = 3axy.$$

The line $x + y + a = 0$ is an asymptote.

(b) Find the corresponding polar equation of the folium of Descartes.

(c) Set $y = tx$ and obtain a parametric representation of the folium of Descartes in terms of t as parameter. Find the ranges for t leading to the loop, the lower arm, and the upper arm.

(d) Find the Cartesian equation of the folium of Descartes when the node is taken as origin and the line of symmetry of the curve as x-axis.

(e) Descartes' solution of a depressed quartic equation employs the method of undetermined coefficients. As an example consider the quartic equation

$$x^4 - 2x^2 + 8x - 3 = 0.$$

Set the left member of the equation equal to the product of two quadratic factors of the forms $x^2 + kx + h$ and $x^2 - kx + m$. Obtain three relations connecting k, h, m by equating corresponding coefficients on the two sides of the equation. Eliminate h and m from the three relations, obtaining a sextic equation in k which can be regarded as a cubic equation in k^2. Thus the solution of the original quartic equation is reduced to the solution of an associated cubic equation. Knowing that one root of the cubic in k^2 is $k^2 = 4$, obtain the four roots of the original quartic equation.

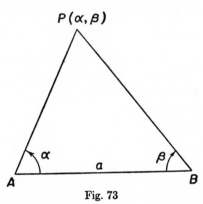

Fig. 73

10.5 Coordinate Systems

Let us designate as a *bipolar coordinate system* one for which the frame of reference is a horizontal line segment AB of length a, with respect to which a point P of the plane is located by recording as coordinates the counterclockwise angle $\alpha = \measuredangle\ BAP$ and the clockwise angle $\beta = \measuredangle\ ABP$ (see Figure 73).

(a) Find the bipolar equation of (1) the perpendicular bisector of AB, (2) an arc of a circle having AB as chord.

(b) Find the equations of transformation connecting the bipolar coordinate system with the rectangular Cartesian coordinate system having x-axis along AB and origin at the mid-point of AB.

(c) Identify the curves: (1) $\cot \alpha \cot \beta = k$, (2) $\cot \alpha / \cot \beta = k$, (3) $\cot \alpha + \cot \beta = k$, where k is a constant.

(d) Find rectangular Cartesian equations of the following curves given by polar equations: (1) lemniscate of Bernoulli, $r^2 = a^2 \cos 2\theta$, (2) cardioid, $r = a(1 - \cos \theta)$, (3) spiral of Archimedes, $r = a\theta$, (4) equiangular spiral, $r = e^{a\theta}$, (5) hyperbolic spiral, $r\theta = a$, (6) four-leaved rose, $r = a \sin 2\theta$.

(e) Describe the latitude and longitude coordinate system on a spherical surface.

(f) A natural extension to space of the polar coordinate system of the plane consists in fixing an origin O and then taking as coordinates of a point P the length r of the radius vector OP and the latitude ϕ and longitude θ of P for the sphere having center O and radius OP. These coordinates are known as *spherical coordinates*. Find equations connecting the spherical coordinates (r,ϕ,θ) of a point P and rectangular Cartesian coordinates (x,y,z) of the point. Essentially such relations are found in the works of Lagrange (1736–1813).

(g) Design a coordinate system to locate points on (1) a circular cylindrical surface, (2) a torus.

10.6 Line Coordinates

(a) Show that on a rectangular Cartesian frame of reference we can use the slope and the y-intercept of a line as its coordinates, or the length of the perpendicular upon the line from the origin and the angle which that perpendicular makes with the x-axis.

(b) The negative reciprocals, u and v, of the x and y-intercepts of a line are known as the line's *Plücker* coordinates. Find the Plücker coordinates of the lines whose Cartesian equations are $5x+3y-6=0$ and $ax+by+1=0$. Write the Cartesian equation of the line having Plücker coordinates $(1,3)$.

(c) Show that the Plücker coordinates u, v of all lines passing through the point with Cartesian coordinates $(2,3)$ satisfy the linear equation $2u + 3v + 1 = 0$. This equation is taken as the Plücker equation of the point $(2,3)$. What are the Cartesian coordinates of the points whose Plücker equations are $5u + 3v - 6 = 0$ and $au + bv + 1 = 0$? Write the Plücker equation of the point having Cartesian coordinates $(1,3)$.

10.7 Dimensionality

(a) Show that the plane is 4 dimensional in directed line segments. What is its dimensionality in directed line segments of a given length?

(b) Show that space is 4 dimensional in lines, 3 dimensional in planes, and 4 dimensional in spheres.

(c) How might we define, analytically, the *straight line* in hyperspace determined by the two points $(x_1, \cdots, x_n)$ and $(y_1, \cdots, y_n)$?

(d) How might we define *direction numbers* and *direction cosines* of the straight line in part (c)?

(e) How might we define the *line segment* determined by the two points in part (c)?

(f) How might we define the *mid-point* of the line segment of part (e)?

10.8 Fermat's Theorems

About 1760, Euler proposed and solved the problem of determining the number of positive integers less than a given positive integer n and prime to n. This number is now usually denoted by $\phi(n)$, and is called *Euler's ϕ-function* of n (also sometimes called the *indicator*, or *totient*, of n). Thus, if $n = 42$, it is found that the 12 integers 1, 5, 11, 13, 17, 19, 23, 25, 29, 31, 37, and 41 are the only positive integers less than and prime to 42. Therefore $\phi(42) = 12$.

(a) Find $\phi(n)$ for $n = 2, 3, \cdots, 12$. A table giving the values of $\phi(n)$ for all $n \leq 10,000$ has been computed by J. W. L. Glaisher (1848–1928).

(b) If p is a prime show that $\phi(p) = p - 1$ and $\phi(p^\alpha) = p^\alpha(1 - 1/p)$.

(c) It can be shown that if $n = ab$, where a and b are relatively prime, then $\phi(n) = \phi(a)\phi(b)$. Using this fact calculate $\phi(42)$ from the results of part (a), and also show that if $n = p_1{}^{\alpha_1}p_2{}^{\alpha_2} \cdots p_r{}^{\alpha_r}$, where $p_1, p_2, \cdots, p_r$ are primes, then

$$\phi(n) = n(1 - 1/p_1)(1 - 1/p_2) \cdots (1 - 1/p_r).$$

Use this last formula to calculate $\phi(360)$.

(d) Euler showed that if a is any positive integer relatively prime to n, then $a^{\phi(n)} - 1$ is divisible by n. Show that the little Fermat theorem is a special case of this.

(e) Show that to establish Fermat's last theorem it is sufficient to consider only prime exponents $p > 2$.

(f) Assuming Fermat's last theorem, show that the curve $x^n + y^n = 1$, where n is a positive integer greater than 2, contains no points with rational coordinates except those points where the curve crosses a coordinate axis.

(g) Assuming item (6) of Section 10-4 (that the area of an integral sided right triangle cannot be a square number), show that the equation

$x^4 - y^4 = z^2$ has no solution in positive integers x, y, z, and then prove Fermat's last theorem for the case $n = 4$.

10.9 Problems from Huygens

(a) A gambler is to win \$300 if a six is thrown with a single die. What is his mathematical expectation?

(b) Suppose a gambler is to win \$300 if he throws a six with a single die but \$600 if he throws a five. What is his mathematical expectation?

Following are some examples of probability problems solved by Huygens:

(1) A and B cast alternately with a pair of ordinary dice. A wins if he throws 6 before B throws 7, and B wins if he throws 7 before A throws 6. If A begins, then his chance of winning is to B's chance of winning as 30 : 31.

(2) A and B each take 12 counters and play with 3 dice as follows: if 11 is thrown, A gives a counter to B; if 14 is thrown, B gives a counter to A; and he wins the game who first obtains all the counters. Then A's chance is to B's as 244,140,625 : 282,429,536,481.

(3) A and B play with 2 dice; if 7 is thrown, A wins; if 10 is thrown, B wins; if any other number is thrown, the game is drawn. Then A's chance of winning is to B's as 13 : 11.

(c) Using the isochronous property of the cycloid, and the fact that the evolute of a cycloid is another cycloid of the same size, show that a pendulum constrained to swing between two successive arches of an inverted cycloid (see Figure 74) must oscillate with a constant period.

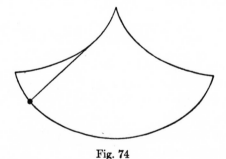

Fig. 74

(d) A ball swings uniformly in a circle at the end of a string, making one revolution per minute. If the length of the string is doubled and the period of revolution halved, how does the centrifugal force compare with that of the first situation?

10.10 Higher Plane Curves

(a) Taking the foci of a Cassinian curve at the points $(-a,0)$ and $(a,0)$ on a rectangular Cartesian frame of reference and denoting the constant product of distances by k^2, find the Cartesian equation of the curve.

(b) Show that the corresponding polar equation of the curve is

$$r^4 - 2r^2a^2 \cos 2\theta + a^4 = k^4.$$

Note that if $k = a$, the curve becomes the lemniscate of Bernoulli,

$$r^2 = 2a^2 \cos 2\theta.$$

(c) Show that the lemniscate of Bernoulli is the cissoid (see Problem Study 4.4) of a circle of radius $a/2$, and itself, for a pole O distant $a\sqrt{2}/2$ units from its center.

(d) Plot carefully a rectangular hyperbola $xy = k^2$ and draw several members of the family of circles with centers on the hyperbola and passing through the origin. The envelope of this family of circles is a lemniscate of Bernoulli.

(e) Using the fact that the normal at a point of the lemniscate of Bernoulli in part (b) makes an angle 2θ with the radius vector to the point, show how we may construct tangents to the lemniscate.

(f) Show that we have the following special cases of the sinusoidal spiral, $r^n = a \cos n\theta$, where n is a rational number.

n	curve
-2	rectangular hyperbola
-1	straight line
$-1/2$	parabola
$-1/3$	Tschirnhausen cubic
$1/2$	cardioid
1	circle
2	lemniscate of Bernoulli

(g) An *epicycloid* is the path traced by a point on a circle rolling externally upon a fixed base circle. The catacaustic of a circle for a light source at infinity is an epicycloid of two cusps whose base circle is concentric with the given circle and whose radius is half the radius of the given circle. An epicycloid of two cusps is called a *nephroid*. The catacaustic of a circle for a light source on the circumference of the circle is an epicycloid of one cusp whose base circle is concentric with the given circle and whose radius is one-third the radius of the given circle. An epicycloid of one cusp is a cardioid. Jakob Bernoulli showed, in 1692, that the catacaustic of a cardioid when the light source is at the cusp of the cardioid is a nephroid. Catacaustics of a circle can be seen as the bright curves on the surface of coffee in a cup or upon the table inside a circular napkin ring. Observe some catacaustics of a circle using a cup of liquid and a movable light source.

10.11   Recreational Problems from Bachet

Following are some arithmetical recreations found in Bachet's *Problèmes plaisants et délectables*. They, and other problems from Bachet, can also be found in Ball-Coxeter, *Mathematical Recreations and Essays*.

(a) (1) Ask a person to choose secretly a number, and then to treble it. (2) Inquire if the product is even or odd. If it is even, ask him to take half of it; if it is odd, ask him to add one and then take half of it. (3) Tell him to multiply the result in (2) by 3, and to announce to you how many times, say n, 9 will divide integrally into the product. (4) Then the number originally chosen was $2n$ or $2n + 1$, according as the result in step (1) was even or odd. Prove this.

(b) Ask a person to choose secretly a number less than 60, and to announce the remainders, say a, b, c, when the number is divided by 3, by 4, and by 5. Then the number originally chosen can be found as the remainder obtained by dividing $40a + 45b + 36c$ by 60. Prove this.

(c) Tell A to choose secretly any number, greater than 5, of counters, and B to take 3 times as many. Ask A to give 5 counters to B, and then ask B to transfer to A 3 times as many counters as A has left. You may now tell B that he has 20 counters. Explain why this is so and generalize to the case where the 3 and 5 are replaced by p and q.

(d) A secretly selects either one of a pair of numbers, one of which is odd and the other even, and the other number is given to B. Ask A to double his number, and B to triple his. Request the sum of the two products. If the sum is even, then A selected the odd number; otherwise A selected the even number. Explain this.

(e) Ask someone to think of an hour, say m, and then to touch on a watch the number that marks some other hour, say n. If, beginning with the number touched, he taps successively in the counterclockwise direction the numbers on the watch, meanwhile mentally counting the taps as m,

$m + 1$, and so on, until he reaches the number $n + 12$, then the last number tapped will be that of the hour he originally thought of. Prove this.

10.12   Some Geometry

(a) Show, by Roberval's method, that the tangent and normal at a point on a central conic bisect the angles between the two focal radii drawn to the point.

(b) A *spherical degree* is defined to be any spherical area which is equivalent to (1/720)th of the entire surface of the sphere. Show that the area of a lune whose angle is $n°$ is equal to $2n$ spherical degrees.

(c) Show that the area of a spherical triangle, in spherical degrees, is equal to the spherical excess of the triangle.

(d) Show that the area A of a spherical triangle of spherical excess E is given by

$$A = \pi r^2 E / 180°,$$

where r is the radius of the sphere.

(e) Find the area of a trirectangular triangle on a sphere whose diameter is 28 inches.

10.13   Computation of Logarithms by Series

The Mercator series

$$\ln (1 + x) = x - x^2/2 + x^3/3 - x^4/4 + \cdots$$

converges for $-1 < x \leq 1$. Replacing x by $-x$, it follows that the series

$$\ln (1 - x) = -x - x^2/2 - x^3/3 - x^4/4 - \cdots$$

must converge for $-1 \leq x < 1$. Since a series whose terms are the differences of the corresponding terms of two given series certainly converges for all values of x for which both of the given series converge, it follows that, for $-1 < x < 1$,

$$\ln [(1 + x)/(1 - x)] = \ln (1 + x) - \ln (1 - x)$$
$$= 2(x + x^3/3 + x^5/5 + x^7/7 + \cdots).$$

If we set $x = 1/(2N + 1)$ we observe that $-1 < x < 1$ for all positive N, and $(1 + x)/(1 - x) = (N + 1)/N$. Substituting in the last equation we find

$$\ln (N + 1) = \ln N + 2[1/(2N + 1) + 1/3(2N + 1)^3 + 1/5(2N + 1)^5 + \cdots],$$

the series converging, and rather rapidly, for all positive N.

(a) By setting $N = 1$, compute $\ln 2$ to four decimal places.

(b) Compute $\ln 3$ to four decimal places.

(c) Compute $\ln 4$ to four decimal places.

Bibliography

ADAMS, O. S., *A Study of Map Projections in General*. Washington: Coast and Geodetic Survey, Special Publication No. 60, Department of Commerce, 1919.

————, AND C. H. DEETZ, *Elements of Map Projection, with Applications to Map and Chart Construction*. Washington: Coast and Geodetic Survey, Special Publication No. 68, Department of Commerce, 1938.

AUBREY, JOHN, *Brief Lives*. New York: Oxford University Press, 1898.

BALL, W. W. R., AND H. S. M. COXETER, *Mathematical Recreations and Essays*, 11th ed. New York: The Macmillan Company, 1939.

BELL, A. E., *Christian* [sic] *Huygens and the Development of Science in the Seventeenth Century*. London: Edward Arnold & Co., 1948.

BELL, E. T., *Men of Mathematics*. New York: Simon and Schuster, Inc., 1937.

BOYER, C. B., *History of Analytic Geometry*. New York: *Scripta Mathematica*, Yeshiva University, 1956.

————, *The History of the Calculus and Its Conceptual Development*. New York: Dover Publications, Inc., 1959.

CONKWRIGHT, N. B., *Introduction to the Theory of Equations*. Boston: Ginn & Company, 1941.

COOLIDGE, J. L., *A History of Geometrical Methods*. New York: Oxford University Press, 1940.

————, *A History of the Conic Sections and Quadric Surfaces*. New York: Oxford University Press, 1947.

————, *The Mathematics of Great Amateurs*. New York: Oxford University Press, 1949.

DAVID, F. N., *Games, Gods and Gambling*. New York: Hafner Publishing Company, 1962.

DESCARTES, RENÉ, *The Geometry of René Descartes*, tr. by D. E. Smith and Marcia L. Latham. New York: Dover Publications, Inc., 1954.

HACKER, S. G., *Arithmetical View Points*. Pullman, Wash.: Mimeographed at Washington State College, 1948.

HALDANE, ELIZABETH S., *Descartes: His Life and Times*. New York: E. P. Dutton & Co., Inc., 1905.

KRAITCHIK, MAURICE, *Mathematical Recreations*. New York: W. W. Norton & Company, Inc., 1942.

MERRIMAN, MANSFIELD, *The Solution of Equations*, 4th ed. New York: John Wiley & Sons, Inc., 1906.

MILLER, G. A., *Historical Introduction to Mathematical Literature*. New York: The Macmillan Company, 1916.

MUIR, JANE, *Of Men and Numbers, the Story of the Great Mathematicians*. New York: Dodd, Mead & Company, Inc., 1961.

ORE, OYSTEIN, *Number Theory and Its History*. New York: McGraw-Hill Book Company, Inc., 1948.

SMITH, D. E., *History of Modern Mathematics*, 4th ed. New York: John Wiley & Sons, Inc., 1906.

———, *A Source Book in Mathematics*. New York: McGraw-Hill Book Company, Inc., 1929.

SULLIVAN, J. W. N., *The History of Mathematics in Europe, from the Fall of Greek Science to the Rise of the Conception of Mathematical Rigour*. New York: Oxford University Press, 1925.

SUMMERSON, JOHN, *Sir Christopher Wren*, No. 9 in a series of *Brief Lives*. New York: The Macmillan Company, 1953.

TODHUNTER, ISAAC, *A History of the Mathematical Theory of Probability from the Time of Pascal to that of Laplace*. New York: Chelsea Publishing Company, 1949.

TURNBULL, H. W., *The Great Mathematicians*. New York: New York University Press, 1961.

WILLIAMSON, BENJAMIN, *An Elementary Treatise on the Differential Calculus*. London: Longmans, Green & Co., Ltd., 1899.

WINGER, R. M., *An Introduction to Projective Geometry*. Boston: D. C. Heath and Company, 1923.

YATES, R. C., *A Handbook on Curves and Their Properties*. Ann Arbor, Mich.: J. W. Edwards, Publisher, Inc., 1947.

11

The Calculus and Related
Concepts

11-1   Introduction

We have seen that many new and extensive fields of mathematical investigation were opened up in the seventeenth century, making that era an outstandingly productive one in the development of mathematics. Unquestionably the most remarkable mathematical achievement of the period was the invention of the calculus, toward the end of the century, by Isaac Newton and Gottfried Wilhelm Leibniz. With this invention, creative mathematics quite generally passed to an advanced level and the history of elementary mathematics essentially terminates. The present chapter will be devoted to a very brief account of the origins and development of the important concepts of the calculus, concepts which are so far-reaching and which have exercised such an impact on the modern world that it is perhaps correct to say that without some knowledge of them a person today can scarcely claim to be well educated.

It is interesting that, contrary to the customary order of presentation found in our sophomore college courses, where we begin with differentiation and later consider integration, the ideas of the integral calculus developed historically before those of the differential calculus. The idea of integration first arose in its role of a summation process in connection with the finding of certain areas, volumes, and arc lengths. Some time later differentiation was created in connection with problems on tangents to curves and with questions about maxima and minima of functions. And still later it was observed that integration and differentiation are related to each other as inverse operations.

Although the major part of our story lies in the seventeenth century, we must, for the beginning, go back to ancient Greece and the fifth century B.C.

11-2   Zeno's Paradoxes

Should we assume that a magnitude is infinitely divisible or that it is made up of a very large number of small indivisible atomic parts? The first assumption appears the more reasonable to most of us, but the utility of the second assumption in the making of discoveries causes it to lose some of its seeming absurdity. There is evidence that in Greek antiquity, schools of mathematical reasoning developed employing each of the above two assumptions.

Some of the logical difficulties encountered in either assumption were strikingly brought out in the fifth century B.C. by four paradoxes devised by the Eleatic philosopher Zeno (*ca.* 450 B.C.). These paradoxes, which have had a profound influence on mathematics, assert that motion is impossible whether we assume a magnitude to be infinitely divisible or to be made up of a large number of atomic parts. We illustrate the nature of the paradoxes by the following two.

The Dichotomy: If a straight line segment is infinitely divisible then motion is impossible, for in order to traverse the line segment it is necessary first to reach the mid-point, and to do this one must first reach the one-quarter point, and to do this one must first reach the one-eighth point, and so on, ad infinitum. It follows that the motion can never even begin.

The Arrow: If time is made up of indivisible atomic instants, then a moving arrow is always at rest, for at any instant the arrow is in a fixed position. Since this is true of every instant it follows that the arrow never moves.

Many explanations of Zeno's paradoxes have been given and it is not difficult to show that they challenge the common intuitive beliefs that the sum of an infinite number of positive quantities is infinitely large, even if each quantity is extremely small ($\sum_{i=1}^{\infty} \epsilon_i = \infty$), and that the sum of either a finite or an infinite number of quantities of dimension zero is zero ($n \times 0 = 0$ and $\infty \times 0 = 0$). Whatever might have been the intended motive of the paradoxes, their effect was to exclude infinitesimals from Greek demonstrative geometry.

11-3 Eudoxus' Method of Exhaustion

The first problems occurring in the history of the calculus were concerned with the computation of areas, volumes, and lengths of arcs, and in their treatment one finds evidence of the two assumptions about the divisibility of magnitudes that we considered above.

One of the earliest important contributions to the problem of squaring the circle was that of Antiphon the Sophist (*ca.* 430 B.C.), a contemporary of Socrates. Antiphon, we are told, advanced the idea that by successively doubling the number of sides of a regular polygon inscribed in a circle, the difference in area between the circle and the polygon would at last be exhausted. Since a square can be constructed equal in area to any given polygon, it will then be possible to construct a square equal to the circle. This argument met immediate criticism on the grounds that it violated the principle that magnitudes are divisible without limit, and that, accordingly, Antiphon's process could never use up the whole area of the circle. Nevertheless, Antiphon's bold pronouncement contained the germ of the famous Greek *method of exhaustion.*

The method of exhaustion is usually credited to Eudoxus (*ca.* 370 B.C.) and can perhaps be considered as the Platonic school's answer to the paradoxes of Zeno. The method assumes the infinite divisibility of magnitudes and has, as a basis, the proposition: *If from any magnitude there be subtracted a part not less than its half, from the remainder another part not less than its half, and so on, there will at length remain a magnitude less than any preassigned magnitude of the same kind.* Let us employ the method of exhaustion to prove that if A_1 and A_2 are the areas of two circles having diameters d_1 and d_2, then

$$A_1 : A_2 = d_1{}^2 : d_2{}^2.$$

We first show, with the aid of the basic proposition, that the difference in area between a circle and an inscribed regular polygon can be made as small as desired. Let AB, in Figure 75, be a side of a regular inscribed polygon, and let M be the mid-point of the arc AB. Since the area of triangle AMB is half that of the rectangle $ARSB$, and hence greater than half the area of the circular segment AMB, it follows that by doubling the number of sides of the inscribed regular polygon we increase the area of the polygon by more than half the difference in area between the polygon and the circle. Consequently, by doubling the number of sides sufficiently often, we can make the difference in area between the polygon and the circle less than any assigned area, however small.

We now return to our theorem, and suppose that instead of equality we have

$$A_1 : A_2 > d_1{}^2 : d_2{}^2.$$

Then we can inscribe in the first circle a regular polygon whose area P_1 differs so little from A_1 that

$$P_1 : A_2 > d_1{}^2 : d_2{}^2.$$

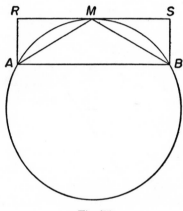

Fig. 75

Let P_2 be a regular polygon similar to P_1, but inscribed in the second circle. Then, from a known theorem about similar regular polygons,

$$P_1 : P_2 = d_1{}^2 : d_2{}^2.$$

It follows that $P_1 : A_2 > P_1 : P_2$, or $P_2 > A_2$, an absurdity, since the area of a regular polygon cannot exceed the area of its circumcircle. In a similar way we can show that we cannot have

$$A_1 : A_2 < d_1{}^2 : d_2{}^2.$$

Consequently, by this double *reductio ad absurdum* process, our theorem is established. Thus, if A is the area and d the diameter of a circle, $A = kd^2$, where k is a constant (actually $\pi/4$) which is the same for all circles.

Archimedes claimed that Democritus (*ca.* 410 B.C.) stated that the volume of a pyramid on any polygonal base is one-third that of a prism with the same base and altitude. Very little is known of Democritus, but he could hardly have given a rigorous demonstration of this theorem. Since a prism can be dissected into a sum of prisms all having triangular bases, and, in turn, a prism of this latter sort can be dissected into three triangular pyramids having, in pairs, equivalent bases and equal altitudes, it follows that the crux of Democritus' problem is to show that two pyramids of the same height and equivalent bases have equal volumes. A demonstration of this was later furnished by Eudoxus, using the method of exhaustion.

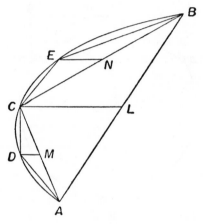

Fig. 76

How, then, might Democritus have arrived at this last result? A clue is furnished by Plutarch, who quotes a dilemma encountered by Democritus on an occasion when he considered a cone as made up of infinitely many plane cross sections parallel to the base. If two "adjacent" sections are of the same size, the solid would be a cylinder and not a cone. On the other hand, if two "adjacent" sections are different in area, the surface of the solid would be broken into a series of small steps, which certainly is not the case. Here we have an assumption concerning the divisibility of magnitudes which is somewhat intermediate to the two assumptions already considered, for here we assume the volume of the cone is infinitely divisible, namely into an infinite number of plane atomic sections, but that these sections are countable in the sense that given one of them there is a next one to it. Now Democritus may have argued that if two pyramids with equivalent bases and equal heights are cut by planes parallel to the bases and dividing the heights in the same ratio, then the corresponding sections formed are equivalent. Therefore the pyramids contain the same infinite number of equivalent plane sections, and hence must be equal in volume. This would be an early instance of Cavalieri's *method of indivisibles*, considered below in Section 11-6.

But of the ancients it was Archimedes who made the most elegant applications of the method of exhaustion and who came the nearest to actual integration. As one of his earliest examples consider his quadrature of a parabolic segment. Let C, D, E be points on the arc of the parabolic segment (see Figure 76) obtained by drawing LC, MD, NE parallel to the axis of the parabola through the mid-points L, M, N of AB, CA, CB. From the geometry of the parabola Archimedes shows that

$$\triangle CDA + \triangle CEB = (\triangle ACB)/4.$$

By repeated applications of this idea it follows that the area of the parabolic segment is given by

$$\triangle ABC + (\triangle ABC)/4 + (\triangle ABC)/4^2 + (\triangle ABC)/4^3 + \cdots$$
$$= \triangle ABC(1 + 1/4 + 1/4^2 + 1/4^3 + \cdots)$$
$$= (4/3)\triangle ABC.$$

Here we have shortened the work by taking the limit of the sum of a geometric progression; Archimedes employs the double *reductio ad absurdum* apparatus of the method of exhaustion.

In his treatment of certain areas and volumes, Archimedes arrived at equivalents of a number of definite integrals found in our elementary calculus textbooks.

11-4   Archimedes' Method of Equilibrium

The method of exhaustion is a rigorous but sterile method. In other words, once a formula is known, the method of exhaustion may furnish an elegant tool for establishing it, but the method does not lend itself to the initial discovery of the result. The method of exhaustion is, in this respect, very much like the process of mathematical induction. How, then, did

Archimedes discover the formulas which he so neatly established by the method of exhaustion?

The above question was finally cleared up only as late as 1906, with the discovery by Heiberg, in Constantinople, of a copy of Archimedes' long lost treatise *Method*, addressed to Eratosthenes. The manuscript was found on a palimpsest (see Section 1-8); that is, it had been written in the tenth century on parchment, and then later, in the thirteenth century, washed off and the parchment reused for a religious text. Fortunately, most of the first text was able to be restored from beneath the later writing.

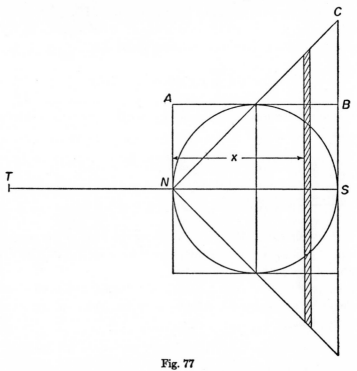

Fig. 77

The fundamental idea of Archimedes' method is this. To find a required area or volume, cut it up into a very large number of thin parallel plane strips, or thin parallel layers, and (mentally) hang these pieces at one end of a given lever in such a way as to be in equilibrium with a figure whose content and centroid are known. Let us illustrate the method by using it to discover the formula for the volume of a sphere.

Let r be the radius of the sphere. Place the sphere with its polar diameter along a horizontal x-axis with the north pole N at the origin (see

Figure 77). Construct the cylinder and the cone of revolution obtained by rotating the rectangle $NABS$ and the triangle NCS about the x-axis. Now cut from the three solids thin vertical slices (assuming that they are flat cylinders) at distance x from N and of thickness Δx. The volumes of these slices are, approximately,

$$\text{sphere:} \quad \pi x(2r - x)\Delta x$$

$$\text{cylinder:} \quad \pi r^2 \Delta x$$

$$\text{cone:} \quad \pi x^2 \Delta x$$

Let us hang at T the slices from the sphere and the cone, where $TN = 2r$. Their combined moment* about N is

$$[\pi x(2r - x)\Delta x + \pi x^2 \Delta x]2r = 4\pi r^2 x \Delta x.$$

This, we observe, is four times the moment of the slice cut from the cylinder when that slice is left where it is. Adding a large number of these slices together we find

$$2r \text{ [vol. of sphere + vol. of cone]} = 4r \text{ [vol. of cylinder],}$$

or

$$2r \text{ [vol. of sphere} + 8\pi r^3/3] = 8\pi r^4,$$

or

$$\text{vol. of sphere} = 4\pi r^3/3.$$

This, we are told in the *Method*, was Archimedes' way of discovering the formula for the volume of a sphere. But his mathematical conscience would not permit him to accept such a method as a proof, and he accordingly supplied a rigorous demonstration by means of the method of exhaustion. In the method of equilibrium we see the fertility of the loosely founded idea of regarding a magnitude as composed of a large number of atomic pieces. Needless to say, with the modern method of limits, Archimedes' method of equilibrium can be made perfectly rigorous, and becomes essentially the same as present-day integration.

* By the *moment* of a volume about a point is meant the product of the volume and the perpendicular distance from the point to the vertical line passing through the centroid of the volume.

11-5   The Beginnings of Integration in Western Europe

The theory of integration received very little stimulus after Archimedes' remarkable achievements until relatively modern times. It was about 1450 that Archimedes' works reached western Europe through a translation of a ninth-century copy of his manuscripts found at Constantinople. This translation was revised by Regiomontanus and was printed in 1540. A few years later a second translation appeared. But it was not until about the beginning of the seventeenth century that we find Archimedes' ideas receiving further development.

Two early writers of modern times who used methods comparable to those of Archimedes were the Flemish engineer Simon Stevin (1548–1620) and the Italian mathematician Luca Valerio (*ca.* 1552–1618). Each of these men tried to avoid the double *reductio ad absurdum* of the method of exhaustion by making a direct passage to the limit, much as we did above (toward the end of Section 11-3) in our treatment of the area of a parabolic segment. Stevin used such a method in his work on hydrostatics where he found the force due to fluid pressure against a vertical rectangular dam by dividing the dam into thin horizontal strips and then rotating these strips about their upper and lower edges until they became parallel to a horizontal plane. This is fundamentally the method we use today in our elementary textbooks on calculus.

Of the early modern Europeans who developed ideas of infinitesimals in connection with integration, particular mention must be made of Johann Kepler. We have already remarked (in Section 9-6) that Kepler had to resort to an integration procedure in order to compute the areas involved in his second law of planetary motion and also the volumes dealt with in his treatise on capacities of wine barrels. But Kepler, like others of the time, had little patience with the careful rigor of the method of exhaustion, and under the temptation to save time and trouble freely adopted processes that Archimedes considered as merely heuristic. Thus Kepler regarded the circumference of a circle as a regular polygon possessing an infinite number of sides. If each of these sides is taken as the base of a triangle whose vertex is at the center of the circle, then the area of the circle is divided into an infinite number of thin triangles, all having an altitude equal to the radius of the circle. Since the area of each thin triangle is equal to half the product of its base and altitude, it turns out that the area of a circle is equal to half the product of its circumference and radius. Similarly, the volume of a sphere was regarded as composed of an infinite number of narrow cones having a common vertex at the center of the sphere. It follows that the volume of a sphere is one-third the product of its surface area and radius. Objectionable as such methods are from the standpoint of mathematical rigor, they produce correct results in a very simple manner. Even today one finds such "atomic" methods used quite regularly by physicists and engineers for setting up a mathematical problem, leaving the rigorous "limit" treatment to the professional mathematician.* And geometers frequently resort to the convenient concept of "consecutive" points and "consecutive" curves and surfaces in a one-parameter family of such entities.†

It probably was Kepler's attempts at integration that led Cavalieri to develop his method of indivisibles.

11-6   Cavalieri's Method of Indivisibles

Bonaventura Cavalieri was born in Milan in 1598, became a Jesuate‡ at an early age, studied under Galileo, and was a professor of mathematics at the University of Bologna from 1629 until his death in 1647. He was one of the most influential mathematicians of his time and wrote a number of works on mathematics, optics, and astronomy. He was largely responsible for the early introduction of logarithms into Italy. But his greatest contribution was a treatise, published in its first form in 1635, devoted to the method of indivisibles—a method which probably can be traced back to Democritus.

Cavalieri's treatise is verbose and not clearly written, and it is difficult to know precisely what is to be understood by an "indivisible." It seems that an indivisible of a given planar piece is a chord of that piece, and an indivisible of a given solid is a plane section of that solid. A planar piece is considered as made up of an infinite set of parallel chords and a solid as

* "Thus, so far as first differentials are concerned, a small part of a curve may be treated as straight and a part of a surface near a point as plane; during a short time dt, a particle may be considered as moving with constant speed and a physical process as occurring at a constant rate."—H. B. Phillips, *Differential Equations*, 3d ed. (New York: John Wiley & Sons, Inc., 1934), p. 28.

†"In other words, *the characteristic of a surface* [of a one-parameter family of surfaces] *is the curve in which a consecutive surface intersects it.*"—E. P. Lane, *Metric Differential Geometry of Curves and Surfaces* (Chicago: University of Chicago Press, 1940), p. 81.

‡Not a Jesuit, as is frequently incorrectly stated.

made up of an infinite set of parallel plane sections. Now, Cavalieri argued, if we slide each member of the set of parallel chords of some given planar piece along its own axis, so that the end points of the chords still trace a continuous boundary, then the area of the new planar piece so formed is the same as that of the original planar piece. A similar sliding of the plane sections of a given solid yields another solid having the same volume as the original one. These results give the so-called *Cavalieri's principle*: (1) *If two planar pieces are included between a pair of parallel lines, and if the two segments cut by them on any line parallel to the including lines are equal in length, then the areas of the planar pieces are equal*; (2) *if two solids are included between a pair of parallel planes, and if the two sections cut by them on any plane parallel to the including planes are equal in area, then the volumes of the solids are equal.*

Cavalieri's principle constitutes a valuable tool in the computation of areas and volumes and can easily be rigorously established. As an illustration of the principle, consider the following application leading to the formula for the volume of a sphere. In Figure 78 we have a hemisphere of radius r on

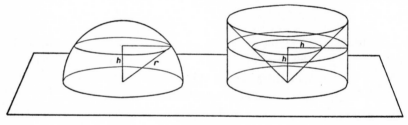

Fig. 78

the left, and on the right a cylinder of radius r and altitude r with a cone removed whose base is the upper base of the cylinder and whose vertex is the center of the lower base of the cylinder. The hemisphere and the gouged out cylinder are resting on a common plane. We now cut both solids by a plane parallel to the base plane and at distance h from it. This plane cuts the one solid in a circular section and the other in an annular, or ring-shaped, section. By elementary geometry we easily show that each of the two sections has an area equal to $\pi(r^2 - h^2)$. It follows, by Cavalieri's principle, that the two solids have equal volumes. Therefore the volume V of a sphere is given by

$$V = 2(\text{vol. of cylinder} - \text{vol. of cone})$$
$$= 2(\pi r^3 - \pi r^3/3) = 4\pi r^3/3.$$

The assumption and consistent use of Cavalieri's principle simplifies the derivation of many formulas encountered in a high school course in solid geometry. This procedure has been adopted by a number of textbook writers and has been advocated on pedagogical grounds.

Cavalieri's conception of indivisibles stimulated considerable discussion, and serious criticisms were leveled by some students of the subject, particularly by the Swiss Paul Guldin. Cavalieri recast his treatment in the vain hope of meeting these objections. The French mathematician Roberval ably handled the method and claimed to be an independent inventor of the conception. The method of indivisibles, or some process very similar to it, was effectively used by Torricelli, Fermat, Pascal, Saint-Vincent, Barrow, and others. In the course of the work of these men results were reached which are equivalent to the integration of expressions like x^n, $\sin \theta$, $\sin^2 \theta$, and $\theta \sin \theta$.

11-7   The Beginning of Differentiation

Differentiation may be said to have originated in the problem of drawing tangents to curves and in finding maximum and minimum values of functions. Although such considerations go back to the ancient Greeks it seems fair to assert that the first really marked anticipation of the method of differentiation stems from ideas set forth by Fermat in 1629.

Kepler had observed that the increment of a function becomes vanishingly small in the neighborhood of an ordinary maximum or minimum value. Fermat translated this fact into a process for determining such a maximum or minimum. The method will be considered in brief. If $f(x)$ has an ordinary maximum or minimum at x, and if e is very small, then the value of $f(x - e)$ is almost equal to that of $f(x)$. Therefore we tentatively set $f(x - e) = f(x)$ and then make the equality correct by letting e assume the value zero. The roots of the resulting equation then give those values of x for which $f(x)$ is a maximum or a minimum.

Let us illustrate the above procedure by considering Fermat's first example—to divide a quantity into two parts such that their product is a maximum. Fermat used Viète's notation, where constants are designated by upper case consonants and variables by upper case vowels. Following the notation to this extent, let B be the given quantity and denote the sought parts by A and $B - A$. Forming

$$(A - E)\,[B - (A - E)]$$

and equating it to $A\,(B - A)$ we have

$$A\,(B - A) = (A - E)\,(B - A + E)$$

or

$$2AE - BE - E^2 = 0.$$

After dividing by E, one obtains

$$2A - B - E = 0.$$

Now setting $E = 0$ we obtain $2A = B$, and thus find the required division.

Although the logic of Fermat's exposition leaves much to be desired, it is seen that his method is equivalent to setting

$$\lim_{h \to 0} \frac{f(x + h) - f(x)}{h} = 0,$$

that is, to setting the derivative of $f(x)$ equal to zero. This is the customary method for finding ordinary maxima and minima of a function $f(x)$, and is sometimes referred to in our elementary textbooks as *Fermat's method*. Fermat, however, did not know that the vanishing of the derivative of $f(x)$ is only a necessary, but not a sufficient, condition for an ordinary maximum or minimum. Also, Fermat's method does not distinguish between a maximum and a minimum value.

Fermat also devised a general procedure for finding the tangent at a point of a curve whose Cartesian equation is given. His idea is to find the *subtangent* for the point, that is, the segment on the x-axis between the foot of the ordinate drawn to the point of contact and the intersection of the tangent line with the x-axis. The method employs the idea of a tangent as the limiting position of a secant when two of its points of intersection with the curve tend to fall together. Using modern notation the method is as follows. Let the equation of the curve (see Figure 79) be $f(x,y) = 0$, and let us seek the subtangent a of the curve for the point (x,y). By similar triangles we easily find the coordinates of a near point on the tangent to be $(x + e, y(1 + e/a))$. This point is tentatively treated as if it were also on the curve, giving us

$$f[x + e, y(1 + e/a)] = 0.$$

The equality is then made correct by letting e assume the value zero. We then solve the resulting equation for the subtangent a in terms of the coordinates x and y of the point of contact. This, of course, is equivalent to setting

$$a = -y(\partial f/\partial y)/(\partial f/\partial x),$$

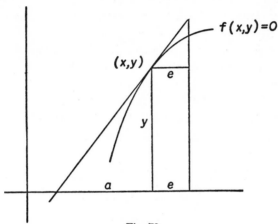

Fig. 79

a general formula that appeared later in the work of Sluze. Fermat in this way found tangents to the ellipse, cycloid, cissoid, conchoid, quadratrix, and folium of Descartes. Let us illustrate the method by finding the subtangent at a general point on the folium of Descartes:

$$x^3 + y^3 = nxy.$$

Here we have

$$(x + e)^3 + y^3(1 + e/a)^3 - ny(x + e)(1 + e/a) = 0,$$

or

$$e(3x^2 + 3y^3/a - nxy/a - ny) + e^2(3x + 3y^3/a^2 - ny/a) + e^3(1 + y^3/a^3) = 0.$$

Now, dividing by e and then setting $e = 0$, we find

$$a = -(3y^3 - nxy)/(3x^2 - ny).$$

11-8   Wallis and Barrow

Isaac Newton's immediate predecessors in England were John Wallis and Isaac Barrow.

John Wallis, who was born in 1616, was one of the ablest and most original mathematicians of his day. He was a voluminous and erudite writer in a number of fields and is said to be one of the first to devise a system for teaching deaf-mutes. In 1649, he was appointed Savilian professor of geometry at Oxford, a position which he held for 54 years until his death in 1703. His work in analysis did much to prepare the way for his great contemporary, Isaac Newton.

Wallis was one of the first to discuss conics as curves of second degree rather than as sections of a cone. In 1656 appeared his *Arithmetica infinitorum* (dedicated to Oughtred), a book which, in spite of some logical blemishes, remained a standard treatise for many years. In this book, the methods of Descartes and Cavalieri are systematized and extended and a number of remarkable results are induced from particular cases. Thus the formula which we would now write as

$$\int_0^1 x^m \, dx = 1/(m+1),$$

where m is a positive integer, is claimed to hold even when m is fractional or negative but different from -1. Wallis was the first to explain with any completeness the significance of zero, negative, and fractional exponents, and he introduced our present symbol (∞) for infinity.

Wallis endeavored to determine π by finding an expression for the area, $\pi/4$, of a quadrant of the circle $x^2 + y^2 = 1$. This is equivalent to evaluating $\int_0^1 (1 - x^2)^{\frac{1}{2}} dx$, which Wallis was unable to do directly since he was not acquainted with the general binomial theorem. He accordingly evaluated $\int_0^1 (1 - x^2)^0 dx$, $\int_0^1 (1 - x^2)^1 dx$, $\int_0^1 (1 - x^2)^2 dx$, and so forth, obtaining the sequence 1, 2/3, 8/15, 16/35, $\cdots$. He then considered the problem of finding the law which for $n = 0, 1, 2, 3, \cdots$ would yield the above sequence. It was the interpolated value of this law for $n = 1/2$ that Wallis was seeking. By a long and complicated process, he finally arrived at his infinite product expression for $\pi/2$ given in Section 4-8. Mathematicians of his day frequently resorted to interpolation processes in order to calculate quantities that they could not evaluate directly.

Wallis accomplished other things in mathematics. He was the mathematician who came nearest to solving Pascal's challenge questions on the cycloid (see Section 9-9). It can be argued fairly that he obtained an equivalent of the formula

$$ds = [1 + (dy/dx)^2]^{\frac{1}{2}}dx$$

for the length of an element of arc of a curve. His *De algebra tractatus; historicus & practicus*, of 1673, is considered as the first serious attempt at a history of mathematics in England. It is in this work that we find the first recorded effort to give a graphical interpretation of the complex roots of a real quadratic equation. Wallis edited parts of the works of a number of the great Greek mathematicians and wrote on a wide variety of physical subjects. He was one of the founders of the Royal Society and for years he assisted the government as a cryptologist.

Whereas Wallis' chief contributions to the development of the calculus lay in the theory of integration, Isaac Barrow's most important contributions were perhaps connected with the theory of differentiation.

Isaac Barrow was born in London in 1630. A story is told that in his early school days he was so troublesome that his father was heard to pray that should God decide to take one of his children he could best spare Isaac. Barrow completed his education at Cambridge and won renown as one of the best Greek scholars of his day. He was a man of high academic caliber, achieving recognition in mathematics, physics, astronomy, and theology. Entertaining stories are told of his physical strength, bravery, ready wit, and scrupulous conscientiousness. He was the first to occupy the Lucasian chair at Cambridge, from which he magnanimously resigned in 1669 in favor of his great pupil, Isaac Newton, whose remarkable abilities he was one of the first to recognize and acknowledge. He died in Cambridge in 1677.

Barrow's most important mathematical work is his *Lectiones opticae et geometricae*, which appeared in the year he resigned his chair at Cambridge. The preface of the treatise acknowledges indebtedness to Newton for some of the material of the book, probably the parts dealing with optics. It is in this book that we find a very near approach to the modern process of differentiation, utilizing the so-called *differential triangle* which we find in our present-day textbooks. Let it be required to find the tangent at a point P on the given curve represented in Figure 80. Let Q be a neighboring point on the curve. Then triangles PTM and PQR are very nearly similar to one another, and, Barrow argued, as the little triangle becomes infinitely small, we have

$$RP/QR = MP/TM.$$

Let us set $QR = e$ and $RP = a$. Then if the coordinates of P are x and y, those of Q are $x - e$ and $y - a$. Substituting these values into the equation of the curve and neglecting squares and higher powers of both e and a, we find the ratio a/e. We then have

$$OT = OM - TM = OM - MP(QR/RP) = x - y(e/a),$$

and the tangent line is determined. Barrow applied this method of constructing tangents to the curves (a) $x^2(x^2 + y^2) = r^2y^2$ (the *kappa curve*), (b) $x^3 + y^3 = r^3$ (a special *Lamé curve*), (c) $x^3 + y^3 = rxy$ (the *folium of*

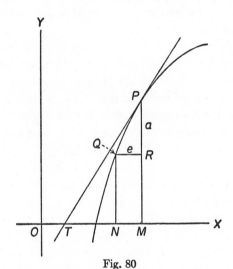

Fig. 80

Descartes, but called *la galande* by Barrow), (d) $y = (r - x) \tan \pi x/2r$ (the *quadratrix*), (e) $y = r \tan \pi x/2r$ (a *tangent curve*). As an illustration let us apply the method to curve (b). Here we have

$$(x - e)^3 + (y - a)^3 = r^3,$$

or

$$x^3 - 3x^2e + 3xe^2 - e^3 + y^3 - 3y^2a + 3ya^2 - a^3 = r^3.$$

Neglecting the square and higher powers of e and a, and using the fact that

$x^3 + y^3 = r^3$, this reduces to

$$3x^2e + 3y^2a = 0,$$

from which we obtain

$$a/e = -x^2/y^2.$$

The ratio a/e is, of course, our modern dy/dx, and Barrow's questionable procedure can easily be made rigorous by the use of the theory of limits.

In spite of tenuous evidence pointing elsewhere, Barrow is generally credited as the first to realize in full generality that differentiation and integration are inverse operations. This capital discovery is the so-called *fundamental theorem of the calculus* and appears to be stated and proved in Barrow's *Lectiones*.

Although Barrow devoted most of the latter part of his life to theology, he did, in 1675, publish an edition (with commentary) of the first four books of Apollonius' *Conic Sections* and of the extant works of Archimedes and Theodosius.

At this stage of the development of differential and integral calculus many integrations had been performed, many cubatures, quadratures, and rectifications effected, a process of differentiation had been evolved and tangents to many curves constructed, the idea of limits had been conceived, and the fundamental theorem recognized. What more remained to be done? There still remained the creation of a general symbolism with a systematic set of formal analytical rules, and also a consistent and rigorous redevelopment of the fundamentals of the subject. It is precisely the first of these, the creation of a suitable and workable *calculus*, that was furnished by Newton and Leibniz, working independently of each other. The redevelopment of the fundamental concepts on an acceptably rigorous basis had to outwait the period of energetic application of the subject, and was the work of the great French analyst Augustin-Louis Cauchy (1789–1857) and his nineteenth-century successors.

11-9   Newton

Isaac Newton was born in Woolsthorpe on Christmas Day, 1642 (old style), the year in which Galileo died. His father, who died before Isaac was born, was a farmer, and it was at first planned that the son also should devote his life to farming. The youngster, however, showed great skill and delight in devising clever mechanical models and in conducting experiments. Thus he made a toy gristmill that ground wheat to flour, with a mouse serving as motive power, and a wooden clock that worked by water. The result was that his schooling was extended, and, when 18, he was allowed to enter Trinity College, Cambridge. It was not until this stage in his schooling that his attention came to be directed to mathematics, by a book on astrology picked up at the Stourbridge Fair. As a consequence he first read Euclid's *Elements*, which he found too obvious, and then Descartes's *La géométrie*, which he found somewhat difficult. He also read Oughtred's *Clavis*, works of Kepler and Vieta, and the *Arithmetica infinitorum* by Wallis. From reading mathematics, he turned to creating it, and early in 1665, when he was 23 years old, he was in possession of the generalized binomial theorem and had created his method of fluxions, as he called what today is known as differential calculus. That year, and part of the next, the university closed because of the rampant bubonic plague, and Newton lived at home. During this period he developed his calculus to the point where he could find the tangent and radius of curvature at an arbitrary point of a curve. He also became interested in various physical questions, performed his first experiments in optics, and formulated the basic principles of his theory of gravitation.

Newton returned to Cambridge in 1667 and for two years occupied himself chiefly with optical researches. In 1669, Barrow resigned the Lucasian professorship in favor of Newton, and the latter began his 18 years of university lecturing. His first lectures, which were on optics, were later communicated in a paper to the Royal Society and aroused considerable interest and discussion. His theory of colors and certain deductions from his optical experiments were vehemently attacked by some scientists. Newton found the ensuing argument so distasteful that he vowed never to publish anything on science again. His tremendous dislike of controversy, which seems to have bordered on the pathological, had an important bearing on the history of mathematics, for the result was that almost all of his findings remained unpublished until many years after their discovery. This postponement of publication later led to the undignified dispute with Leibniz concerning priority of discovery of the calculus. It was owing to this controversy that the English mathematicians, backing Isaac Newton as their leader, cut themselves off from continental developments, and mathematical progress in England was detrimentally retarded for practically a hundred years.

Newton continued his work in optics, and in 1675 communicated his work on the emission, or corpuscular, theory of light to the Royal Society. His reputation and his ingenious handling of the theory led to its general adoption, and it was not until many years later that the wave theory was shown to be a better hypothesis for research. Newton's university lectures from 1673 to 1683 were devoted to algebra and the theory of equations. It was in this period, in 1679, that he verified his law of gravitation* by using a new measurement of the earth's radius in conjunction with a study of the motion of the moon. He also established the compatibility of his law of gravitation with Kepler's laws of planetary motion, on the assumption that the sun and the planets may be regarded as heavy particles. But these important findings were not communicated to anyone until five years later, in 1684, when Halley visited Newton at Cambridge to discuss the law of force that causes the planets to move in elliptical orbits about the sun. With his interest in celestial mechanics rearoused in this way, Newton proceeded to work out many of the propositions later to become fundamental in the first book of his *Principia*. When Halley, somewhat later, saw Newton's manuscript he realized its tremendous importance, and secured the author's promise to send the results to the Royal Society. This Newton did, and at about the same time he finally solved a problem that had been bothering him for some years, namely that a spherical body whose density at any point depends only on its distance from the center of the sphere attracts an external particle as if its whole mass were concentrated at the center. This theorem completed his justification of Kepler's laws of planetary motion, for the slight departure of the sun and the planets from true sphericity is here negligible. Newton now worked in earnest on his theory and by a gigantic intellectual effort wrote the first book of the *Principia* by the summer of 1685. A year later the second book was completed and a third begun. Jealous accusations by Hooke, and the resulting unpleasantness of the matter to Newton, almost led to the abandonment of the third book, but Halley finally persuaded Newton to finish the task. The complete treatise, entitled *Philosophiae naturalis principia mathematica*, was published, at Halley's expense, in the middle of 1687 and immediately made an enormous impression throughout Europe.

In 1689, Newton represented the university in parliament. In 1692, he suffered a curious illness which lasted about two years and which involved some form of mental derangement. Most of his later life was devoted to

*Any two particles in the universe attract one another with a force which is directly proportional to the product of their masses and inversely proportional to the square of the distance between them.

chemistry, alchemy, and theology. As a matter of fact, even during the earlier part of his life, he probably spent about as much time on these pursuits as he did on mathematics and natural philosophy. Although his creative work in mathematics practically ceased, he did not lose his remarkable powers, for he masterfully solved numerous challenge problems that were submitted to him and which were quite beyond the powers of the other mathematicians in England. In 1696, he was appointed Warden of the Mint, and in 1699 he was promoted to be Master of the Mint. In 1703, he was elected president of the Royal Society, a position to which he was annually re-elected until his death, and in 1705 he was knighted. The last part of his life was made unhappy by the unfortunate controversy with Leibniz. He died in 1727 when 84 years old, after a lingering and painful illness, and was buried in Westminster Abbey.

As remarked above, all of Newton's important published works, except the *Principia*, appeared years after the author had discovered their contents, and almost all of them finally appeared only because of pressure from friends. The dates of these works, in order of publication, are as follows: *Principia*, 1687; *Opticks*, with two appendices on *Cubic Curves* and *Quadrature and Rectification of Curves by the Use of Infinite Series*, 1704; *Arithmetica universalis*, 1707; *Analysis per Series, Fluxiones, etc.*, and *Methodus differentialis*, 1711; *Lectiones opticae*, 1729; and *The Method of Fluxions and Infinite Series*, translated from Newton's Latin by J. Colson, 1736. One should also mention two important letters written in 1676 to H. Oldenburg, secretary of the Royal Society, in which Newton describes some of his mathematical methods.

It is in the letters to Oldenburg that Newton describes his early induction of the generalized binomial theorem, which he enunciates in the form

$$(P + PQ)^{m/n} = P^{m/n} + \frac{m}{n} AQ + \frac{m-n}{2n} BQ + \frac{m-2n}{3n} CQ + \cdots ,$$

where A represents the first term, namely $P^{m/n}$, B represents the second term, namely $(m/n)AQ$, C represents the third term, and so forth. The correctness, under proper restrictions, of the binomial expansion for all complex values of the exponent was established over 150 years later by the Norwegian mathematician N. H. Abel (1802–1829).

A more important mathematical discovery made by Newton at about the same time was his method of fluxions, the essentials of which he communicated to Barrow in 1669. His *Method of Fluxions* was written in 1671, but was not published until 1736. In this work, Newton considers a curve as generated by the continuous motion of a point. Under this conception the abscissa and the ordinate of the generating point are, in general, changing quantities. A changing quantity is called a *fluent* (a flowing quantity), and its rate of change is called the *fluxion* of the fluent. If a fluent, such as the ordinate of the point generating a curve, be represented by y, then the fluxion of this fluent is represented by $\dot{y}$. In modern notation we see that this is equivalent to dy/dt, where t represents time. In spite of this introduction of time into geometry, the idea of time can be evaded by supposing that some quantity, say the abscissa of the moving point, increases constantly. This constant rate of increase of some fluent is called the *principal fluxion*, and the fluxion of any other fluent can be compared with this principal fluxion. The fluxion of $\dot{y}$ is denoted by $\ddot{y}$, and so on for higher ordered fluxions. On the other hand, the fluent of y is denoted by the symbol y with a small square drawn about it, or sometimes by $\overset{\shortmid}{y}$. Newton also introduces another concept, which he calls the *moment* of a fluent; it is the infinitely small amount by which a fluent such as x increases in an infinitely small interval of time o. Thus the moment of the fluent x is given by the product $\dot{x}o$. Newton remarks that we may, in any problem, neglect all terms that are multiplied by the second or higher power of o, and thus obtain an equation between the coordinates x and y of the generating point of a curve and their fluxions $\dot{x}$ and $\dot{y}$. As an example he considers the cubic curve $x^3 - ax^2 + axy - y^3 = 0$. Replacing x by $x + \dot{x}o$ and y by $y + \dot{y}o$, we get

$$\begin{aligned}
& x^3 + 3x^2(\dot{x}o) + 3x(\dot{x}o)^2 + (\dot{x}o)^3 \\
& - ax^2 - 2ax(\dot{x}o) - a(\dot{x}o)^2 \\
& \cdot + axy + ay(\dot{x}o) + a(\dot{x}o)(\dot{y}o) + ax(\dot{y}o) \\
& - y^3 - 3y^2(\dot{y}o) - 3y(\dot{y}o)^2 - (\dot{y}o)^3 = 0.
\end{aligned}$$

Now, using the fact that $x^3 - ax^2 + axy - y^3 = 0$, dividing the remaining terms by o, and then rejecting all terms containing the second or higher power of o, we find .

$$3x^2\dot{x} - 2ax\dot{x} + ay\dot{x} + ax\dot{y} - 3y^2\dot{y} = 0.$$

Newton considers two types of problems. In the first type, we are given a relation connecting some fluents and we are asked to find a relation connecting these fluents and their fluxions. This is what we did above, and is, of course, equivalent to differentiation. In the second type, we are given a relation connecting some fluents and their fluxions and we are asked to find a relation connecting the fluents alone. This is the inverse problem and is equivalent to solving a differential equation. The idea of discarding terms containing the second and higher powers of o was later justified by Newton by the use of limit notions. Newton made numerous and remarkable applications of his method of fluxions. He determined maxima and minima, tangents to curves, curvature of curves, points of inflection, and convexity and concavity of curves, and he applied his theory to numerous quadratures and to the rectification of curves. In the integration of some differential equations he showed extraordinary ability. In this work is found a method (a modification of which is now known by Newton's name) for approximating the values of the real roots of either an algebraic or a transcendental numerical equation.

The *Arithmetica universalis* contains the substance of Newton's lectures of 1673 to 1683. In it are found many important results in the theory of equations, such as the fact that imaginary roots of a real polynomial must occur in pairs, rules for finding an upper bound to the roots of a polynomial, his formulas expressing the sum of the nth powers of the roots of a polynomial in terms of the coefficients of the polynomial, an extension of Descartes's rule of signs to give limits to the number of imaginary roots of a real polynomial, and many other things.

Cubic Curves, which appeared as an appendix to the work on *Optics*, investigates the properties of cubic curves by analytic geometry. In his classification of cubic curves Newton enumerates 72 out of the possible 78 forms which a cubic may assume. Many of his theorems are stated without proof. The most attractive of these, as well as the most baffling, was his assertion that just as all conics can be obtained as central projections of a circle, so all cubics can be obtained as central projections of the curves

$$y^2 = ax^3 + bx^2 + cx + d.$$

This theorem remained a puzzle until a proof was discovered in 1731.

Of course, Newton's greatest work is his *Principia*, in which there appears for the first time a complete system of dynamics and a complete mathematical formulation of the principal terrestrial and celestial phenomena of motion. It proved to be the most influential and most admired work in the history of science. It is interesting that the theorems, although perhaps discovered by fluxional methods, are all masterfully established by classical Greek geometry aided, here and there, with some simple notions of limits. Until the development of the theory of relativity, all physics and astronomy rested on the assumption, made by Newton in this work, of a privileged frame of reference. In the *Principia* are found many results concerning higher plane curves, and proofs of such attractive geometric theorems as

(1) The locus of the centers of all conics tangent to the sides of a quadrilateral is the line (*Newton's line*) through the mid-points of its diagonals.

(2) If a point P moving along a straight line is joined to two fixed points O and O', and if lines OQ and $O'Q$ make fixed angles with OP and $O'P$, then the locus of Q is a conic.

Newton was never beaten by any of the various challenge problems that circulated among the mathematicians of his time. In one of these, proposed by Leibniz, he solved the problem of finding the orthogonal trajectories of a family of curves.

Newton was a skilled experimentalist and a superb analyst. As a mathematician, he is ranked almost universally as the greatest the world has yet produced. His insight into physical problems and his ability to treat them mathematically has probably never been excelled. One can find many testimonials by competent judges as to his greatness, such as the noble tribute paid by Leibniz, who said, "Taking mathematics from the beginning of the world to the time when Newton lived, what he did was much the better half." And there is the remark by Lagrange to the effect that Newton was the greatest genius that ever lived, and the most fortunate, for we can find only once a system of the universe to be established. His accomplishments were poetically expressed by Pope in the lines,

> Nature and Nature's laws lay hid in night;
> God said, 'Let Newton be,' and all was light.

In contrast to these eulogies is Newton's own modest estimate of his work: "I do not know what I may appear to the world; but to myself I seem to have been only like a boy playing on the seashore, and diverting myself in now and then finding a smoother pebble or a prettier shell than ordinary, whilst the great ocean of truth lay all undiscovered before me." In generosity to his predecessors he once explained that if he had seen farther than other men, it was only because he had stood on the shoulders of giants.

It has been reported that Newton often spent 18 or 19 hours of the 24 in writing, and that he possessed remarkable powers of concentration. Amusing tales are told in support of his absent-mindedness when engaged in an investigation. Thus there is the story which relates that, when giving a dinner to some friends, he left the table for a bottle of wine, and forgetting his errand went to his room, donned his surplice, and ended up in chapel. On another occasion he dismounted from his horse to lead the animal up a hill, but when he reached the top and endeavored to remount he discovered that in the ascent the horse had slipped away leaving only the empty bridle in his hands.

11-10   Leibniz

Gottfried Wilhelm Leibniz, the great universal genius of the seventeenth century and Newton's rival in the invention of the calculus, was born in Leipzig in 1646 (old style). Having taught himself to read Latin and Greek when he was a mere child, he had, before he was 20, mastered the ordinary textbook knowledge of mathematics, philosophy, theology, and law. At this young age he began to develop the first ideas of his "characteristica generalis," which involved a universal mathematics that later blossomed into the symbolic logic of George Boole (1815–1864), and still later, in 1910, into the great *Principia mathematica* of Whitehead and Russell. When, ostensibly because of his youth, he was refused the degree of doctor of laws at the University of Leipzig, he moved to Nuremberg. There he wrote a brilliant essay on teaching law by the historical method and dedicated it to the Elector of Mainz. This led to his appointment by the Elector to a commission for the recodification of some statutes. The rest of Leibniz's life from this point on was spent in diplomatic service, first for the Elector of Mainz and then, from about 1676 until his death, for the estate of the Duke of Brunswick at Hanover.

In 1672, while in Paris on a diplomatic mission, Leibniz met Huygens, who was then residing there, and the young diplomat prevailed upon the scientist to give him lessons in mathematics. The following year Leibniz was sent on a political mission to London, where he made the acquaintance of Oldenburg and others and where he exhibited his calculating machine (see Section 9-10) to the Royal Society. Before he left Paris to take up his lucrative post as librarian for the Duke of Brunswick, Leibniz had already discovered the fundamental theorem of the calculus, developed much of his notation in this subject, and worked out a number of the elementary formulas of differentiation.

Leibniz' appointment in the Hanoverian service gave him leisure time to-pursue his favorite studies, with the result that he left behind him a mountain of papers on all sorts of subjects. He was a particularly gifted linguist, winning some fame as a Sanskrit scholar, and his writings on philosophy have ranked him high in that field. He entertained various grand projects that came to nought, such as that of reuniting the Protestant and Catholic churches, and then later, just the two Protestant sects of his day. In 1682, he and Otto Mencke founded a journal called the *Acta eruditorum*, of which he became editor-in-chief. Most of his mathematical papers, which were largely written in the ten-year period from 1682 to 1692, appeared in this journal. The journal had a wide circulation in continental Europe. In 1700, Leibniz founded the Berlin Academy of Science, and endeavored to create similar academies in Dresden, Vienna, and St. Petersburg.

The closing seven years of Leibniz' life were embittered by the controversy which others had brought upon him and Newton concerning whether he had discovered the calculus independently of Newton. In 1714, his employer became the first German King of England, and Leibniz was left, neglected, at Hanover. They say that two years later, in 1716, when he died, his funeral was attended only by his faithful secretary.

Leibniz' search for his "characteristica generalis" led to plans for a theory of mathematical logic and a symbolic method with formal rules that would obviate the necessity of thinking. Although this dream has only today reached a noticeable stage of realization, Leibniz had, in current terminology, stated the principal properties of logical addition, multiplication, and negation, had considered the null class and class inclusion, and had noted the similarity between some properties of the inclusion of classes and the implication of propositions (see Problem Study 11.10).

Leibniz invented his calculus sometime between 1673 and 1676. It was on October 29, 1675, that he first used the modern integral sign, as a long letter S derived from the first letter of the Latin word *summa* (sum), to indicate the sum of Cavalieri's indivisibles. A few weeks later he was writing differentials and derivatives as we do today, as well as integrals like $\int y\,dy$ and $\int y\,dx$. His first published paper on differential calculus did not appear until 1684. In this paper he introduces dx as an arbitrary finite interval and then defines dy by the proportion

$$dy : dx = y : \text{subtangent}.$$

Many of the elementary rules for differentiation, which a student learns early in one of our college courses in the calculus, were derived by Leibniz. The rule for finding the nth derivative of the product of two functions (see Problem Study 11.6) is still referred to as *Leibniz' rule.*

Leibniz had a remarkable feeling for mathematical form and was very sensitive to the potentialities of a well-devised symbolism. His notation in the calculus proved to be very fortunate, and is unquestionably more convenient and flexible than the fluxional notation of Newton. The English mathematicians, though, clung long to the notation of their leader. It was as late as the nineteenth century that there was formed, at Cambridge, the Analytical Society, as it was named by one of its founders, Charles Babbage (see Section 9-10), for the purpose of advocating "the principles of pure d-ism as opposed to the *dot*-age of the university."

The theory of determinants is usually said to have originated with Leibniz, in 1693, when he considered these forms with reference to systems of simultaneous linear equations, although a similar consideration had been made ten years earlier in Japan by Seki Kōwa. The generalization of the binomial theorem into the multinomial theorem, which concerns itself with the expansion of

$$(a + b + \cdots + n)^r,$$

is due to Leibniz. He also did much in laying the foundation of the theory of envelopes, and he defined the osculating circle and showed its importance in the study of curves.

We shall not enter here into a discussion of the unfortunate Newton-Leibniz controversy. The universal opinion today is that each discovered the calculus independently of the other. While Newton's discovery was made first, Leibniz was the earlier in publishing results. If Leibniz was not as penetrating a mathematician as Newton, he was perhaps a broader one, and while inferior to his English rival as an analyst and mathematical physicist, he probably had a keener mathematical imagination and a superior instinct for mathematical form.

For some time after Newton and Leibniz, the foundations of the calculus remained obscure and little heeded, for it was the remarkable applicability of the subject that attracted the early researchers. By 1700, most of our undergraduate college calculus had been founded, along with sections of more advanced fields, such as the calculus of variations. The first textbook of the subject appeared in 1696, written by the Marquis de l'Hospital (1661–1704), when he published the lectures of his teacher, Johann Bernoulli. In this book is found the so-called *l'Hospital's rule* for finding the limiting value of a fraction whose numerator and denominator tend simultaneously to zero.

Problem Studies

11.1 The Method of Exhaustion

(a) Assuming the so-called *axiom of Archimedes*: *If we are given two magnitudes of the same kind, then we can find a multiple of the smaller which exceeds the larger*, establish the basic proposition of the method of exhaustion: *If from any magnitude there be subtracted a part not less than its half, from the remainder another part not less than its half, and so on, there will at length remain a magnitude less than any preassigned magnitude of the same kind.* (The axiom of Archimedes is implied in the fourth definition of Book V of Euclid's *Elements*, and the basic proposition of the method of exhaustion is found as Proposition 1 of Book X of the *Elements*.)

(b) Show, with the aid of the basic proposition of the method of exhaustion, that the difference in area between a circle and a circumscribed regular polygon can be made as small as desired.

11.2 The Method of Equilibrium

Figure 81 represents a parabolic segment having AC as chord. CF is tangent to the parabola at C and AF is parallel to the axis of the parabola. OPM is also parallel to the axis of the parabola. K is the mid-point of FA and $HK = KC$. Take K as a fulcrum, place OP with its center at H, and leave OM where it is. Using the geometrical fact that $OM/OP = AC/AO$ show, by Archimedes' method of equilibrium, that the area of the parabolic segment is one-third the area of triangle AFC.

11.3 Some Archimedean Problems

Archimedes devoted a number of tracts to solving volume and area problems. He established his results by the "method of exhaustion." By modern methods solve the following Archimedean problems.

(a) Find the area of a spherical zone of height h and radius r.

(b) Find the centroid of a spherical segment.

(c) Find the volume of a *cylindrical wedge* or *hoof*, cut from a right circular cylinder by a plane passing through a diameter of the base of the cylinder.

(d) Find the volume common to two right circular cylinders of equal radii and having their axes intersecting perpendicularly.

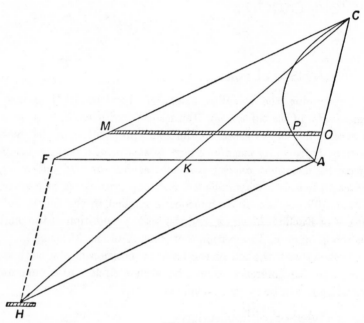

Fig. 81

11.4 The Method of Indivisibles

(a) Show that a triangular prism can be dissected into three triangular pyramids having, in pairs, equivalent bases and equal altitudes.

(b) Establish Cavalieri's principle by modern integration.

(c) Find, by Cavalieri's principle, the volume of a *cylindrical wedge*, or *hoof* (see Problem Study 11.3 (c)), whose height is twice the radius of the base. (Use for a comparison solid a rectangular parallelepiped $a \times a \times 2a$, with two square pyramids removed having the $a \times a$ ends of the parallelepiped as bases and having a common vertex at the center of the parallelepiped. Stand the parallelepiped upon one of its square ends and tip the hoof over so that the bounding diameter of its base is vertical.)

(d) Find, by Cavalieri's principle, the volume of the *spherical ring* obtained by removing from a solid sphere a cylindrical boring which is coaxal with the polar axis of the sphere. (Use for a comparison solid a sphere with diameter equal to the altitude h of the ring.)

(e) Show that all spherical rings of the same altitude have the same volume, irrespective of the radii of the spheres of the rings.

11.5   The Prismoidal Formula

A *prismatoid* is a polyhedron all of whose vertices lie in two parallel planes. The faces in these two parallel planes are called the *bases* of the prismatoid. If the two bases have the same number of sides, the prismatoid is called a *prismoid*. A *generalized prismoid* is any solid having two parallel base planes and having the areas of its sections parallel to the bases given by a quadratic function of their distances from one base.

(a) Show that the volumes of a prism, a wedge (a right triangular prism turned so as to rest on one of its lateral faces as a base), and a pyramid are given by the *prismoidal formula*:

$$V = h(U + 4M + L)/6,$$

where h is the altitude, and U, L, and M are the areas of the upper and lower bases and mid-section, respectively.

(b) Show that the volume of any convex prismatoid is given by the prismoidal formula.

(c) Show, by Cavalieri's principle, that the volume of any generalized prismoid is given by the prismoidal formula.

(d) Establish part (c) by integral calculus.

(e) Show by integral calculus that the prismoidal formula gives the volume of any solid having two parallel base planes and having the areas of its sections parallel to the bases given by a *cubic* function of their distances from one base.

(f) Using the prismoidal formula find the volumes of (1) a sphere, (2) an ellipsoid, (3) a cylindrical wedge, (4) the solid in 11.3 (d).

11.6   Differentiation

Find the slope of the tangent at the point (3,4) on the circle $x^2 + y^2 = 25$ by

(a) Fermat's method,

(b) Barrow's method,

(c) Newton's method of fluxions,

(d) the modern method.

(e) If $y = uv$, where u and v are functions of x, show that the nth derivative of y with respect to x is given by

$$y^{(n)} = uv^{(n)} + nu'v^{(n-1)} + n(n-1)u''v^{(n-2)}/2\,!$$
$$+ n(n-1)(n-2)u'''v^{(n-3)}/3\,! + \cdots + u^{(n)}v.$$

This is known as *Leibniz's rule*.

11.7 The Binomial Theorem

(a) Show that Newton's enunciation of the binomial theorem as given in Section 11-9 is equivalent to the familiar expansion

$$(a+b)^r = a^r + ra^{r-1}b + r(r-1)a^{r-2}b^2/2\,!$$
$$+ r(r-1)(r-2)a^{r-3}b^3/3\,! + \cdots.$$

(b) Show by the binomial theorem that if $(a + ib)^k = p + iq$, where a, b, p, q are real, k is a positive integer, and $i = \sqrt{-1}$, then $(a - ib)^k = p - iq$.

(c) Show by using part (b) that imaginary roots of a polynomial with real coefficients occur in conjugate pairs. (This result was given by Newton.)

11.8 An Upper Bound for the Roots of a Polynomial

(a) By using the binomial theorem, or otherwise, show that if $f(x)$ is a polynomial of degree n, then

$$f(y + h) \equiv f(h) + f'(h)y + f''(h)y^2/2\,! + \cdots + f^{(n)}(h)y^n/n\,!.$$

(b) Show that any number which makes a polynomial $f(x)$, and all of its derivatives $f'(x), f''(x), \cdots, f^{(n)}(x)$ positive, is an upper bound for the roots of $f(x) = 0$. (This result was given by Newton.)

(c) Show that if for $x = a$ we have $f^{(n-k)}(x), f^{(n-k+1)}(x), \cdots, f^{(n)}(x)$ all positive, then these functions will also all be positive for any number $x > a$.

(d) Parts (b) and (c) may be used to find a close upper bound for the roots of a polynomial. The general procedure is as follows: *Take the smallest integer that will make $f^{(n-1)}(x)$ positive. Substitute this integer in $f^{(n-2)}(x)$. If we obtain a negative result, increase the integer successively by units until* *an integer is found that makes this function positive. Now proceed with the new integer as before. Continue in this way until an integer is found that makes all of the functions* $f(x), f'(x), \cdots, f^{(n-1)}(x)$ *positive.* Find, by this procedure, an upper bound for the roots of

$$x^4 - 3x^3 - 4x^2 - 2x + 9.$$

11.9 Approximate Solution of Equations

(a) Newton devised a method for approximating the values of the real roots of a numerical equation which applies equally well to either an algebraic or a transcendental equation. The modification of this method, now known as Newton's method, says: *If $f(x) = 0$ has only one root in the interval $[a,b]$, and if neither $f'(x)$ nor $f''(x)$ vanishes in this interval, and if x_0 be chosen as that one of the two numbers a and b for which $f(x_0)$ and $f''(x_0)$ have the same sign, then*

$$x_1 = x_0 - f(x_0)/f'(x_0)$$

is nearer to the root than is x_0.

Establish this result.

(b) Solve by Newton's method the cubic $x^3 - 2x - 5 = 0$ for the root lying between 2 and 3.

(c) Solve by Newton's method the equation $x = \tan x$ for the root lying between 4.4 and 4.5.

(d) Find by Newton's method $\sqrt{12}$ correct to three decimal places.

11.10 Algebra of Classes

The concept of a "class of objects" is fundamental in logic. Leibniz developed some of the elementary algebra of classes. Using modern notation, if A and B are classes of objects, then $A \frown B$ (called the *intersection*, or *product*, of A and B) represents the class of all objects belonging to both A and B, and $A \smile B$ (called the *union*, or *sum*, of A and B) represents the class of all objects belonging to either A or B.

The algebra of classes can be illustrated graphically by means of so-called *Venn diagrams*, where a class A is represented by all points inside of a given area. Thus if we represent classes A and B by the points inside the two circles A and B, as indicated in Figure 82, the set $A \frown B$ is represented by the area common to these two circles, and the set $A \smile B$ is represented by the area containing all the points in either one or the other of the two circles. If we represent all our classes inside a surrounding rectangle, then by A', called the *complement* of A, we mean the area inside the rectangle but outside the area which represents A.

(a) On a Venn diagram shade each of the following areas: $A \frown (B' \smile C)$, $(A' \frown B) \smile (A \frown C')$, $(A \smile B') \smile C'$.

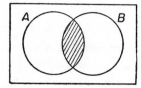 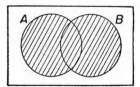

Fig. 82

(b) By shading the appropriate areas on a Venn diagram verify the following equations in the algebra of classes: $A \frown (B \frown C) = (A \frown B) \frown C$, $A \frown (B \smile C) = (A \frown B) \smile (A \frown C)$, $(A \smile B)' = A' \frown B'$.

(c) By shading the appropriate areas on a Venn diagram determine which of the following equations are valid: $(A' \smile B)' = A \frown B'$, $A' \smile B' = (A \smile B)'$, $A \smile (B \frown C)' = (A \smile B') \frown C'$.

Bibliography

BELL, E. T., *Men of Mathematics*. New York: Simon and Schuster, Inc., 1937.

BOYER, C. B., *The History of the Calculus and Its Conceptual Development*. New York: Dover Publications, Inc., 1959.

BREWSTER, SIR DAVID, *Life of Newton*. London: John Murray, 1831.

BRODETSKY, SELIG, *Sir Isaac Newton: A Brief Account of His Life and Work*. London: Methuen & Co., Ltd., 1927.

CAJORI, FLORIAN, *A History of the Conceptions of Limits and Fluxions in Great Britain, from Newton to Woodhouse*. Chicago: The Open Court Publishing Company, 1919.

———, *A History of Mathematical Notations*, 2 vols. Chicago: The Open Court Publishing Company, 1928–1929.

———, *Sir Isaac Newton's Mathematical Principles of Natural Philosophy and His System of the World*, a revision of the translation of 1729 by Andrew Motte. Berkeley, Calif.: University of California Press, 1934.

CHILD, J. M., *The Early Mathematical Manuscripts of Leibniz*. Chicago: The Open Court Publishing Company, 1920.

———, *The Geometrical Lectures of Isaac Barrow*. Chicago: The Open Court Publishing Company, 1916.

COOLIDGE, J. L., *Geometry of the Complex Domain*. New York: Oxford University Press, 1924.

———, *A History of Geometrical Methods*. New York: Oxford University Press, 1940.

DE MORGAN, AUGUSTUS, *Essays on the Life and Work of Newton*. Chicago: The Open Court Publishing Company, 1914.

HEATH, T. L., *History of Greek Mathematics*, Vol. 2. New York: Oxford University Press, 1921.

———, *A Manual of Greek Mathematics*. New York: Oxford University Press, 1931.

———, *The Method of Archimedes Recently Discovered by Heiberg*. New York: Cambridge University Press, 1912.

———, *The Works of Archimedes*. New York: Cambridge University Press, 1897.

KERN, W. F., AND J. R. BLAND, *Solid Mensuration; with Proofs*, 2d ed. New York: John Wiley & Sons, Inc., 1938.

MACFARLANE, ALEXANDER, *Lectures on Ten British Mathematicians of the Nineteenth Century*, Mathematical Monographs, No. 17. New York: John Wiley & Sons, Inc., 1916.

MORE, L. T., *Isaac Newton, a Biography*. New York: Dover Publications, Inc., 1962.

MUIR, JANE, *Of Men and Numbers, the Story of the Great Mathematicians*. New York: Dodd, Mead & Company, Inc., 1961.

ROSENTHAL, A., "The History of Calculus," *American Mathematical Monthly*, 58 (1951), 75–86.

ROYAL SOCIETY OF LONDON, *Newton Tercentenary Celebrations, 15–19 July, 1946*. New York: The Macmillan Company, 1947.

SCOTT, J. F., *The Mathematical Work of John Wallis (1616–1703)*. London: Taylor & Francis, Ltd., 1938.

SMITH, D. E., *History of Modern Mathematics*, 4th ed. New York: John Wiley & Sons, Inc., 1906.

———, *A Source Book in Mathematics*. New York: McGraw-Hill Book Company, Inc., 1929.

SULLIVAN, J. W. N., *The History of Mathematics in Europe, from the Fall of Greek Science to the Rise of the Conception of Mathematical Rigour*. New York: Oxford University Press, 1925.

TOEPLITZ, OTTO, *The Calculus, a Genetic Approach*. Chicago: University of Chicago Press, 1963.

TURNBULL, H. W., *Mathematical Discoveries of Newton*. Glasgow: Blackie & Son, Ltd., 1945.

———, *The Great Mathematicians*. New York: New York University Press, 1961.

WALKER, EVELYN, *A Study of the Traité des Indivisibles of Gilles Persone de Roberval*. New York: Teachers College, Columbia University, 1932.

12

Transition to the Twentieth Century

12-1   Introduction

The preceding three chapters have treated the remarkable contributions to elementary mathematics* of the very able mathematicians of the seventeenth century; moreover, in conjunction with the practice adopted in the rest of the book, many recent but elementary modifications and extensions of their work have also been presented. At this point, then, we have virtually concluded the historical treatment of elementary mathematics in the form that we have it today. It is interesting to note, without carrying the generalization too far, that the sequence of mathematics courses studied in the classroom follows quite closely the evolutionary trend of the subject.

For one who would study, with genuine understanding, what has happened in mathematics during the past two and a half centuries, an extensive study of advanced courses beyond the calculus is requisite. When the student possesses such a background, the excellent book, *The Development of Mathematics* of E. T. Bell, is recommended. Nevertheless, it seems advisable to add the pages of this final chapter in an attempt to provide a few additional highlights of the mathematics of the eighteenth, nineteenth, and twentieth centuries that are within the student's comprehension and which show very briefly the more recent trend of mathematical development from the elementary basis. The field of elementary mathematics will then appear in its proper setting, as a prelude to the more remarkable achievements of modern times.

One cannot point out too strongly the extreme sketchiness and incompleteness of what follows. Moritz Cantor's great history of mathematics,

---

*Essentially the arithmetic, elementary algebra, geometry, trigonometry, college algebra, analytic geometry, and beginning calculus ordinarily taught in lower schools and during the freshman or sophomore year in college.

which terminates with the end of the eighteenth century, consists of four large volumes averaging almost a thousand pages each. It has been conservatively estimated that if the history of the mathematics of the nineteenth century should be written with the same detail it would require at least fourteen more such volumes! No one has yet hazarded an estimate of the number of such volumes needed for a similar treatment of the history of the mathematics of the first half of the twentieth century, which is by far the most active era of all. And, as indicated above, little of this additional material could be properly appreciated by the ordinary undergraduate; indeed, an understanding of much of the material would require the deep background of a mathematical expert.

The almost explosive growth of mathematical research in modern times is further illustrated by the fact that prior to 1700 there were only, by one count, 17 periodicals containing mathematical articles, in the eighteenth century there were 210 such periodicals, in the nineteenth century 950 of them, and the number has increased enormously during the first half of the twentieth century. Furthermore, it was not until the nineteenth century that there appeared journals devoted either primarily or exclusively to mathematics. It has, probably quite properly, been remarked that the articles in these research journals constitute the true history of modern mathematics, and it must be confessed that very few of the present-day articles can be read by anyone but the specialist.

The calculus, aided by analytic geometry, was the greatest mathematical tool discovered in the seventeenth century. It proved to be remarkably powerful and capable of attacking problems quite unassailable in earlier days. It was the wide and astonishing applicability of the discipline that attracted the bulk of the mathematical researchers of the day, with the result that papers were turned out in great profusion with little concern regarding the very unsatisfactory foundations of the subject. The processes employed were justified largely on the ground that they worked, and it was not until the eighteenth century had almost elapsed, after a number of absurdities and contradictions had crept into mathematics, that mathematicians felt it was essential that the basis of their work be logically examined and rigorously established. The painstaking effort to place analysis on a logically rigorous foundation was a natural reaction to the pell-mell employment of intuition and formalism of the previous century. The task proved to be a difficult one, its various ramifications occupying the better part of the next hundred years. A result of this careful work in the foundations of analysis was that it led to equally careful work in the foundations of all branches of mathematics and to the refinement of many important concepts. Thus the function idea itself had to be clarified, and such notions as limit, continuity, differentiability, and integrability had to be very carefully and clearly defined. This task of refining the basic concepts of mathematics led, in turn, to intricate generalizations. Such concepts as space, dimension, convergence, and integrability, to name only a few, underwent remarkable generalization and abstraction. A good part of the mathematics of the first half of the twentieth century has been devoted to this sort of thing, until now generalization and abstraction have become striking features of present-day mathematics. But some of these developments have, in turn, brought about a fresh batch of paradoxical situations. The generalization to transfinite numbers and the abstract study of sets have widened and deepened many branches of mathematics, but, at the same time, they have revealed some very disturbing paradoxes which appear to lie in the innermost depths of mathematics. Here is where we seem to be today, and it may be that the second half of the twentieth century will witness the resolution of these critical problems.

In summarizing the last paragraph we may say, with a fair element of truth, that the eighteenth century was largely spent in exploiting the new and powerful methods of the calculus, that the nineteenth century was largely devoted to the effort of establishing on a firm logical foundation the enormous but shaky superstructure erected in the preceding century, and that the first half of the twentieth century has, in large part, been spent in generalizing as far as possible the gains already made, and that at present many mathematicians are becoming concerned with even deeper foundational problems. This general picture is complicated by the various sociological factors that affect the development of any science. Such matters as the growth of life insurance, the construction of the large navies of the eighteenth century, the economic and technological problems brought about by the opening of the industrial revolution on continental Europe, the present world-wide war atmosphere, and today's concentrated effort to conquer outer space, have led to many practical developments in the field of mathematics. A division of mathematics into "pure" and "applied" has come about, research in the former being carried on to a great extent by those specialists who have become interested in the subject for its own sake, and in the latter by those who remain attached to immediately practical uses.

We shall now fill in some of the details of the general picture sketched above.

12-2   The Bernoulli Family

As already noted in Section 10-3, the early part of the eighteenth century saw, with the work of Antoine Parent and Alexis Clairaut, the beginnings of solid analytic geometry. The first half of the eighteenth century also saw the significant work of Girolamo Saccheri, the forerunner of non-Euclidean geometry; this work was considered in Section 5-7. It will be recalled that Lambert's work in this area occurred toward the end of the century and that the actual discovery of non-Euclidean geometry by Lobachevsky, Janós Bolyai, and Gauss took place in the early nineteenth century. But the bulk of the mathematics of the eighteenth century found its genesis and its goal in the fields of mechanics and astronomy, and it was not until well into the nineteenth century that mathematical research generally emancipated itself from this viewpoint.

The principal contributions to mathematics in the eighteenth century were made by members of the Bernoulli family, Abraham De Moivre, Brook Taylor, Colin Maclaurin, Leonhard Euler, Alexis Claude Clairaut, Jean-le-Rond d'Alembert, Johann Heinrich Lambert, Joseph Louis Lagrange, and Gaspard Monge.*

One of the most distinguished families in the history of mathematics and science is the Bernoulli family of Switzerland, which, from the late seventeenth century on, has produced a remarkable number of capable mathematicians and scientists. The family record starts with the two brothers, Jakob Bernoulli (1654–1705) and Johann Bernoulli (1667–1748), some of whose mathematical accomplishments have already been mentioned in our book. These two men gave up earlier vocational interests and became mathematicians when Leibniz' papers began to appear in the *Acta eruditorum*. They were among the first mathematicians to realize the surprising power of the calculus and to apply the tool to a great diversity of problems. From 1687 until his death, Jakob occupied the mathematical chair at Basel University. Johann, in 1697, became a professor at Groningen University, and then, on Jakob's death in 1705, succeeded his brother in the chair at Basel University, to remain there for the rest of his life. The two brothers, often bitter rivals, maintained an almost constant exchange of ideas with Leibniz and with each other.

*The reader will find, by going through the index, that about 150 mathematicians who flourished after 1700 have already been mentioned, in one context or another, in the first eleven chapters of the book.

Among Jakob Bernoulli's contributions to mathematics are the early use of polar coordinates (already mentioned in Section 10-3), the derivation in both rectangular and polar coordinates of the formula for the radius of curvature of a plane curve, the study of the catenary curve with extensions to strings of variable density and strings under the action of a central force, the study of a number of other higher plane curves, the discovery of the so-called *isochrone*—or curve along which a body will fall with uniform vertical velocity (it turned out to be a semicubical parabola with a vertical cusptangent), the determination of the form taken by an elastic rod fixed at one end and carrying a weight at the other, the form assumed by a flexible rectangular sheet having two opposite edges held horizontally fixed at the same height and loaded with a heavy liquid, and the shape of a rectangular sail filled with wind. He also proposed and discussed the problem of isoperimetric figures (planar closed paths of given species and fixed perimeter which include a maximum area), and was thus one of the first mathematicians to work in the calculus of variations. He was also (as was pointed out in Section 10-5) one of the early students of mathematical probability; his book in this field, the *Ars conjectandi*, was posthumously published in 1713. There are several things in mathematics which now bear Jakob Bernoulli's name; among these are the *Bernoulli distribution* and *Bernoulli theorem* of statistics and probability theory, the *Bernoulli equation* met by every student of a first course in differential equations, the *Bernoulli numbers* and *Bernoulli polynomials* of number-theory interest, and the *lemniscate of Bernoulli* encountered in any first course in the calculus. In Jakob Bernoulli's solution to the problem of the isochrone curve, which was published in the *Acta eruditorum* in 1690, we meet for the first time the word integral in a calculus sense. Leibniz had called the integral calculus *calculus summatorius*; in 1696 Leibniz and Johann Bernoulli agreed to call it *calculus integralis*. Jakob Bernoulli was struck by the way the equiangular spiral reproduces itself under a variety of transformations and asked, in imitation of Archimedes, that such a spiral be engraved on his tombstone, along with the inscription "Eadem mutata resurgo" ("I shall arise the same, though changed").

Johann Bernoulli was an even more prolific contributor to mathematics than was his brother Jakob. Though he was a jealous and cantankerous man, he was one of the most successful teachers of his time. He greatly enriched the calculus and was very influential in making the power of the new subject appreciated in continental Europe. As we have seen (in Section 11-10), it was his material that the Marquis de l'Hospital (1661–1704), under a curious financial agreement with Johann, assembled in 1696 into the first calculus textbook. It was in this way that the familiar method of evaluating the indeterminate form 0/0 became incorrectly known, in later calculus texts, as *l'Hospital's rule*. Johann Bernoulli wrote on a wide variety of topics, including optical phenomena connected with reflection and refraction, the determination of the orthogonal trajectories of families of curves, rectification of curves and quadrature of areas by series, analytical trigonometry, the exponential calculus, and other subjects. One of his more noted pieces of work is his contribution to the problem of the *brachystochrone* —the determination of the curve of quickest descent of a weighted particle moving between two given points in a gravitational field; the curve turned out to be an arc of an appropriate cycloid curve. This problem was also discussed by Jakob Bernoulli. The cycloid curve is also the solution to the problem of the *tautochrone*—the determination of the curve along which a weighted particle will arrive at a given point of the curve in the same time interval no matter from what initial point of the curve it starts. This latter problem, which was more generally discussed by Johann Bernoulli, Euler, and Lagrange, had earlier been solved by Huygens (1673) and Newton (1687), and applied by Huygens in the construction of pendulum clocks (see Problem Study 10.9 (c)).

Johann Bernoulli had three sons, Nicolaus (1695–1726), Daniel (1700–1782), and Johann (II) (1710–1790), all of whom won renown as eighteenth-century mathematicians and scientists. Nicolaus, who showed great promise in the field of mathematics, was called to the St. Petersburg Academy, where he unfortunately died, by drowning, only eight months later. He wrote on curves, differential equations, and probability. A problem in probability, which he proposed from St. Petersburg, later became known as the *Petersburg paradox*. The problem is: if A receives a penny should head appear on the first toss of a coin, 2 pennies if head does not appear until the second toss, 4 pennies if head does not appear until the third toss, and so on, what is A's expectation? Mathematical theory shows that A's expectation is infinite, which seems a paradoxical result. The problem was investigated by Nicolaus' brother Daniel, who succeeded Nicolaus at St. Petersburg. Daniel returned to Basel seven years later. He was the most famous of Johann's three sons, and devoted most of his energies to probability, astronomy, physics, and hydrodynamics. In probability he devised the concept of *moral expectation*, and in his *Hydrodynamica*, of 1738, appears the principle of hydrodynamics that bears his name in all present-day elementary physics texts. He wrote on tides, established the kinetic theory of gases, studied the vibrating string, and pioneered in partial differential equations. Johann (II), the youngest of the three sons, studied law but spent his later years as a professor of mathematics at the University of Basel. He was particularly interested in the mathematical theory of heat and light.

There was another eighteenth-century Nicolaus Bernoulli (1687–1759), a nephew of Jakob and Johann, who achieved some fame in mathematics. This Nicolaus held, for a time, the chair of mathematics at Padua once filled by Galileo. He wrote extensively on geometry and differential equations. Later in life he taught logic and law.

Johann Bernoulli (II) had a son Johann (III) (1744–1807) who, like his father, studied law but then turned to mathematics. When barely 19 years old, he was called as a professor of mathematics to the Berlin Academy. He wrote on astronomy, the doctrine of chance, recurring decimals, and indeterminate equations.

Lesser Bernoulli descendants are Daniel (II) (1751–1834) and Jakob (II) (1759–1789), two other sons of Johann (II), Christoph (1782–1863), a son of Daniel (II), and Johann Gustav (1811–1863), a son of Christoph.

12-3   De Moivre, Taylor, Maclaurin

Abraham De Moivre (1667–1754) was born in France but lived most of his life in England, becoming an intimate friend of Isaac Newton. He is particularly noted for his work *Annuities upon Lives*, which played an important role in the history of actuarial mathematics, his *Doctrine of Chances*, which contained much new material on the theory of probability, and his *Miscellanea analytica*, which contributed to recurrent series, probability, and analytic trigonometry. De Moivre is credited with the first treatment of the probability integral,

$$\int_0^\infty e^{-x^2}\, dx = \sqrt{\pi}/2,$$

and of (essentially) the normal frequency curve

$$y = ce^{-hx^2}, \qquad c \text{ and } h \text{ constants},$$

so important in the study of statistics. The misnamed *Stirling's formula*, which says that for very large n

$$n! \approx (2\pi n)^{\frac{1}{2}} e^{-n} n^n,$$

is due to De Moivre and is highly useful for approximating factorials of large numbers. The familiar formula

$$(\cos x + i \sin x)^n = \cos nx + i \sin nx, \qquad i = \sqrt{-1},$$

known by De Moivre's name and found in every theory of equations textbook, was familiar to De Moivre for the case where n is a positive integer. This formula has become the keystone of analytic trigonometry.

Rather interesting is the fable often told of De Moivre's death. According to the story De Moivre noticed that each day he required a quarter of an hour more sleep than on the preceding day. When the arithmetic progression reached 24 hours De Moivre passed away.

Every student of the calculus is familiar with the name of the Englishman Brook Taylor (1685–1731) and the name of the Scotsman Colin Maclaurin (1698–1746), through the very useful Taylor's expansion and Maclaurin's expansion of a function. It was in 1715 that Taylor published (with no consideration of convergence) his well-known expansion theorem,

$$f(a + h) = f(a) + hf'(a) + \frac{h^2}{2!} f''(a) + \cdots.$$

In 1717 Taylor applied his series to the solution of numerical equations as follows: Let a be an approximation to a root of $f(x) = 0$; set $f(a) = k$, $f'(a) = k'$, $f''(a) = k''$, and $x = a + h$; expand $0 = f(a + h)$ by the series; discard all powers of h above the second; substitute the values of k, k', k'', and then solve for h. By successive applications of this process, closer and closer approximations can be obtained. Some work done by Taylor in the theory of perspective has found recent application in the mathematical treatment of photogrammetry, the science of surveying by means of photographs taken from an airplane.

Maclaurin was one of the ablest mathematicians of the eighteenth century. The so-called Maclaurin expansion is nothing but the case where $a = 0$ in the Taylor expansion above and was actually given by James

Stirling 25 years before Maclaurin used it in 1742. Maclaurin did very notable work in geometry, particularly in the study of higher plane curves, and he showed great power in applying classical geometry to physical problems. Among his many papers in applied mathematics is a prize-winning memoir on the mathematical theory of tides.

12-4   Euler, Clairaut, d'Alembert

The name Leonhard Euler has already been referred to many times in this book. Euler was born in Basel, Switzerland, in 1707, and he studied mathematics there under Johann Bernoulli. In 1727 he accepted the chair of mathematics at the new St. Petersburg Academy formed by Peter the Great. Fourteen years later he accepted the invitation of Frederick the Great to go to Berlin to head the Prussian Academy. After 25 years in this post Euler returned to St. Petersburg, remaining there until his death in 1783 when he was 76 years old.

Euler was a voluminous writer on mathematics, indeed, the most prolific writer in the history of the subject; his name is attached to every branch of the study. It is of interest to note that his amazing productivity was not in the least impaired when, about 1768, he had the misfortune to become totally blind.

Euler's work represents the outstanding example of eighteenth-century formalism, or the manipulation, without proper attention to matters of convergence and mathematical existence, of formulas involving infinite processes. For example, if the binomial theorem is applied formally to $(1 - 2)^{-1}$ we find

$$-1 = 1 + 2 + 4 + 8 + 16 + \cdots,$$

a result which caused Euler no wonderment! Also, by adding the two series

$$x + x^2 + \cdots = x/(1 - x) \qquad \text{and} \qquad 1 + 1/x + 1/x^2 + \cdots = x/(x - 1),$$

Euler found that

$$\cdots + 1/x^2 + 1/x + 1 + x + x^2 + \cdots = 0.$$

The nineteenth-century effort to inject rigor into mathematics was brought about by an accumulation of absurdities such as these.

Euler's contributions to mathematics are too numerous and, in general, too advanced to expound here, but we may note some of his contributions to the elementary field. First of all, we owe to Euler the conventionalization of the following notations:

$f(x)$ for functional notation,
e for the base of natural logarithms,
a, b, c for the sides of a triangle ABC,
s for the semiperimeter of triangle ABC,
$\sum$ for the summation sign,
i for the imaginary unit, $\sqrt{-1}$.

To Euler is also due the very remarkable formula

$$e^{ix} = \cos x + i \sin x,$$

which, for $x = \pi$, becomes

$$e^{i\pi} + 1 = 0,$$

a relation connecting five of the most important numbers in mathematics. By purely formal processes, Euler arrived at an enormous number of curious relations, like

$$i^i = e^{-\pi/2},$$

and he succeeded in showing that any nonzero real number r has an infinite number of logarithms (for a given base), all imaginary if $r < 0$ and all imaginary but one if $r > 0$. In college geometry we find the *Euler line* of a triangle (see Problem Study 5.12), in college courses in the theory of equations the student sometimes encounters *Euler's method* for solving quartic equations, and in even the most elementary course in number theory one meets *Euler's theorem* and the *Euler ϕ-function* (see Problem Study 10.8). The beta and gamma functions of advanced calculus are credited to Euler, though they were adumbrated by Wallis. Euler employed the idea of an integrating factor in the solution of differential equations, was one of the first to develop a theory of continued fractions, contributed to the fields of differential geometry and the calculus of variations, and considerably enriched number theory. In one of his smaller papers occurs the relation

$$V - E + F = 2,$$

already known to Descartes, connecting the number of vertices V, edges E, and faces F of any simple closed polyhedron. In another paper he investigates *orbiform curves*, or curves which, like the circle, are convex ovals of constant width. Several of his papers are devoted to mathematical recreations, such as unicursal and multicursal graphs (inspired by the seven bridges of Königsberg), the re-entrant knight's path on a chess board, and Graeco-Latin squares. He also published extensively in areas of applied mathematics, in particular to lunar theory, the three-body problem of celestial mechanics, the attraction of ellipsoids, hydraulics, ship-building, artillery, and a theory of music.

Euler was a masterful writer of textbooks, in which he presented his material with great clarity, detail, and completeness. These texts enjoyed a marked and a long popularity, and to this day make very interesting and profitable reading. One cannot but be surprised at Euler's enormous fertility of ideas, and it is no wonder that so many of the great mathematicians coming after him have admitted their indebtedness to him.

Claude Alexis Clairaut was born in Paris in 1713, and died there in 1765. He was a youthful mathematical prodigy, composing in his eleventh year a treatise on curves of the third order. This early paper, and a singularly elegant subsequent one on the differential geometry of twisted curves in space, won him a seat in the French Academy of Sciences at the illegal age of 18. In 1736 he accompanied Pierre Louis Moreau de Maupertuis (1698–1759) on an expedition to Lapland to measure the length of a degree of one of the earth's meridians. The expedition was undertaken to settle a dispute as to the shape of the earth. Newton and Huygens had concluded, from mathematical theory, that the earth is flattened at the poles. But about 1712, the Italian astronomer and mathematician Giovanni Domenico Cassini (1625–1712), and his French-born son Jacques Cassini (1677–1756), measured an arc of longitude extending from Dunkirk to Perpignan, and obtained a result that seemed to support the Cartesian contention that the earth is elongated at the poles. The measurement made in Lapland unquestionably confirmed the Newton-Huygens belief, and earned Maupertuis the title of "earth flattener." In 1743, after his return to France, Clairaut published his definitive work, *Théorie de la figure de la Terre*. In 1752 he won a prize from the St. Petersburg Academy for his paper *Théorie de la Lune*, a mathematical study of lunar motion which cleared up some, to then, unanswered questions. He applied the process of differentiation to the differential equation

$$y = px + f(p), \qquad p = dy/dx,$$

now known in elementary textbooks on differential equations as *Clairaut's equation*, and he found the singular solution, but this process had been used earlier by Brook Taylor. In 1759 he calculated, with an error of about a month, the 1759 return of Halley's comet.

Clairaut had a brother who died when only 16, but who at 14 read a paper on geometry before the French Academy and at 15 published a work on geometry. The father of the Clairaut children was a teacher of mathematics, a correspondent of the Berlin Academy, and a writer on geometry.

Jean-le-Rond d'Alembert (1717–1783), like Alexis Clairaut, was born in Paris and died in Paris. As a newly-born infant he was abandoned near the church of Saint Jean-le-Rond and was discovered there by a gendarme who had him hurriedly christened with the name of the place where he was found. Later, for reasons not known, the name d'Alembert was added.

There existed a scientific rivalry, often not friendly, between d'Alembert and Clairaut. At the age of 24, d'Alembert was admitted to the French Academy. In 1743 he published his *Traité de dynamique*, based upon the great principle of kinetics that now bears his name. In 1744 he applied his principle in a treatise on the equilibrium and motion of fluids, and in 1746 in a treatise on the causes of winds. In each of these works, and also in one of 1747 devoted to vibrating strings, he was led to partial differential equations, and he became a pioneer in the study of such equations. With the aid of his principle he was able to obtain a complete solution of the baffling problem of the precession of the equinoxes. D'Alembert showed interest in the foundations of analysis, and in 1754 he made the important suggestion that a sound theory of limits was needed to put analysis on a firm foundation, but his contemporaries paid little heed to his suggestion. It was in 1754 that d'Alembert became permanent secretary of the French Academy. During his later years he worked on the great French *Encyclopédie*, which had been begun by Denis Diderot and himself.

12-5   Lambert, Lagrange, Monge

A little younger than Clairaut and d'Alembert was Johann Heinrich Lambert (1728–1777), born in Mulhouse (Alsace), then part of Swiss territory. It was in 1766 that Lambert wrote his investigation of the parallel postulate entitled *Die Theorie der Parallellinien* (see Section 5-7). Lambert was a mathematician of high quality. As the son of a poor tailor he was largely self-taught. He possessed a fine imagination and he established his results with great attention to rigor. In fact, Lambert was the first to prove rigorously that the number π is irrational. He showed that if x is rational, but not zero, then tan x cannot be rational; since tan $\pi/4 = 1$, it follows that $\pi/4$, or π, cannot be rational. We also owe to Lambert the first systematic development of the theory of hyperbolic functions, and, indeed, our present notation for these functions. Lambert was a many-sided scholar and contributed noteworthily to the mathematics of numerous other topics, such as descriptive geometry, the determination of comet orbits, and the theory of projections employed in the making of maps (a much-used one of these projections is now named after him). At one time he considered plans for a mathematical logic of the sort once outlined by Leibniz.

The two greatest mathematicians of the eighteenth century were Euler and Joseph Louis Lagrange (1736–1813), and which of the two is to be accorded first place is a matter of debate that often reflects the varying mathematical sensitivities of the debaters. Lagrange was born in Turin, Italy. In 1766, when Euler left Berlin, Frederick the Great wrote to Lagrange that "the greatest king in Europe" wished to have at his court "the greatest mathematician of Europe." Lagrange accepted the invitation and for twenty years held the post vacated by Euler. A few years after leaving Berlin, and in spite of the chaotic political situation in France, Lagrange accepted a professorship at the newly established École Normale, and then at the École Polytechnique, schools which became famous in the history of mathematics inasmuch as many of the great mathematicians of modern France have been trained there and many held professorships there. Lagrange did much to develop the high degree of scholarship in mathematics that has been associated with these institutions.

Lagrange's work had a very deep influence on later mathematical research, for he was the earliest mathematician of the first rank to recognize the thoroughly unsatisfactory state of the foundations of analysis and accordingly to attempt a rigorization of the calculus. The attempt, which was far from successful, was made in 1797 in his great publication *Théorie des fonctions analytiques contenant les principes du calcul différentiel*. The cardinal idea here was the representation of a function $f(x)$ by a Taylor's series. The derivatives $f'(x)$, $f''(x)$, $\cdots$ were then defined as the coefficients of h, $h^2/2!$, $\cdots$ in the Taylor expansion of $f(x + h)$ in terms of h. The notation $f'(x)$, $f''(x)$, $\cdots$, very commonly used today, is due to Lagrange. But Lagrange failed to give sufficient attention to matters of convergence and divergence. Nevertheless, we have here the first "theory of functions of a real variable." Two other great works of Lagrange are his *Traité de la résolution des équations numériques de tous degrés* and his monumental *Mécanique analytique*; the former was written late in the century and gives a method of approximating the real roots of an equation by means of continued fractions, the latter (which has been described as a "scientific poem") dates from Lagrange's Berlin period and contains the general equations of motion of a dynamical system known today as *Lagrange's equations*. His work in differential equations (for example, the method of variation of parameters), and particularly in partial differential equations, is very notable, and his contributions to the calculus of variations did much for the development of that subject. Lagrange had a penchant for number theory and wrote important papers in this field also, such as the first published proof of the theorem that every positive integer can be expressed as the sum of not more than four squares. Some of his early work on the theory of equations later led Galois to his theory of groups. In fact, the important theorem of group theory that states that the order of a subgroup of a group G is a factor of the order of G, is called *Lagrange's theorem*. Lagrange has been mentioned a number of times in earlier parts of our book.

Whereas Euler wrote with a profusion of detail and a free employment of intuition, Lagrange wrote concisely and with attempted rigor. Lagrange was "modern" in style and can be characterized as the first real analyst.

The last outstanding mathematician of the eighteenth century whom we shall consider is Gaspard Monge (1746–1818), who in 1794 became a professor of mathematics at the École Polytechnique, which he had vigorously helped to establish. He is particularly noted as the elaborator of descriptive geometry, the science of representing three dimensional objects by appropriate projections on the two dimensional plane. A work of his entitled *Application de l'analyse à la géométrie* ran through five editions and was one of the most important of the early treatments of the differential geometry of surfaces. Monge was a gifted teacher and his lectures inspired a large following of able geometers.

We conclude our very brief survey of eighteenth-century mathematics by noting that while the century witnessed considerable further development in such subjects as trigonometry, analytic geometry, calculus, theory of numbers, theory of equations, probability, differential equations, and analytic mechanics, it witnessed also the creation of a number of new subjects, such as actuarial science, the calculus of variations, higher functions, descriptive geometry, and differential geometry.

12-6   Three Significant Events of the Nineteenth Century

In the nineteenth century the superstructure of mathematics continued to rise, but only on ever-deepening foundations. The evolution from intuition to rigor made marked progress, mathematics slowly obtained a freedom from traditional bonds, and generalization and abstraction began to become the order of the day. The teaching responsibility of professional mathematicians increased. Research became more and more centered about the universities rather than around royal courts or membership in learned academies. National languages gradually replaced scientific Latin, and mathematicians began to work in ever-more-specialized fields. A number of important philosophies of mathematics emerged.

Three profoundly significant mathematical events occurred during the nineteenth century—one in the field of geometry, one in the field of algebra, and one in the field of analysis. The event in geometry was the first one of the three to occur; it was the discovery, about 1829, of a self-consistent geometry different from the geometry of Euclid. The story of this event has been told in Section 5-7, and the heroes were Nicolai Ivanovitch Lobachevsky (1793–1856), János Bolyai (1802–1860), and Carl Friedrich Gauss (1777–1855). The immediate consequence of the discovery of this first non-Euclidean geometry was, of course, the final settlement of the ages-old problem of the parallel postulate—the parallel postulate was shown to be independent of the other assumptions of Euclidean geometry. But a much more far-reaching consequence than this was the liberation of geometry from its traditional mold. A deep-rooted and centuries-old conviction that there could be only the one possible geometry was shattered, and the way was opened for the creation of many different systems of geometry. With the possibility of creating such purely "artificial" geometries, it became apparent that geometry is not necessarily tied to actual physical space. The postulates of geometry became, for the mathematician, mere hypotheses whose physical truth or falsity need not concern him. The mathematician may take his postulates to suit his pleasure, just so long as they are consistent with one another.

The second of the three events to occur was the one in algebra, and it occurred not many years after the first event; it was the creation, in 1843, of a noncommutative algebra. In the early nineteenth century, algebra was considered to be simply generalized arithmetic.* In other words, instead of working with specific numbers, as we do in arithmetic, in algebra we employ letters as symbols to represent arbitrary numbers. It was the early nineteenth-century British school of algebraists—George Peacock (1791–1858), Duncan Farquharson Gregory (1813–1844), Augustus De Morgan (1806–1871), and others—who first noticed the presence of structure in algebra, such as the commutative and associative laws of addition and multiplication, and the distributive law of multiplication over addition. It seemed inconceivable, in the early nineteenth century, that there could exist a consistent algebra with a structure contrary to that of the common algebra of arithmetic. Such was the feeling about algebra when, in 1843, the Irish mathematician William Rowan Hamilton (1805–1865) was forced, by physical considerations and after years of cogitation on a particular problem, to invent his quaternion algebra in which the commutative law of multiplication does not hold. The following year, the German mathematician Hermann Grassmann (1809–1877) published the first edition of his remarkable *Ausdehnungslehre*, in which were developed whole classes of algebras with a structure differing from that of the familiar algebra of arithmetic. In 1857, the English mathematician Arthur Cayley (1821–1895) devised his matrix algebra, which is another example of a noncommutative algebra. By developing algebras satisfying structural laws different from those obeyed by common algebra, these men opened the flood gates of modern abstract algebra.

By weakening or deleting various postulates of common algebra, or by replacing one or more of the postulates by others, which are consistent with the remaining postulates, an enormous variety of systems can be studied. As some of these systems we have groupoids, quasigroups, loops, semigroups, monoids, groups, rings, integral domains, lattices, division rings, Boolean rings, Boolean algebras, fields, vector spaces, Jordan algebras, and Lie algebras, the last two being examples of nonassociative algebras. It is probably correct to say that mathematicians have, to date, studied well over 200 such algebraic structures. Most of this work belongs to the twentieth century and reflects the spirit of generalization and abstraction so prevalent in mathematics today. Abstract algebra has become the vocabulary of much of present-day mathematics.

The third of the three profound mathematical events of the nineteenth century was in the field of analysis and was slow in materializing; it was the

*This is still the view of algebra as commonly taught in the high school and frequently in the freshman year at college.

so-called *arithmetization of analysis*. We have seen that some mathematicians of the eighteenth century became alarmed over the deepening crisis in the foundations of analysis. D'Alembert observed, in 1754, that a theory of limits was needed, and Lagrange, in 1797, attempted a rigorization of analysis. A great forward step was made in 1821, when the French mathematician Augustin-Louis Cauchy successfully executed d'Alembert's suggestion, by developing an acceptable theory of limits and then defining convergence, continuity, differentiability, and the definite integral in terms of the limit concept. It is essentially these definitions that we find in the more carefully written of today's elementary textbooks on the calculus.

But the demand for an even deeper understanding of the foundations of analysis was strikingly brought out in 1874 with the publicizing of an example, due to the German mathematician Karl Weierstrass, of a continuous function having no derivative, or, what is the same thing, a continuous curve possessing no tangent at any of its points. Georg Bernhard Riemann produced a function which is continuous for all irrational values of the variable but discontinuous for all rational values. Examples such as these seemed to contradict human intuition and made it increasingly apparent that Cauchy had not struck the true bottom of the difficulties in the way of a sound foundation of analysis. The theory of limits had been built upon a simple intuitive notion of the real number system. Indeed, the real number system was taken more or less for granted, as it still is in most of our elementary calculus texts. It became clear that the theory of limits, continuity, and differentiability depend upon more recondite properties of the real number system than had been supposed. Accordingly, Weierstrass advocated a program wherein the real number system itself should first be rigorized, then all the basic concepts of analysis should be derived from this number system. This remarkable program, known as the *arithmetization of analysis*, proved to be difficult and intricate, but was ultimately realized by Weierstrass and his followers, so that today all of analysis can be logically derived from a postulate set characterizing the real number system.

Mathematicians have gone considerably beyond the establishment of the real number system as the foundation of analysis. Euclidean geometry, through its analytical interpretation, can also be made to rest upon the real number system, and mathematicians have shown that most branches of geometry are consistent if Euclidean geometry is consistent. Again, since the real number system, or some part of it, can serve for interpreting so many branches of algebra, it appears that the consistency of a good deal of algebra can also be made to depend upon that of the real number system. In fact, today it can be stated that essentially all of existing mathematics is consistent if the real number system is consistent. Herein lies the tremendous importance of the real number system for the foundations of mathematics.

Since the great bulk of existing mathematics can be made to rest on the real number system, one naturally wonders if the foundations can be pushed even deeper. In the late nineteenth century, with the work of Richard Dedekind (1831–1916), Georg Cantor (1845–1918), and Giuseppi Peano (1858–1932), these foundations were established in the much simpler and more basic system of natural numbers. That is, these men showed how the real number system, and thence the great bulk of mathematics, can be derived from a postulate set for the natural number system. Then, in the early twentieth century, it was shown that the natural numbers can be defined in terms of concepts of set theory, and thus that the great bulk of mathematics can be made to rest on a platform in set theory. Logicians, led by Bertrand Russell (1872–   ) and Alfred North Whitehead (1861–1947), have endeavored to push the foundations even deeper, by deriving the theory of sets from a foundation in the calculus of propositions of logic, though not all mathematicians feel this step has been successfully executed.

Since the great bulk of mathematics can be made to rest on set theory, the discovery of paradoxes in the fringes of set theory has precipitated another crisis in the foundations of mathematics, and many mathematicians of the twentieth century are concerning themselves with the resolution of this crisis. Three important schools of philosophy of mathematics have emerged, the logistic school, which received its definitive expression in the monumental *Principia mathematica* of Whitehead and Russell, the intuitionist school, originated by the Dutch mathematician L. E. J. Brouwer (1881–   ), and the formalist school, founded by the great German mathematician David Hilbert (1862–1943).

Because of the feasibility of the logistic thesis, and because a branch of mathematics is the resultant of a given postulate set and a given logic, the subject of symbolic logic is being intensively pursued by many present-day mathematicians, and a number of special journals have come into being to publicize the work of these mathematicians.

12-7   The Evolution of Some Basic Concepts

Following the development of the theory of sets by Georg Cantor toward the end of the nineteenth century, interest in that theory developed rapidly until today virtually every field of mathematics has felt its impact. Notions of space and the geometry of a space, for example, have been completely revolutionized by the theory of sets. Also, the basic concepts in analysis, such as those of limit, function, continuity, derivative, and integral, are now most aptly described in terms of set-theory ideas. Most important, however, has been the opportunity for new mathematical developments undreamed of fifty years ago. Thus, in companionship with the new appreciation of postulational procedures in mathematics, abstract spaces have been born, general theories of dimension and measure have been created, and the branch of mathematics called *topology* has undergone a spectacular growth. In short, under the influence of set theory, a considerable unification of traditional mathematics has occurred, and new mathematics has been created at an explosive rate.

To illustrate the historical evolution of basic mathematical concepts, let us first consider notions of space and the geometry of a space. These concepts have undergone marked changes since the days of the ancient Greeks. For the Greeks there was only one space and one geometry; these were absolute concepts. The space was not thought of as a collection of points, but rather as a realm, or locus, in which objects could be freely moved about and compared with one another. From this point of view, the basic relation in geometry was that of congruence or superposability.

With the development of analytic geometry in the seventeenth century, space came to be regarded as a collection of points, and with the creation of the classical non-Euclidean geometries in the nineteenth century, mathematicians accepted the situation that there is more than one geometry. But space was still regarded as a locus in which figures could be compared with one another. The central idea became that of a group of congruent transformations of space into itself, and a geometry came to be regarded as the study of those properties of configurations of points which remain unchanged when the enclosing space is subjected to these transformations. We have seen, in Section 9-8, how this point of view was expanded by Felix Klein (1849–1929) in his *Erlanger Programm* of 1872. In the *Erlanger Programm*, a geometry was defined as the invariant theory of a transformation group. This concept synthesized and generalized all earlier concepts of geometry, and supplied a singularly neat classification of a large number of important geometries.

At the end of the nineteenth century, with the development of the idea of a branch of mathematics as an abstract body of theorems deduced from a set of postulates, each geometry became, from this point of view, a particular branch of mathematics. Postulate sets for a large variety of geometries were studied, but the *Erlanger Programm* was in no way upset, for a geometry could be regarded as a branch of mathematics which is the invariant theory of a transformation group.

In 1906, however, Maurice Fréchet (1878–   ) inaugurated the study of abstract spaces, and very general geometries came into being which no longer necessarily fit into the neat Kleinian classification. A space became merely a set of objects, usually called *points*, together with a set of relations in which these points are involved, and a geometry became simply the theory of such a space. The set of relations to which the points are subjected is called *the structure of the space*, and this structure may or may not be explainable in terms of the invariant theory of a transformation group. Thus, through set theory, geometry received a further generalization. Although abstract spaces were first formally introduced in 1906, the idea of a geometry as the study of a set of points with some superimposed structure was really already contained in remarks made by Riemann in his great lecture of 1854. It is interesting that some of these new geometries have found valuable application in the Einstein theory of relativity, and in other developments of modern physics.

The concept of function, like the notions of space and geometry, has undergone a marked evolution, and every student of mathematics encounters various refinements of this evolution as his studies progress from the elementary courses of high school into the more advanced and sophisticated courses of the college postgraduate level.

The history of the term "function" furnishes another interesting example of the tendency of mathematicians to generalize and extend their concepts. The word "function" seems to have been introduced in 1637 by Descartes, who employed the term simply to mean some positive integral power, x^n, of a variable x. Somewhat later, Leibniz employed the term to denote any quantity connected with a curve, such as the coordinates of a point on the curve, the slope of the curve, and so on. Johann Bernoulli regarded a function as any expression made up of a variable and some constants, and Euler regarded a function as any equation or formula involving variables and constants. This latter idea is the notion of a function formed by most students of elementary mathematics courses. The Euler concept remained unchanged until Joseph Fourier (1768–1830) was led, in his investigations of heat flow, to consider so-called trigonometric series. These series involve a more general type of relationship between variables than had previously been studied, and, in an attempt to furnish a definition of function broad enough to encompass such relationships, Lejeune Dirichlet (1805–1859) arrived at the following formulation: A *variable* is a symbol which represents any one of a set of numbers; if two variables x and y are so related that whenever a value is assigned to x there is automatically assigned, by some rule or correspondence, a value to y, then we say y is a (single-valued) *function* of x. The variable x, to which values are assigned at will, is called the *independent variable*, and the variable y, whose values depend upon those of x, is called the *dependent variable*. The permissible values that x may assume constitute the *domain of definition* of the function, and the values taken on by y constitute the *range of values* of the function.

The student of mathematics usually meets the Dirichlet definition of function in his introductory course in calculus. The definition is a very broad one and does not imply anything regarding the possibility of expressing the relationship between x and y by some kind of analytic expression; it stresses the basic idea of a relationship between two sets of numbers.

Set theory has extended the concept of function to embrace relationships between any two sets of elements, be the elements numbers or anything else. Thus, in set theory, a function f is defined to be any set of ordered pairs of elements such that if $(a_1,b_1) \in f$, $(a_2,b_2) \in f$, and $a_1 = a_2$, then $b_1 = b_2$. The set A of all first elements of the ordered pairs is called the *domain* (*of definition*) of the function, and the set B of all second elements of the ordered pairs is called the *range* (*of values*) of the function. A functional relationship is thus nothing but a special kind of subset of the Cartesian product set $A \times B$. A one-to-one correspondence is, in its turn, a special kind of function, namely, a function f such that if $(a_1,b_1) \in f$, $(a_2,b_2) \in f$, and $b_1 = b_2$, then $a_1 = a_2$. If, for a functional relationship f, $(a,b) \in f$, we write $b = f(a)$.

The notion of function pervades much of mathematics, and since the early part of the present century various influential mathematicians have advocated the employment of this concept as the unifying and central principle in the organization of elementary mathematics courses. The concept seems to form a natural and effective guide for the selection and development of textual material. There is no doubt of the value of a mathematics student's early acquaintance with the function concept.

12-8   Laplace, Legendre, Gauss

There were many mathematical giants in the nineteenth century, a large number of whom, along with lesser men, have already been mentioned, in one connection or another, either in this chapter or in earlier chapters. We shall close the present chapter by very briefly considering Pierre-Simon Laplace, Adrien-Marie Legendre, Carl Friedrich Gauss, some geometers of the nineteenth century, Augustin-Louis Cauchy, Karl Weierstrass, Georg Bernhard Riemann, and Georg Cantor.

Laplace and Legendre were contemporaries of Lagrange, but they published their principal works in the nineteenth century. Pierre-Simon Laplace was born of poor parents in 1749. His mathematical ability early won him good teaching posts, and as a political opportunist he ingratiated himself with whichever party happened to be in power during the uncertain days of the French Revolution. His most outstanding work was done in the fields of celestial mechanics, probability, differential equations, and geodesy. He published two monumental works, *Traité de mécanique céleste* (five volumes, 1799–1825) and *Théorie analytique des probabilités* (1812), each of which was preceded by an extensive nontechnical exposition. The five-volume *Traité de mécanique céleste*, which earned him the title of "the Newton of France," embraced all previous discoveries in this field along with Laplace's own contributions, and marked the author as the unrivaled master in the subject. It may be of interest to repeat a couple of anecdotes often told in connection with this work. When Napoleon teasingly remarked that God was not mentioned in his treatise, Laplace replied, "Sir, I did not need that hypothesis." And the American astronomer, Nathaniel Bowditch, when he translated Laplace's treatise into English remarked, "I never come across one of Laplace's 'Thus it plainly appears' without feeling sure that I have hours of hard work before me to fill up the chasm and find out and show how it plainly appears." Laplace's name is connected with the *nebular hypothesis* of cosmology and with the so-called *Laplace equation* of potential theory (though neither of these contributions originated with Laplace), with the so-called *Laplace transform* which later became the key to the operational calculus of Heaviside, and with the *Laplace expansion* of a determinant. Laplace died in 1827, exactly one hundred years after the death of Isaac Newton.

Adrien-Marie Legendre (1752–1833) is known in the history of elementary mathematics principally for his very popular *Éléments de géométrie*, in which he attempted a pedagogical improvement of Euclid's *Elements* by considerably rearranging and simplifying many of the propositions. This work was very favorably received in America and became the prototype of the geometry textbooks in this country. In fact, the first English translation of Legendre's geometry was made in 1819 by John Farrar of Harvard University. Three years later another English translation was made, by the famous Scottish littérateur, Thomas Carlyle, who early in life was a teacher of mathematics. Carlyle's translation, as later revised by Charles Davies, and later still by J. H. Van Amringe, ran through 33 American editions. In later editions of his geometry, Legendre attempted to prove the parallel postulate. Legendre's chief work in higher mathematics centered about number theory, elliptic functions, the method of least squares, and integrals; this work is too advanced to be discussed here. He was also an assiduous computer of mathematical tables. Legendre's name is today connected with the second-order differential equation

$$(1 - x^2)y'' - 2xy' + n(n + 1)y = 0,$$

which is of considerable importance in applied mathematics. Functions satisfying this differential equation are called *Legendre functions* (of order n). When n is a nonnegative integer, the equation has polynomial solutions of special interest called *Legendre polynomials*. Legendre's name is also associated with the symbol $(c|p)$ of number theory. The *Legendre symbol* $(c|p)$ is equal to ± 1 according as the integer c, which is prime to p, is or is not a quadratic residue of the odd prime p. (For example, $(6|19) = 1$ since the congruence $x^2 \equiv 6$ (mod 19) has a solution, and $(39|47) = -1$ since the congruence $x^2 \equiv 39$ (mod 47) has no solution.)

The greatest mathematician of the nineteenth century, and usually ranked with Archimedes and Isaac Newton as one of the three greatest mathematicians of all time, was Carl Friedrich Gauss (1777–1855). Gauss was one of those remarkable infant prodigies who appear from time to time. They tell of him the incredible story that at the age of three he detected an error in his father's bookkeeping. We have already, in Section 5-4, considered Gauss's surprising contribution to the theory of the Euclidean construction of regular polygons, and, in Section 5-7, his anticipation of the discovery of non-Euclidean geometry. In his doctoral dissertation, written at the age of 20, he gave the first wholly satisfactory proof of the *fundamental theorem of algebra* (that a polynomial equation with complex coefficients and of degree n has at least one complex root). Unsuccessful attempts to prove this theorem had been made by Newton, Euler, d'Alembert, and Lagrange. Gauss's greatest single publication is his *Disquisitiones arithmeticae*, a work of fundamental importance in the modern theory of numbers. Gauss's findings on the construction of regular polygons appear in this work, as does his facile notation for congruence, and the first proof of the beautiful quadratic reciprocity law, which says, in terms of the Legendre symbol defined above, that if $p = 2P + 1$ and $q = 2Q + 1$ are unequal primes, then

$$(p|q)(q|p) = (-1)^{PQ}$$

Gauss contributed notably to astronomy, geodesy, electricity, differential geometry, and the method of least squares. It was in 1812, in a paper on hypergeometric series, that Gauss made the first systematic investigation of the convergence of a series. Famous is Gauss's assertion that "mathematics is the queen of the sciences, and the theory of numbers is the queen of mathematics." Gauss has been described as "the mathematical giant who from his lofty heights embraces in one view the stars and the abysses." He worked from 1807 until his death as director of the observatory and professor of mathematics at the University of Göttingen.

12-9   Geometry in the Nineteenth Century

Quite apart from the discovery of non-Euclidean geometry, the field of geometry made enormous strides in the nineteenth century. Much of this work, in both its synthetic and analytic forms, can be traced back to the inspiring teaching of Gaspard Monge. Projective geometry, as an individual branch of mathematics, began in 1822 with the publication of the *Traité des propriétés des figures* of Jean Victor Poncelet (1788–1867), perhaps Monge's most outstanding student. This work of Poncelet gave a tremendous impetus to the study of the subject and inaugurated the "great period" in the history of projective geometry. As we have seen, in Section 9-8, the idea of poles and polars in projective geometry was elaborated by Poncelet and Joseph-Diez Gergonne (1771–1859) into a regular method out of which grew the elegant principle of duality. Many of Poncelet's ideas were further developed by the Swiss geometer Jacob Steiner (1796–1867), one of the greatest synthetic geometers the world has ever known. Projective geometry was finally completely freed of any metrical basis by Karl Georg Christian von Staudt (1798–1867) in his *Geometrie der Lage* of 1847. The analytical side of geometry made spectacular gains in the work of Augustus Ferdinand Möbius (1790–1868), Michel Chasles (1793–1880), and, particularly, Julius Plücker (1801–1868). Michel Chasles was also an outstanding synthetic geometer, and his *Aperçu historique sur l'origine et le développement des méthodes en géométrie* (1837) is still a standard historical work. Felix Klein (1849–1929) introduced his *Erlanger Programm* for the codification of geometries in 1872. It was shown how, by the adoption of a suitable projective definition of a metric, we can study metric geometry in the framework of projective geometry, and by the adjunction of an invariant conic to a projective geometry in the plane we can obtain the classical non-Euclidean geometries. In the late nineteenth and early twentieth centuries, projective geometry received a number of postulational treatments, and finite geometries were discovered. It was shown that, by gradually adding and altering postulates, one can pass from projective geometry to Euclidean geometry, encountering a number of other important geometries on the way. Finally, differential geometry, which also started in an essential way with Gaspard Monge, developed deeply in the nineteenth and early twentieth centuries. This field of geometry received highly directive contributions from Carl Friedrich Gauss and Georg Bernhard Riemann. The last half of the nineteenth century saw the birth of the great Italian school of geometers.

12-10　Cauchy, Weierstrass, Riemann, Cantor

With Lagrange and Gauss the nineteenth century rigorization of analysis got under way. Later nineteenth-century mathematicians who played outstanding roles in this work were the Frenchman, Augustin-Louis Cauchy (1789–1857), and the German, Karl Weierstrass (1815–1897). As noted above, our present definitions of limit, continuity, the derivative as the limit of a difference quotient, and the definite integral as the limit of a sum are substantially those given by Cauchy. Cauchy wrote extensively and profoundly in both pure and applied mathematics, and he can probably be ranked next to Euler in volume of output. His numerous contributions to advanced mathematics include researches in convergence and divergence of infinite series, real and complex function theory, differential equations, determinants, probability, and mathematical physics. His name is met by the student of calculus in the so-called *Cauchy root test* and *Cauchy ratio test* for convergence or divergence of a series of positive terms, and in the *Cauchy product* of two given series. In even a first course in complex function theory one encounters the *Cauchy inequality, Cauchy's integral formula, Cauchy's integral theorem,* and the basic *Cauchy-Riemann differential equations.*

Cauchy's rigor inspired other mathematicians to attempt the banishment of formal manipulation and intuition from analysis. Cauchy was an ardent partisan of the Bourbons and, after the revolution of 1830, was forced to give up his professorship at the École Polytechnique and was excluded from public employment for eighteen years. Part of this time he spent in exile in Turin and Prague, and part in teaching in some church schools in Paris. In 1848 he was allowed to return to a professorship at the École Polytechnique without having to take the oath of allegiance to the new government. Throughout his life he was an indefatigable worker and it is regrettable that he possessed a narrow conceit and often ignored the meritorious efforts of younger men.

It is generally thought that a potential mathematician of the first rank, in order to succeed in his field, must start serious mathematical studies at an early age and must not be dulled by an inordinate amount of elementary teaching. Karl Weierstrass is an outstanding exception to these two general rules. A misdirected youth spent in studying the law and finance gave Weierstrass a late start in mathematics, and it was not until he was forty that he finally emancipated himself from secondary teaching by obtaining an instructorship at the University of Berlin, and another eight years passed before, in 1864, he was awarded a full professorship at the university and could finally devote all his time to advanced mathematics. Weierstrass never regretted the years he spent in elementary teaching, and he later carried over his remarkable pedagogical abilities into his university work, becoming probably the greatest teacher of advanced mathematics that the world has yet known.

Weierstrass wrote a number of early papers on hyperelliptic integrals, Abelian functions, and algebraic differential equations, but his widest known contribution to mathematics is his foundation of the theory of complex functions on power series. This, in a sense, was an extension to the complex plane of the idea earlier attempted by Lagrange, but Weierstrass carried it through with absolute rigor. Weierstrass showed particular interest in entire functions and in functions defined by infinite products. He discovered uniform convergence and, as we have seen above, started the so-called arithmetization of analysis, or the reduction of the principles of analysis to real number concepts. A large number of his mathematical findings became possessions of the mathematical world, not through publication by him, but through notes taken of his lectures. He was very generous in allowing students and others to carry out, and receive credit for, investigations of many of his mathematical gems. As an illustration, somewhat in point, it was in his lectures of 1861 that he first discussed his example of a continuous nondifferentiable function, which was finally published in 1874 by Paul du Bois-Reymond (1831–1889).

Weierstrass was a very influential teacher, and his meticulously prepared lectures established an ideal for many future mathematicians; "Weierstrassian rigor" became synonymous with "extremely careful reasoning." Weierstrass was "the mathematical conscience par excellence," and he became known as "the father of modern analysis." He died in 1897, just one hundred years after the first publication, in 1797 by Lagrange, of an attempt to rigorize the calculus.

Along with this rigorization of mathematics there appeared a tendency toward abstract generalization, a process which has become very pronounced in present-day mathematics. Perhaps the German mathematician Georg Bernhard Riemann (1826–1866) influenced this feature of modern mathematics more than any other nineteenth-century mathematician. His doctoral dissertation of 1851 led to the concept of *Riemann surfaces*, which, in turn, introduced so-called topological considerations into analysis. Riemann clarified the concept of integrability by the definition of what we now know as the *Riemann integral*, which led, in the twentieth century, to the more general *Lebesgue integral*, and thence to further generalizations of the integral. Riemann's famous probationary lecture of 1854 on the hypotheses which lie at the foundation of geometry generalized the idea of space and led, in more recent times, to the extensive and important theory of abstract spaces. Riemann died of tuberculosis when only forty, but he left to the mathematical world, in his small collection of published papers, a singularly rich legacy of ideas not yet exhausted by later mathematicians.

Georg Cantor, with a life span astride the nineteenth and twentieth centuries, is the last mathematician to be considered in detail here. He was born in St. Petersburg in 1845, moved with his parents to Germany in 1856, studied at the University of Berlin (where he came under the influence of Weierstrass) from 1863 to 1869, taught in the University of Halle from 1869 until 1905, and died in a mental hospital in Halle in 1918. His early interests were in number theory, indeterminate equations, and trigonometric series. The subtle theory of trigonometric series seems to have inspired him to look into the foundations of analysis. He produced his beautiful treatment of irrational numbers—which utilizes convergent sequences of rational numbers and differs radically from the geometrically inspired treatment of Dedekind —and commenced, in 1874, his revolutionary work on set theory and the theory of the infinite. With this latter work, Cantor created a whole new field of mathematical research. In his papers he developed a theory of transfinite numbers, based on a mathematical treatment of the actual infinite, and created an arithmetic of transfinite numbers analogous to the arithmetic of finite numbers. Cantor was deeply religious, and his work, which in a sense is a continuation of the arguments connected with the paradoxes of Zeno, reflects his sympathetic respect for medieval scholastic speculation on the nature of the infinite. His views met considerable opposition, chiefly from Leopold Kronecker (1823–1891) of the University of Berlin. Today, Cantor's set theory has penetrated into almost every branch of mathematics, and it has proved to be of particular importance in topology and in the foundations of real function theory. There are logical difficulties, and paradoxes have appeared. The twentieth-century controversy between the formalists, led by Hilbert, and the intuitionists, led by Brouwer, is essentially a continuation of the controversy between Cantor and Kronecker.

Here we terminate our work, with an admission of the dissatisfaction that is inevitable when so many fundamental and interesting topics must be omitted. An acceptable treatment of the history of mathematics beyond 1700 requires an entire volume of its own—with a semester of class meetings to cover it and a fair number of advanced mathematics courses as a prerequisite. The best that can be hoped is that the interested student may be stimulated to take up the story elsewhere, after he becomes sufficiently prepared mathematically.

General Bibliography

ARCHIBALD, R. C., "Outline of the History of Mathematics," Herbert Ellsworth Slaught Memorial Paper No. 2. Buffalo, N. Y.: The Mathematical Association of America, Inc., 1949.

BALL, W. W. R., *A Primer of the History of Mathematics*, 4th ed. New York: The Macmillan Company, 1895.

————, *A Short Account of the History of Mathematics*, 5th ed. New York: The Macmillan Company, 1912.

BELL, E. T., *The Development of Mathematics*, 2d ed. New York: McGraw-Hill Book Company, Inc., 1945.

————, *Mathematics, Queen and Servant of Science*. New York: McGraw-Hill Book Company, Inc., 1951.

CAJORI, FLORIAN, *A History of Elementary Mathematics*, rev. ed. New York: The Macmillan Company, 1924.

————, *A History of Mathematics*, 2d ed. New York: The Macmillan Company, 1919.

FINK, KARL, *A Brief History of Mathematics*, tr. by W. W. Beman and D. E. Smith. Chicago: The Open Court Publishing Company, 1900.

FREEBURY, H. A., *A History of Mathematics*. New York: The Macmillan Company, 1961.

HOFMANN, J. E., *The History of Mathematics*. New York: Philosophical Library, Inc., 1957.

————, *Classical Mathematics, a Concise History of the Classical Era in Mathematics*. New York: Philosophical Library, Inc., 1959.

HOGBEN, L. T., *Mathematics for the Million*. New York: W. W. Norton & Company, Inc., 1937.

HOOPER, ALFRED, *Makers of Mathematics*. New York: Random House, Inc., 1948.

KLINE, MORRIS, *Mathematics in Western Culture*. New York: Oxford University Press, 1953.

————, *Mathematics and the Physical World*. New York: Thomas Y. Crowell Company, 1959.

————, *Mathematics, a Cultural Approach*. Reading, Mass.: Addison-Wesley Publishing Company, Inc., 1962.

LARRETT, DENHAM, *The Story of Mathematics*. New York: Greenberg: Publisher, Inc., 1926.

SANFORD, VERA, *A Short History of Mathematics*. Boston: Houghton Mifflin Company, 1930.

SARTON, GEORGE, *The Study of the History of Mathematics*. New York: Dover Publications, Inc., 1954.

SCOTT, J. F., *A History of Mathematics from Antiquity to the Beginning of the Nineteenth Century*. London: Taylor & Francis, Ltd., 1958.

SMITH, D. E., *History of Mathematics*, 2 vols. Boston: Ginn & Company, 1923–1925.

————, AND JEKUTHIEL GINSBURG, *A History of Mathematics in America Before 1900* (Carus Mathematical Monograph No. 5). Chicago: The Open Court Publishing Company, 1934.

STRUIK, D. J., *A Concise History of Mathematics*, 2 vols. New York: Dover Publications, Inc., 1948.

Journal literature devoted to the history of mathematics is very vast. For an excellent beginning one may consult:

READ, C. B., "Articles on the History of Mathematics: A Bibliography of Articles Appearing in Six Periodicals," *School Science and Mathematics*, Dec. 1959, pp. 689–717. This is a bibliography of over 1000 articles devoted to the history of mathematics and appearing prior to July 1, 1959 in the six journals: *The American Mathematical Monthly*, *The Mathematical Gazette* (New Series), *The Mathematics Teacher*, *National Mathematics Magazine* (volumes 1–8 were published as *Mathematics News Letter;* starting with volume 21 the title became *Mathematics Magazine*), *Scripta Mathematica*, and *School Science and Mathematics*. The articles are classified into some thirty convenient categories.

For portraits of eminent mathematicians, suitable for framing, there are:

SMITH, D. E., *Portraits of Eminent Mathematicians, with Brief Biographical Sketches*, Portfolio I (Archimedes, Copernicus, Viète, Galileo, Napier, Descartes, Newton, Leibniz, Lagrange, Gauss, Lobachevsky, Sylvester). New York: *Scripta Mathematica*, Yeshiva University, 1946.

SMITH, D. E., *Portraits of Eminent Mathematicians, with Brief Biographical Sketches*, Portfolio II (Euclid, Cardan, Kepler, Fermat, Pascal, Euler, Laplace, Cauchy, Jacobi, Hamilton, Cayley, Chebyshev, Poincaré). New York: *Scripta Mathematica*, Yeshiva University, 1938.

KEYSER, C. J., *Portraits of Famous Philosophers who were also Mathematicians, with Biographical Accounts* (Pythagoras, Plato, Aristotle, Epicurus, Roger Bacon, Descartes, Pascal, Spinoza, Leibniz, Berkeley, Kant, Peirce). New York: *Scripta Mathematica*, Yeshiva University, 1939.

SINGLE PORTRAITS, Series I (10 by 15 inches) (Archimedes, Aristotle, Berkeley, Cardan, Cauchy, Cayley, Chebyshev, Clausius, Copernicus,

Descartes, Einstein, Epicurus, Euclid, Euler, Fermat, Fresnel, Galileo, Gauss, Gibbs, Hamilton, Jacobi, Joule, Kant, Kepler, Lagrange, Laplace, Leibniz, Lobachevsky, Napier, Newton, Pascal, Plato, Plato and Aristotle, Pythagoras, Poincaré, Rowland, Spinoza, Sylvester, Viète, Weierstrass). New York: *Scripta Mathematica*, Yeshiva University, 1963.

SINGLE PORTRAITS, Series II ($6\frac{1}{2}$ by $9\frac{1}{2}$ inches) (Airey, Archibald, Aristotle, Byerly, Cardan, Clausius, Comrie, Davis, Epicurus, Felber, Fresnel, Galois, Germain, Hamilton, Heaviside, Herschel, Huygens, Jonquieres, Kant, Karapetoff, Keyser, Kruse, Lamé, D. H. Lehmer, D. N. Lehmer, Liouville, Loewy, Loria, Lowan, Newton, Pareto, Peacock, Benjamin Peirce, Charles Peirce, Pupin, Pythagoras, David E. Smith, William B. Smith, Spinoza, Steiner, Talquist, Tannery, Zaviska). New York: *Scripta Mathematica*, Yeshiva University, 1963.

A Chronological Table*

(up to the year 1700, with a number of important dates beyond)

−4700   Possible beginning of Babylonian calendar.

−4228   Probable introduction of Egyptian calendar.

−4000   Discovery of metal.

−3500   Writing in use.

−3100   Approximate date of a royal Egyptian mace in a museum at Oxford.

−2900   Great pyramid of Gizeh erected.

−2400   Babylonian tablets of Ur.

−2200   Date of many mathematical tablets found at Nippur; mythical date of the *lo-shu*, the oldest known example of a magic square.

−1950   Hammurabi, King of Babylonia.

−1850   Moscow papyrus (25 numerical problems, "greatest Egyptian pyramid"); oldest extant astronomical instrument.

−1750   Plimpton 322 dates somewhere from −1900 to −1600.

−1650   Rhind, or Ahmes, papyrus (85 numerical problems).

−1600   Approximate date of many of the Babylonian tablets in the Yale collection.

−1500   Largest existing obelisk; oldest extant Egyptian sundial.

−1350   Date of later mathematical tablets found at Nippur; Rollin papyrus (elaborate bread problems).

−1167   Harris papyrus (list of temple wealth).

−1105   Possible date of the *Chóu-peï*, oldest Chinese mathematical work.

−753   Rome founded.

−650   Papyrus had been introduced into Greece by this date.

−600   Tha'les; beginning of demonstrative geometry.

−540   Pythag'oras (geometry, arithmetic, music).

−500   Possible date of the *Śulvasūtras* (religious writings showing acquaintance with Pythagorean numbers and with geometric constructions).

−480   Battle of Thermop'ylae.

−460   Parmen'ides (sphericity of the earth).

−450   Ze'no (paradoxes of motion).

−440   Hippoc'rates of Chi'os (reduction of the duplication problem, lunes, arrangement of the propositions of geometry in a scientific fashion).

*A much fuller chronological table may be found in D. E. Smith, *History of Mathematics*, Vol. I (Boston: Ginn & Company, 1923–1925), 549–570. A minus sign before a date indicates that the date is B.C.

−430 An'tiphon (method of exhaustion).

−425 Hip'pias of Elis (trisection with quadratrix); Theodo'rus of Cyrene (irrational numbers); Soc'rates.

−410 Democ'ritus (atomic theory).

−404 Athens finally defeated by Sparta.

−400 Archy'tas (leader of Pythagorean school at Tarentum, application of mathematics to mechanics).

−380 Pla'to (mathematics in the training of the mind, Plato's Academy).

−375 Theaete'tus (incommensurables, regular solids).

−370 Eudox'us (incommensurables, method of exhaustion, astronomy).

−350 Menaech'mus (conics); Dinos'tratus (quadrature with quadratrix, brother of Menaechmus); Xenoc'rates (history of geometry); Thymar'idas (solution of systems of simple equations).

−340 Ar'istotle (systematizer of deductive logic).

−336 Alexander the Great began his reign.

−335 Eude'mus (history of mathematics).

−332 Alexandria founded.

−323 Alexander the Great died.

−320 Aristae'us (conics, regular solids).

−300 Eu'clid (*Elements*, perfect numbers, optics, data).

−280 Aristar'chus (Copernican system).

−260 Co'non (astronomy, spiral of Archimedes); Dosi'theus (recipient of several papers by Archimedes).

−250 Stone columns erected by King Ásoka and containing earliest preserved examples of our present number.symbols.

−240 Nicome'des (trisection with conchoid).

−230 Eratos'thenes (prime number sieve, size of the earth).

−225 Apollo'nius (conic sections, plane loci, tangencies, circle of Apollonius); Archime'des (greatest mathematician of antiquity, circle and sphere, computation of π, area of parabolic segment, spiral of Archimedes, infinite series, method of equilibrium, mechanics, hydrostatics).

−180 Hyp'sicles (astronomy, number theory); Di'ocles (duplication with cissoid).

−140 Hippar'chus (trigonometry, astronomy, star catalogue).

−100 Probable date of carvings on the walls of a cave near Poona.

 −75 Cicero discovered the tomb of Archimedes.

  75 Possible date of Her'on (machines, plane and solid mensuration, root extraction, surveying) and of Diophan'tus (number theory, syncopation of Greek algebra).

 100 Nicom'achus (number theory); Menela'us (spherical trigonometry); Plu'tarch.

150 Ptol'emy (trigonometry, table of chords, planetary theory, star catalogue, geodesy, *Almagest*).

200 Probable date of inscriptions carved in the caves at Nasik.

300 Pap'pus (*Mathematical Collection*, commentaries, isoperimetry, projective invariance of cross ratio, Castillon-Cramer problem, arbelos theorem, generalization of Pythagorean theorem, centroid theorems, Pappus' theorem).

320 Iam'blichus (number theory).

390 The'on of Alexandria (commentator, edited Euclid's *Elements*).

410 Hypa'tia of Alexandria (commentator, first woman mentioned in the history of mathematics, daughter of Theon of Alexandria).

460 Pro'clus (commentator).

480 Chinese value of π as 355/113.

500 Metrodo'rus and the *Greek Anthology*.

505 Varāhamihira (Hindu astronomy).

510 Boethius (writings on geometry and arithmetic became standard texts in the monastic schools); Āryabhata the Elder (astronomy and arithmetic).

530 Simpli'cius (commentator).

560 Euto'cius (commentator).

622 Flight of Mohammed from Mecca.

628 Brahmagupta (algebra, cyclic quadrilaterals).

641 Last library at Alexandria burned.

710 Bede (calendar, finger reckoning).

711 Saracens invade Spain.

766 Brahmagupta's works brought to Bagdad.

775 Alcuin called to the court of Charlemagne.

790 Harun al-Rashid (caliph patron of learning).

820 Mohammed ibn Mûsâ al-Khowârizmî (wrote influential treatise on algebra and a book on the Hindu numerals, astronomy, "algebra," "algorism"); al-Mâmûn (caliph patron of learning).

850 Mahāvīra (arithmetic, algebra).

870 Tâbit ibn Qorra (translator of Greek works, conics, algebra, magic squares, amicable numbers).

871 Alfred the Great began his reign.

900 Abû Kâmil (algebra).

920 Al-Battânî, or Albategnius (astronomy).

980 Abû'l-Wefâ (geometric constructions with compasses of fixed opening, trigonometric tables).

1000 Alhazen (optics, geometric algebra); Gerbert, or Pope Sylvester II (arithmetic, globes).

1020 Al-Karkhî (algebra).

1042   Edward the Confessor became king.

1066   Norman Conquest.

1095   First Crusade.

1100   Omar Khayyam (geometric solution of cubic equations, calendar).

1120   Plato of Tivoli (translator from the Arabic); Adelard of Bath (translator from the Arabic).

1130   Jabir ibn Aflah, or Geber (trigonometry).

1140   Johannes Hispalensis (translator from the Arabic); Robert of Chester (translator from the Arabic).

1146   Second Crusade.

1150   Gherardo of Cremona (translator from the Arabic); Bhāskara (algebra, indeterminate equations).

1202   Fibonacci (arithmetic, algebra, geometry, Fibonacci sequence, *Liber abaci*).

1225   Jordanus Nemorarius (algebra).

1250   Sacrobosco (Hindu-Arabic numerals, sphere); Nasîr ed-dîn (trigonometry); Roger Bacon (eulogized mathematics); rise of European universities.

1260   Campanus (translation of Euclid's *Elements*, geometry).

1271   Marco Polo began his travels.

1303   Chu Shĭ-kié (algebra, numerical solution of equations).

1325   Thomas Bradwardine (arithmetic, geometry, star polygons).

1349   Black Death destroyed a large part of the European population.

1360   Nicole Oresme (coordinates, fractional exponents).

1431   Joan of Arc burned.

1435   Ulugh Beg (trigonometric tables).

1450   Nicholas Cusa (geometry, calendar reform); printing from movable type.

1453   Fall of Constantinople.

1460   Georg von Peurbach (arithmetic, astronomy, table of sines).

1470   Regiomontanus, or Johann Müller (trigonometry).

1478   First printed arithmetic, in Treviso, Italy.

1482   First printed edition of Euclid's *Elements*.

1484   Nicolas Chuquet (arithmetic, algebra); Borghi's arithmetic.

1489   Johann Widman (arithmetic, algebra, + and − signs).

1491   Calandri's arithmetic.

1492   Columbus discovered America.

1494   Pacioli (*Sūma*, arithmetic, algebra, double entry bookkeeping).

1500   Leonardo da Vinci (optics, geometry).

1506   Scipione del Ferro (cubic equation); Antonio Maria Fior (cubic equation).

1510   Albrecht Dürer (curves, perspective, approximate trisection, patterns for folding the regular polyhedra).

1514   Jakob Köbel (arithmetic).
1518   Adam Riese (arithmetic).
1521   Luther excommunicated.
1522   Tonstall's arithmetic.
1525   Rudolff (algebra, decimals); Stifel (algebra, number mysticism); Buteo (arithmetic).
1530   Da Coi (cubic equation); Copernicus (trigonometry, planetary theory).
1545   Ferrari (quartic equation); Tartaglia (cubic equation, arithmetic, science of artillery); Cardano (algebra).
1550   Rhaeticus (tables of trigonometric functions); Scheubel (algebra); Commandino (translator, geometry).
1556   First work on mathematics printed in the New World.
1557   Robert Recorde (arithmetic, algebra, geometry, = sign).
1558   Elizabeth became Queen of England.
1570   Billingsley and Dee (first English translation of the *Elements*).
1572   Bombelli (algebra, irreducible case of cubic equations).
1573   Valentin Otho found early Chinese value of π, namely 355/113.
1575   Xylander, or Wilhelm Holzmann (translator).
1580   François Viète, or Vieta (algebra, geometry, trigonometry, notation, numerical solution of equations, theory of equations, infinite product converging to $2/\pi$).
1583   Clavius (arithmetic, algebra, geometry, calendar).
1590   Cataldi (continued fractions); Stevin (decimal fractions, compound interest table, statics, hydrostatics).
1593   Adrianus Romanus (value of π, problem of Apollonius).
1595   Pitiscus (trigonometry).
1600   Thomas Harriot (algebra, symbolism); Jobst Bürgi (logarithms); Galileo (falling bodies, pendulum, projectiles, astronomy, telescopes, cycloid); Shakespeare.
1603   Accademia dei Lincei founded (Rome).
1608   Telescope invented.
1610   Kepler (laws of planetary motion, volumes, star polyhedra, principle of continuity); Ludolf van Ceulen (computation of π).
1612   Bachet de Méziriac (mathematical recreations, edited Diophantus' *Arithmetica*).
1614   Napier (logarithms, rule of circular parts, computing rods).
1615   Henry Briggs (common logarithms, tables)..
1619   Savilian professorships at Oxford established.
1620   Gunter (logarithmic scale, Gunter's chain in surveying); Paul Guldin (centroid theorems of Pappus); Snell (geometry, trigonometry, refinement of classical method of computing π, loxodromes).
1630   Mersenne (number theory, Mersenne numbers, clearinghouse for

mathematical ideas); Oughtred (algebra, symbolism, slide rule, first table of natural logarithms); Mydorge (optics, geometry); Albert Girard (algebra, spherical geometry).

1635   Fermat (number theory, maxima and minima, probability, analytic geometry, Fermat's last theorem); Cavalieri (method of indivisibles).

1637   Descartes (analytic geometry, $F + V = E + 2$, folium, ovals, rule of signs).

1640   Desargues (projective geometry); de Beaune (Cartesian geometry); Torricelli (physics, geometry, isogonic center); Frénicle de Bessy (geometry); Roberval (geometry, tangents, indivisibles); de la Loubère (curves, magic squares).

1643   Louis XIV crowned.

1649   Charles I executed.

1650   Blaise Pascal (conics, cycloid, probability, Pascal triangle, computing machines); John Wallis (algebra, imaginary numbers, arc length, exponents, symbol for infinity, infinite product converging to $\pi/2$, early integration); Frans van Schooten (edited Descartes and Viète); Grégoire de Saint-Vincent (circle squarer, other quadratures); Wingate (arithmetic); Nicolaus Mercator (trigonometry, astronomy, series computation of logarithms); John Pell (algebra, incorrectly credited with the so-called Pell equation).

1660   Sluze (spirals, points of inflection); Viviani (geometry); Brouncker (first president of Royal Society, rectification of parabola and cycloid, infinite series, continued fractions).

1662   Royal Society founded (London).

1663   Lucasian professorships at Cambridge established.

1666   French Academy founded (Paris).

1670   Barrow (tangents, fundamental theorem of the calculus); James Gregory (optics, binomial theorem, expansion of functions into series, astronomy); Huygens (circle quadrature, probability, evolutes, pendulum clocks, optics); Sir Christopher Wren (architecture, astronomy, physics, rulings on hyperboloid of one sheet, arc length of cycloid).

1671   Giovanni Domenico Cassini (astronomy, Cassinian curves).

1675   Greenwich observatory founded.

1680   Sir Isaac Newton (fluxions, dynamics, hydrostatics, hydrodynamics, gravitation, cubic curves, series, numerical solution of equations, challenge problems); Johann Hudde (theory of equations); Robert Hooke (physics, spring-balance watches); Seki Kōwa (determinants, calculus).

1682   Leibniz (calculus, determinants, multinomial theorem, symbolic logic, notation, computing machines).

1685 Kochanski (approximate rectification of circle).
1690 Marquis de l'Hospital (applied calculus, indeterminate forms); Halley (astronomy, mortality tables and life insurance, translator); Jakob (James, Jacques) Bernoulli (isochronous curves, clothoid, logarithmic spiral, probability); de la Hire (curves, magic squares, maps); Tschirnhausen (optics, curves, theory of equations).
1700 Johann (John, Jean) Bernoulli (applied calculus); Giovanni Ceva (geometry); David Gregory (optics, geometry); Parent (solid analytic geometry).

     * * * * *

1706 William Jones (first use of π for circle ratio).
1715 Taylor (expansion in series, geometry).
1720 De Moivre (actuarial mathematics, probability, complex numbers, Stirling's formula).
1731 Alexis Clairaut (solid analytic geometry).
1733 Saccheri (forerunner of non-Euclidean geometry).
1740 Maclaurin (higher plane curves, physics); Frederick the Great became King of Prussia.
1750 Euler (notation, $e^{i\pi} = -1$, Euler line, quartic equation, ϕ-function, beta and gamma functions, applied mathematics).
1760 Comte de Buffon (calculation of π by probability).
1770 Lambert (non-Euclidean geometry, hyperbolic functions, map projection, irrationality of π).
1776 United States independence.
1780 Lagrange (calculus of variations, differential equations, mechanics, numerical solution of equations, attempted rigorization of calculus (1797), theory of numbers).
1789 French Revolution.
1794 École Normale Supérieure, and École Polytechnique founded; Monge (descriptive geometry, differential geometry of surfaces).
1797 Mascheroni (geometry of compasses).
1804 Napoleon made emperor.
1805 Laplace (celestial mechanics, probability, differential equations); Legendre (*Éléments de géométrie*, theory of numbers, elliptic functions, method of least squares, integrals).
1810 Gergonne (geometry, editor of *Annales*).
1820 Gauss (polygon construction, number theory, differential geometry, non-Euclidean geometry, fundamental theorem of algebra, astronomy, geodesy); Poinsot (geometry).
1822 Feuerbach (geometry of the triangle).
1824 **Thomas Carlyle (English translation of Legendre's *Géométrie*).**
1825 Bolyai and Lobachevsky (non-Euclidean geometry); Abel (elliptic functions).

1826 *Crelle's Journal.*

1830 Cauchy (rigorization of analysis, functions of a complex variable); Poncelet (projective geometry, ruler constructions); Galois (groups, theory of equations); Babbage (computing machines).

1836 *Liouville's Journal.*

1837 Trisection of an angle and duplication of cube proved impossible.

1839 *Cambridge Mathematical Journal,* which in 1855 became *Quarterly Journal of Pure and Applied Mathematics.*

1840 Steiner (geometry).

1841 *Archiv der Mathematik und Physik.*

1842 *Nouvelles annales de mathématiques.*

1847 Staudt (freed projective geometry of metrical basis).

1850 Chasles (higher geometry, history of geometry); Cayley (invariants, matrices and determinants, hyperspace); H. G. Grassmann (calculus of extension); Plücker (higher analytic geometry); Mannheim (standardized the modern slide rule).

1854 Riemann (Riemann surfaces, Riemann integral, Riemannian geometry).

1855 Zacharias Dase (lightning calculator).

1865 London Mathematical Society founded; *Proceedings of London Mathematical Society.*

1872 Société Mathématique de France founded; Klein's *Erlanger Programm;* Dedekind (irrational numbers).

1873 Brocard (geometry of the triangle).

1878 *American Journal of Mathematics.*

1880 Georg Cantor (irrational numbers, transcendental numbers, transfinite numbers).

1882 Lindemann (transcendence of π, squaring of circle proved impossible).

1884 Circolo Matematico di Palermo founded.

1887 *Rendiconti.*

1888 Lemoine (geometry of the triangle, geometrography); American Mathematical Society founded (at first under a different name); *Bulletin of the American Mathematical Society.*

1890 Weierstrass (arithmetization of mathematics); Deutsche Mathematiker-Vereinigung organized.

1892 *Jahresbericht.*

1900 *Transactions of American Mathematical Society.*

Answers and Suggestions for the Solution of the Problem Studies

1.2 (a). 27, 3, 2.

1.2 (b). $5780 = \epsilon'\psi\pi$, $72,803 = \zeta M\beta'\omega\gamma$, $450,082 = \mu M\epsilon M\pi\beta$, $3,257,888 = \tau M\kappa M\epsilon M\zeta'\omega\pi\eta$.

1.3 (c). $360 = 2(5^3) + 4(5^2) + 2(5) = (()))^{**}$,
$$252 = 2(5^3) + \cdot 2(1) = ((//,$$
$$78 = 3(5^2) + 3(1) =)))///,$$
$$33 = 1(5^2) + 1(5) + 3(1) =)*///.$$

1.3 (d). $360 = *(*\#, 252 = *\#\#*, 78 =)\#), 33 = //).$

1.4 (a). Note that $ab = [(a - 5) + (b - 5)]10 + (10 - a)(10 - b)$.

1.5 (b). Multiply the decimal fraction by b, then the decimal part of this product by b, and so forth.

1.5 (c). $(.3012)_4 = 99/128 = .7734375$.

1.7 (a). First express to base 10, and then to base 8.

1.7 (b). 9, 8, 7.

1.7 (c). no, yes, yes, no.

1.7 (d). In the first case, we have $79 = b^2 + 4b + 2$.

1.7 (e). Denoting the digits by a, b, c we have $49a + 7b + c = 81c + 9b + a$, where a, b, c are less than 7.

1.7 (f). We must have $3b^2 + 1 = t^2$, t and b positive integers, $b > 3$.

1.8 (a). Express w in the binary scale.

1.9 (a). Let t be the tens digit and u the units digit. Following instructions we have

$$2(5t + 7) + u = (10t + u) + 14$$

as the announced final result. The trick is now obvious.

2.1 (a). Suppose n is regular. Then

$$1/n = a_0 + a_1/60 + \cdots + a_r/60^r$$
$$= (a_0 60^r + a_1 60^{r-1} + \cdots + a_r)/60^r$$
$$= m/60^r, \text{ say.}$$

It follows that $mn = 60^r$ and n can have no prime factors other than those of 60.

2.2 (a).   We have $(1.2)^x = 2$, whence $x = (\log 2)/(\log 1.2)$.

2.3 (a).   We have $x^2 + y^2 = 1000$, $y = 2x/3 - 10$.

2.4 (b).   Set $x = 2y$.

2.4 (c).   Eliminate x and y, obtaining a cubic equation in z.

2.4 (d).   Take the cubic in x with unit leading coefficient and subject it to a linear transformation of the type $x = y + m$. Determine m so that the resulting cubic in y lacks the linear term.

2.5 (a).   Find the first two terms of the binomial expansion of $(a^2 + h)^{1/2}$.

2.6 (b).   Express, in the binary scale, the factor that is successively halved.

2.8 (a).   By *fraction*, Ahmes means *unit fraction*. Only the denominators of unit fractions were written.

2.8 (c).   Let x be the largest share and d the common difference in the arithmetic progression. Then we find $5x - 10d = 100$ and $11x - 46d = 0$.

2.9 (a).   256/81, or, approximately, 3.16.

2.9 (b).   Consider the right triangle T_1 with legs a and b, and any other triangle T_2 with sides a and b. Place T_2 on T_1 so that one pair of equal sides coincide. Or use the formula $K = \frac{1}{2}ab \sin C$.

2.9 (c).   Draw the diagonal DB and use 2.9 (b).

2.9 (d). $(a + c)(b + d)/4 = [(ad + bc)/2 + (ab + cd)/2]/2$. Now use 2.9 (c).

2.10 (b).   Start with $\sqrt{m} - \sqrt{n} \geqq 0$.

2.10 (c).   Complete the pyramid of which the frustum is a part and express the volume of the frustum as the difference between the volumes of the completed and added pyramids.

2.11 (a).   The magic constant $= (1 + 2 + 3 + \cdots + n^2)/n$.

2.11 (c).   Denote the numbers of the magic square by letters and then add together the letters of the middle row, the middle column, and the two diagonals.

2.11 (d).   Use 2.11 (c) and an indirect argument.

2.12   A simple solution is indicated by a figure which appears in the *Chôu-peï*, the oldest known Chinese mathematical work, and which may date back to the second millennium B.C. Show that four right triangles having legs of lengths 3 and 4 may be placed, as shown in Figure 83, to form a square whose area is 25. Then the hypotenuse of the right triangle must be 5. Since a triangle is determined by its three sides, it follows that a 3, 4, 5 triangle is a right triangle.

3.2 (a)   Show that $2^{mn} - 1$ contains the factor $2^m - 1$.

3.2 (b).   8128.

3.2 (c).   If $a_1, a_2, \cdots, a_n$ represent all the divisors of N, then N/a_1, $N/a_2, \cdots, N/a_n$ also represent all the divisors of N.

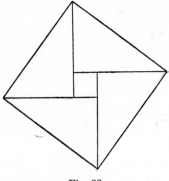

Fig. 83

3.2 (d).   The sum of the proper divisors of p^n is $(p^n - 1)/(p - 1)$.

3.3 (a).   1, 6, 15, 28.

3.3 (b).   An oblong number is a number of the form $a(a + 1)$.

3.3 (d).

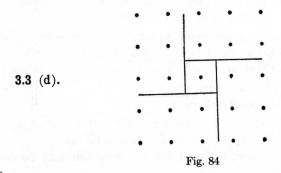

Fig. 84

3.4 (a).   Use the fact that $(a - b)^2 \geqq 0$.

3.4 (e).   A cube has 8 vertices, 12 edges, and 6 faces.

3.6 (c).   If there were an isosceles right triangle with integral sides then $\sqrt{2}$ would be rational.

3.6 (d).   If there are positive integers a, b, c $(a \neq 1)$ such that $a^2 + b^2 = c^2$ and $b^2 = ac$, then a, b, c cannot be relatively prime. But if there is a Pythagorean triple in which one integer is a mean proportional between the other two, there must be a primitive Pythagorean triple of this sort.

3.7 (c).   Assume $\sqrt{p} = a/b$, where a and b are relatively prime.

3.7 (d). Assume $\log_{10} 2 = a/b$, where a and b are integers. Then we must have $10^a = 2^b$, which is impossible.

3.9 (b). ab is the fourth proportional to $1, a, b$.

3.9 (c). a/b is the fourth proportional to $b, 1, a$.

3.9 (d). $\sqrt{a}$ is a mean proportional between 1 and a.

3.10 (a). Obtain $\sqrt{12}$ as in 3.9 (d).

3.10 (c). Denote the parts by x and $a - x$. Then $x^2 - (a - x)^2 = x(a - x)$, or $x^2 + ax - a^2 = 0$.

3.10 (e). Show that $OM + ON = g$ and $(OM)(ON) = h$.

3.11 (b). First trisect the diagonal BD by points E and F. Then the broken lines AEC and AFC divide the figure into three equivalent parts. Transform these parts so as to fulfill the conditions by drawing parallels to AC through E and F.

3.11 (d). Through B draw BD parallel to MN to cut AC in D. Then, if the required triangle is $AB'C'$, AC' is a mean proportional between AC and AD.

3.11 (e). Let ABC be the given triangle. Draw AB' making the given vertex angle with AC and let it cut the parallel to AC through B in B'. Now use 3.11 (d).

3.12 (a). A convex polyhedral angle must contain at least three faces, and the sum of its face angles must be less than $360°$.

4.1 (b). Let A be the given point and BC the given line segment. Construct by Proposition 1 an equilateral triangle ABD. Draw circle $B(C)$ and let DB produced cut this circle in G. Now draw circle $D(G)$ to cut DA produced in L. Then AL is the sought segment.

4.1 (c). Use Proposition 2 of Book I.

4.2 (a). The equations of the parabolas may be taken as $x^2 = sy$ and $y^2 = 2sx$, where s and $2s$ are the latera recta of the parabolas.

4.2 (b). The equations of the parabola and hyperbola may be taken as $x^2 = sy$ and $xy = 2s^2$.

4.3 (a). Let M be the mid-point of OA and let E be the center of the rectangle $OADB$. Then, by Proposition 6, Bk. II (see Section 3-6), $(OA')(AA') + (MA)^2 = (MA')^2$. Adding $(ME)^2$ to both sides we find $(OA')(AA') + (EA)^2 = (EA')^2$. Similarly, $(OB')(BB') + (EB)^2 = (EB')^2$. Therefore $(OA')(AA') = (OB')(BB')$.

4.4 (a). We have $r = P_1P_2 = AP_1 \tan \theta = 2a \sin \theta \tan \theta$. It follows that $r = 2a(y/r)(y/x)$, or $r^2x = 2ay^2$.

4.4 (b). Denote the coordinates of P by (x,y). Then $(AQ)^3/(OA)^3 = y^3/x^3 = y/(2a - x) = RP/RA = OD/OA = n$.

4.4 (c). Let S be the foot of the perpendicular from R on MN, and let T be the mid-point of RS. Draw the circle $S(T)$ to cut TP in U. Then $SCPU$ is a parallelogram. Let TP cut MN in V and the tangent to $S(T)$ at the point Q diametrically opposite to T in W. Triangles SUV and APV are congruent, and $UV = VP$. It is now easy to show that $TP = UW$. Thus P lies on the cissoid of $S(T)$ and QW for the pole T.

4.5 (a). The equation of the hyperbola, referred to its asymptotes as coordinate axes, is $xy = ab$, where $(b/2, a/2)$ is the center of the rectangle. The equation of the circumcircle of the rectangle is $x^2 + y^2 - ay - bx = 0$. The point of intersection of the hyperbola and circle, other than the point (b,a), is $(\sqrt[3]{a^2b}, \sqrt[3]{ab^2})$. But $\sqrt[3]{a^2b}$ and $\sqrt[3]{ab^2}$ are the mean proportionals between a and b.

4.6 (a). Denote AB by a, AC by b, BC by c, and angle ADB by θ. Then, by the law of sines, applied first to triangle BCD and then to triangle ABD, $\sin 30°/\sin\theta = a/c$, $\sin\theta/\sin 120° = a/(b + a)$. Consequently $1/\sqrt{3} = \tan 30° = a^2/c(b + a)$. Squaring both sides and recalling that $c^2 = b^2 - a^2$, we find $2a^3(2a + b) = b^3(2a + b)$, or $b^3 = 2a^3$.

4.6 (b). Draw CO and use the fact that an exterior angle of a triangle is equal to the sum of the two remote interior angles.

4.7 (a). Let R be the foot of the perpendicular from Q on the x-axis and let RQ cut c in S. Then $OQ/RQ = PQ/SQ$.

4.7 (b). See 4.6 (a).

4.7 (c). See 4.6 (b).

4.7 (d). See 4.4.

4.8 (a). Let Q and N be the feet of the perpendiculars from P and M on OA, and let QP cut OM in S. Since P and R are on the hyperbola we have $(OQ)(QP) = (ON)(NR)$, or $NR = (OQ)(NM)/ON$. But, from similar triangles OQS and ONM, $QS = (OQ)(NM)/ON$. It follows that $SRMP$ is a rectangle. If T is the center of this rectangle, $OP = PT = TM$.

4.8 (b). Take radius $OA = 1$ and denote angle AOB by 3θ. Take P on arc AB such that angle $AOP = 1/3$ angle AOB, and let Q be the foot of the perpendicular from P on OC. Then $AP = 2\sin\theta/2 = 2\,PQ$.

4.9 (a). Use the fact that the sum of the infinite geometric series $1/2 - 1/4 + 1/8 - 1/16 + \cdots$ is $1/3$. For another asymptotic Euclidean solution of the trisection problem see *American Mathematical Monthly*, Dec. 1945, Problem 4134, pp. 587–589.

4.10 (a). We have angle $AOP = k\pi/2$ when $OM = k(OA) = k$. Therefore, if we denote the coordinates of P by (x,y), $y = k = x\tan(k\pi/2) = x\tan(\pi y/2)$.

4.10 (c).   Let the quadratrix cut OA in Q. Then

$$OQ = \lim_{y \to 0} \{y/\tan(\pi y/2)\} = 2/\pi,$$

by l'Hospital's rule in calculus. It is now easy to show that

$$\overset{\frown}{AC} : OA = OA : AQ.$$

4.11 (a).   3.1414.

4.11 (b).   3.14153.

4.11 (c).   We have $GB/BA = EF/FA = (DE)^2/(DA)^2 = (DE)^2/[(BA)^2 + (BC)^2]$. Therefore $GB = 4^2/(7^2 + 8^2) = 16/113 = 0.1415929\cdots$. This leads to 355/113 as an approximation of π.

4.13 (b).   Let $\alpha = \tan^{-1}(1/5)$ and $\beta = \tan^{-1}(1/239)$. Then show that $4\alpha - \beta = \pi/4$ by showing that $\tan(4\alpha - \beta) = 1$.

4.13 (c).   Consider a circle of unit radius. Then the side of an inscribed square is given by $\sec\theta$, where $\theta = 45°$. The sum of 2 sides of a regular inscribed 8-gon is given by $\sec\theta \sec\theta/2$; the sum of 4 sides of a regular inscribed 16-gon is given by $\sec\theta \sec\theta/2 \sec\theta/4$; and so forth. It follows that

$$\sec\theta \sec\theta/2 \sec\theta/4 \cdots \to \pi/2,$$

the length of a quadrant of the circle. Therefore

$$2/\pi = \cos\theta \cos\theta/2 \cos\theta/4 \cdots.$$

Now use the fact that $\cos\theta = \sqrt{2}/2$ and $\cos\theta/2 = [(1 + \cos\theta)/2]^{1/2}$, $\cos\theta/4 = [(1 + \cos\theta/2)/2]^{1/2}$, and so forth.

4.13 (d).   Set $x = \sqrt{1/3}$ in Gregory's series.

4.15 (a). $\cos\theta = \cos(2\theta/3 + \theta/3)$.

4.15 (d).   Let $7\theta = 360°$. Then $\cos 3\theta = \cos 4\theta$, or, setting $x = \cos\theta$, $8x^3 + 4x^2 - 4x - 1 = 0$.

4.15 (e).   Take $AOB = 90°$ and let M and N be the feet of the perpendiculars from P on OA and OB. Let R be the center of the rectangle $OMPN$. Now if CD is Philon's line for angle AOB and point P, show that $RE = RP$, and hence that $RD = RC$. We now have Apollonius' solution of the duplication problem (see 4.3).

5.1 (c). Suppose $a > b$. Then the algorism may be summarized as follows:

$$a = q_1 b + r_1 \qquad\qquad 0 < r_1 < b$$

$$b = q_2 r_1 + r_2 \qquad\qquad 0 < r_2 < r_1$$

$$r_1 = q_3 r_2 + r_3 \qquad\qquad 0 < r_3 < r_2$$

$$\cdot\ \cdot\ \cdot\ \cdot\ \cdot \qquad\qquad \cdot\ \cdot\ \cdot\ \cdot\ \cdot$$

$$r_{n-2} = q_n r_{n-1} + r_n \qquad\qquad 0 < r_n < r_{n-1}$$

$$r_{n-1} = q_{n+1} r_n$$

Now, from the last step, r_n divides r_{n-1}. From next to the last step r_n divides r_{n-2}, since it divides both terms on the right. Similarly r_n divides r_{n-3}. Successively, r_n divides each r_i, and finally a and b.

On the other hand, from the first step, any common divisor c of a and b divides r_1. From the second step, c then divides r_2. Successively, c divides each r_i. Thus c divides r_n.

5.1 (d). From the next to the last step in the algorism we can express r_n in terms of r_{n-1} and r_{n-2}. From the preceding step we can then express r_n in terms of r_{n-2} and r_{n-3}. Continuing this way we finally obtain r_n in terms of a and b.

5.2 (a). If p does not divide u, then integers P and Q exist such that $Pp + Qu = 1$, or $Ppv + Quv = v$.

5.2 (b). Suppose there are two prime factorizations of the integer n. If p is one of the prime factors in the first factorization it must, by 5.2 (a), divide one of the factors in the second factorization, that is, coincide with one of the factors.

5.2 (c). We note that $273 = (13)(21)$. Find [see 5.1 (e)] integers p and q such that $13p + 21q = 1$. Dividing by 273 we then have $p/21 + q/13 = 1/273$. Similarly find integers r and s such that $1/21 = r/3 + s/7$.

5.3 (c). For [see 5.1 (f)] there exist positive integers p and q such that $pr - qs = \pm 1$. Then the difference between the angle subtended at the center of the r-gon by q of its sides is

$$p(360°/s) - q(360°/r) = (pr - qs)(360°/rs) = \pm 360°/rs.$$

5.4 (c). Let ABC be the given triangle and let XY, parallel to BC, cut AB in X and AC in Y. Draw BY and CX. Show that $\triangle BXY : \triangle AXY = \triangle CXY : \triangle AXY$. But, by VI 1, $\triangle BXY : \triangle AXY = BX : XA$ and $\triangle CXY : \triangle AXY = CY : YA$.

5.5 (c). For each b_i in 5.5 (b) may have $a_i + 1$ values.

5.5 (f). Since b divides ac, we have $b_i \leqq a_i + c_i$. Also, since a and b are relatively prime, we have $a_i = 0$ or $b_i = 0$. In either case $b_i \leqq c_i$.

5.5 (h). Suppose $\sqrt{2} = a/b$, where a and b are positive integers. Then, since $a^2 = 2b^2$, we have $(2a_1, 2a_2, \cdots) = (1 + 2b_1, 2b_2, \cdots)$, whence $2a_1 = 1 + 2b_1$, which is impossible.

5.6 (a). Let M be the mid-point of the base AB. Draw DM and CM.

5.6 (c). Drop a perpendicular from the vertex of the triangle upon the line joining the mid-points of the two sides of the triangle.

5.9. T 1. Suppose we have both $R(a,b)$ and $R(b,a)$. Then, by P 3, we have $R(a,a)$. But this is impossible by P 2. Hence the theorem by *reductio ad absurdum*.

T 2. Since $c \neq a$ we have, by P 1, either $R(a,c)$ or $R(c,a)$. If we have $R(c,a)$, since we also have $R(a,b)$, we have, by P 3, $R(c,b)$. Hence the theorem.

T 3. Suppose the theorem is false and let a be any element of K. Then there exists an element b of K such that we have $R(a,b)$. By P 2, $a \neq b$. Thus a and b are distinct elements of K.

By our supposition there exists an element c of K such that we have $R(b,c)$. By P 2, $b \neq c$. By P 3 we also have $R(a,c)$. By P 2, $a \neq c$. Thus a, b, c are distinct elements of K.

By our supposition there exists an element d of K such that we have $R(c,d)$. By P 2, $c \neq d$. By P 3 we also have $R(b,d)$ and $R(a,d)$. By P 2, $b \neq d$, $a \neq d$. Thus a, b, c, d are distinct elements of K.

By our supposition there exists an element e of K such that we have $R(d,e)$. By P 2, $d \neq e$. By P 3 we also have $R(c,e)$, $R(b,e)$, $R(a,e)$. By P 2, $c \neq e$, $b \neq e$, $a \neq e$. Thus a, b, c, d, e are distinct elements of K.

We now have a contradiction of P 4. Hence the theorem by *reductio ad absurdum*.

T 4. By T 3 there is at least one such element, say a. Let $b \neq a$ be any other element of K. By P 1 we have either $R(a,b)$ or $R(b,a)$. But, by hypothesis, we do not have $R(a,b)$. Therefore we must have $R(b,a)$, and the theorem is proved.

T 5. By Definition 1 we have $R(b,a)$ and $R(c,b)$. By P 3 we then have $R(c,a)$, or, by Definition 1, we have $D(a,c)$.

T 6. Suppose $a \neq b$. Then, by P 1, we have either $R(a,b)$ or $R(b,a)$. Suppose we have $R(a,b)$. Since we have $F(b,c)$ we also have, by Definition 2, $R(b,c)$. This is impossible since we are given that we have $F(a,c)$. Suppose we have $R(b,a)$. Since we have $F(a,c)$ we also have, by Definition 2, $R(a,c)$.

This is impossible since we are given that we have $F(b,c)$. Thus in either case we are led to a contradiction of our hypothesis. Hence the theorem by *reductio ad absurdum*.

T 7. For, by Definition 2, we have $R(a,b)$ and $R(b,c)$. Hence, by Definition 2, we cannot have $F(a,c)$.

5.10 (b). T 1. If a is an ancestor of b, then b is not an ancestor of a.

T 2. If a is an ancestor of b and if c is some third member of K distinct from a and b, then either a is an ancestor of c or c is an ancestor of b.

T 3. There is some man in K who is not an ancestor of anyone in K.

T 4. There is only one man in K who is not an ancestor of anyone in K.

Definition 1. If b is an ancestor of a we say that a is a *descendent* of b.

T 5. If a is a descendent of b, and b is a descendent of c, then a is a descendent of c.

Definition 2. If a is an ancestor of b and there is no individual c of K such that a is an ancestor of c and c is an ancestor of b, then we say that a is a *father* of b.

T 6. A man in K has at most one father in K.

T 7. If a is the father of b and b is the father of c, then a is not the father of c.

Definition 3. If a is the father of b and b is the father of c, we say that a is a *grandfather* of c.

5.10 (d). Since T 1 has been deduced from P 1, P 2, P 3, P 4, all that remains is to deduce P 2 from P 1, T 1, P 3, P 4.

5.11 (b). The *converse* of "If A then B" is "If B then A."

5.11 (c). The *opposite* of "If A then B" is "If not A then not B."

5.11 (d). The *contradictory* of "If A then B" is "If A then not B."

5.13 (c). By 5.13 (b) the problem is reduced to that of constructing a triangle given a, angle A, and $b + c$. This may be accomplished by constructing a triangle DBC where $BC = a$, $BD = b + c$, and angle $D =$ (angle A)/2. The perpendicular bisector of CD will then cut BD in point A such that triangle ABC is the sought triangle.

6.2 (e). The volume of the segment is equal to the volume of a spherical sector minus the volume of a cone. Also, $a^2 = h(2R - h)$.

6.2 (f). The segment is the difference of two segments, each of one base, and having, say, altitudes u and v. Then

$$V = \pi R(u^2 - v^2) - \pi(u^3 - v^3)/3$$

$$= \pi h[(Ru + Rv) - (u^2 + uv + v^2)/3].$$

But $u^2 + uv + v^2 = h^2 + 3uv$ and $(2R - u)u = a^2$, $(2R - v)v = b^2$. Therefore

$$V = \pi h[(a^2 + b^2)/2 + (u^2 + v^2)/2 - h^2/3 - uv]$$
$$= \pi h[(a^2 + b^2)/2 + h^2/2 + uv - h^2/3 - uv], \text{and so forth.}$$

6.4 (a). $(GC)^2 = (TW)^2 = 4\, r_1 r_2$.

6.5 (b). If p is composite, then $p = ab$, where $a \leq b$ and, consequently, $a^2 \leq p$.

6.5 (c). For $n = 10^9$ we have $(A_n \log_e n)/n = 1.053 \cdots$.

6.5 (d). Consider $(n + 1)\,! + 2$, $(n + 1)\,! + 3, \cdots$, $(n + 1)\,! + (n + 1)$.

6.7 (b). Let A and B be points, and C a straight line. Produce AB to cut line C in S. Now find T on line C such that $(ST)^2 = (SA)(SB)$. In general there are two solutions.

6.7 (c). Reflect the given point in a bisector of the angle determined by the two given lines.

6.7 (d). Reflect the focus F in the line m, obtaining a point F'. Now, by 6.7 (b), find the centers of the circles passing through F and F' and touching the given directrix.

6.8 (b). For problem (1) take A and B on the x-axis and reflections of one another in the origin.

6.8 (c). (1) Let the interior and exterior bisectors of angle APB cut AB in M and N. Then M and N are on the required locus and angle MPN is a right angle.

(2) Let A and B be the fixed points, P the moving point, and O the mid-point of AB. Add the expressions for $(PA)^2$ and $(PB)^2$ as given by the law of cosines applied to triangles PAO and PBO.

6.9 (a). Let $ABCD$ be the cyclic quadrilateral. Find E on diagonal AC such that $\angle ABE = \angle DBC$. From similar triangles ABE and DBC obtain $(AB)(DC) = (AE)(BD)$. From similar triangles ABD and EBC obtain $(AD)(BC) = (EC)(BD)$.

6.9 (b). In part (a) take AC as a diameter, $BC = a$, and $CD = b$.

6.9 (c). In part (a) take AB as a diameter, $BD = a$, and $BC = b$.

6.9 (d). In part (a) take AC as a diameter, $BD = t$ and perpendicular to AC.

6.11 (b). Let a ray of light emanating from a point A hit the mirror at point M and reflect toward a point B. If B' is the image of B in the mirror, then BB' is perpendicularly bisected by the plane of the mirror, and we must have AMB' a straight line.

6.11 (c).   Apply 6.11 (b).

6.12 (b).   Show, from a figure, that $ab = 2rs$ and $a + b = r + s$, and then solve simultaneously.

6.13 (a).   120 apples.

6.13 (b).   60 years old.

6.13 (c).   960 talents.

6.13 (d).   Each Grace had $4n$ apples, gave away $3n$, and had n left.

6.14 (a).   2/5 of a day.

6.14 (b).   144/37 hours.

6.14 (c).   30.5 minae of gold, 9.5 minae of copper, 14.5 minae of tin, and 5.5 minae of iron.

6.15 (a).   84 years old.

6.15 (b).   7, 4, 11, 9.

6.15 (c).   Set $CD = 3x$, $AC = 4x$, $AD = 5x$, $CB = 3y$. Then, since $AB/DB = AC/CD$, we find $AB = 4(y - x)$. By the Pythagorean theorem we are led to $7y = 32x$. We finally get $AB = 100$, $AD = 35$, $AC = 28$, $BD = 75$, $DC = 21$.

6.15 (d).   1806.

6.16 (a). $481 = 20^2 + 9^2 = 16^2 + 15^2$.

6.16 (b).   We have $5 = 2^2 + 1^2$, $13 = 3^2 + 2^2$, $17 = 4^2 + 1^2$. By using the identities of 6.16 (a) we find

$$(5)(13) = 8^2 + 1^2 = 7^2 + 4^2,$$

$$(5)(17) = 9^2 + 2^2 = 7^2 + 6^2,$$

$$(13)(17) = 14^2 + 5^2 = 11^2 + 10^2.$$

Again, by the identities of 6.16 (a), we find

$$1105 = 33^2 + 4^2 = 32^2 + 9^2 = 31^2 + 12^2 = 24^2 + 23^2.$$

6.17 (a).   From the similar triangles DFB and DBO, $FD/DB = DB/OD$. Therefore $FD = (DB)^2/OD = 2(AB)(BC)/(AB + BC)$.

6.17 (b).   From similar right triangles, $OA/OB = AF/BD = AF/BE = AC/CB = (OC - OA)/(OB - OC)$. Now solve for OC.

6.17 (c).   Let HA cut BC in R and LM in S, and let LB cut DH in U and MC cut FH in V. Then $\square ABDE = \square ABUH = \square BRSL$, and $\square ACFG = \square ACVH = \square RCMS$.

6.17 (d).   An analytic solution is easy if we recall that the coordinates of a point dividing the segment joining points (a,b) and (c,d) in the ratio

m/n are $(na + mc)/(m + n)$ and $(nb + md)/(m + n)$ and that the co-ordinates of the centroid of the triangle determined by the points (a,b), (c,d), (e,f) are $(a + c + e)/3$ and $(b + d + f)/3$.

A synthetic solution is not so easy. One, due to Fuhrmann, is given in R. A. Johnson, *Modern Geometry* (Boston: Houghton Mifflin Company, 1929), Section 276, p. 175.

6.18 (a). $V = 2\pi^2 r^2 R$, $S = 4\pi^2 rR$.

6.18 (b).   The centroid of the semicircular arc lies on the bisecting radius of the semicircle and at the distance $2r/\pi$, where r is the radius of the semicircle, from the diameter of the semicircle.

6.18 (c).   The centroid of the semicircular area lies on the bisecting radius of the semicircle and at the distance $4r/3\pi$, where r is the radius of the semicircle, from the diameter of the semicircle.

7.1 (a). $x = hd/(2h + d)$.

7.1 (b).   8 cubits and 10 cubits.

7.1 (c).   40.

7.2 (a).   8 days.

7.2 (b).   18 mangoes.

7.2 (c).   8 for a citron and 5 for a wood apple.

7.2 (d).   36 camels.

7.3 (a).   72 bees.

7.3 (b).   20 cubits.

7.3 (c).   22/7 yojanas.

7.3 (d).   10 !, 4 !.

7.3 (e).   100 arrows.

7.4 (a).   Suppose $\sqrt{a} = b + \sqrt{c}$. Then $\sqrt{c} = (a - b^2 - c)/2b$.

7.4 (b).   If $a + \sqrt{b} = c + \sqrt{d}$, then $\sqrt{b} = (c - a) + \sqrt{d}$. Now use 7.4 (a).

7.5 (b).   It is easily shown that $x = x_1 + mb$ and $y = y_1 - ma$ constitute a solution. Conversely, assume x and y form a solution. Then $a(x - x_1) = b(y_1 - y)$, or $x - x_1 = mb$ and $y_1 - y = ma$.

7.5 (c).   Dividing by 7 we find

$$x + 2y + 2y/7 = 29 + 6/7.$$

Therefore there exists an integer z such that

$$2y/7 + z = 6/7,$$

or

$$2y + 7z = 6.$$

This can be solved by inspection to give $z_1 = 0$, $y_1 = 3$. Then $x_1 = 23$. The general solution of the original equation is then, by 7.5 (b),

$$x = 23 + 16m, \qquad y = 3 - 7m.$$

Since, by requirement, $x > 0$, $y > 0$, we must have $m \geq -1$ and $m \leq 0$. The only permissible values for m are 0 and -1. We thus get two solutions

$$x = 23,\, y = 3 \quad \text{and} \quad x = 7,\, y = 10.$$

Or find, as in 5.1 (f), p and q such that $7p + 16q = 1$. Then we may take $x_1 = 209p$ and $y_1 = 209q$.

7.5 (d). There are the four solutions $x = 124,\, y = 4$; $x = 87,\, y = 27$; $x = 50,\, y = 50$; $x = 13,\, y = 73$.

7.5 (e). Let x represent the number of dimes and y the number of quarters. Then we must have $10x + 25y = 500$.

7.5 (f). Let x denote the number of fruits in a pile and y the number of fruits each traveler receives. Then we have $63x + 7 = 23y$. The smallest permissible value for x is 5.

7.6 (a). Draw the circumdiameter through the vertex through which the altitude passes, and use similar triangles.

7.6 (b). Apply 7.6 (a) to triangles DAB and DCB.

7.6 (c). Use the result of 7.6 (b) along with Ptolemy's relation, $mn = ac + bd$.

7.6 (d). Here $\theta = 0°$ and $\cos \theta = 1$. Now use 7.6 (b) and (c).

7.7 (b). Since the quadrilateral has an incircle we have $a + c = b + d = s$. Therefore $s - a = c$, $s - b = d$, $s - c = a$, $s - d = b$.

7.7 (c). In Figure 85 we have

$$a^2 + c^2 = r^2 + s^2 + m^2 + n^2 - 2(rn + sm) \cos \theta,$$

$$b^2 + d^2 = r^2 + s^2 + m^2 + n^2 + 2(sn + rm) \cos \theta.$$

Therefore $a^2 + c^2 = b^2 + d^2$ if and only if $\cos \theta = 0$, or $\theta = 90°$.

7.7 (d). Use 7.7 (c).

7.7 (e). The consecutive sides of the quadrilateral are 39, 60, 52, 25; the diagonals are 56 and 63; the circumdiameter is 65; the area is 1764.

7.9 (a). We shall indicate a proof of the theorem for the four-digit number N having a, b, c, d for its thousands, hundreds, tens, and units digits; the proof is easily generalized. Now

$$N = 1000a + 100b + 10c + d.$$

Let $S = a + b + c + d$. Then

$$N = 999a + 99b + 9c + S = 9(111a + 11b + c) + S,$$

and so forth.

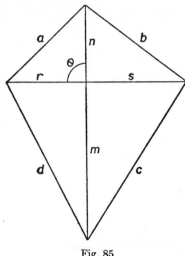

Fig. 85

7.9 (b).   Let M and N be any two numbers with excesses e and f. Then there exist integers m and n such that

$$M = 9m + e, \qquad N = 9n + f.$$

Now

$$M + N = 9(m + n) + (e + f)$$

and

$$MN = 9(9mn + ne + mf) + ef,$$

and so forth.

7.9 (d).   Let M be the given number and N that obtained by some permutation of the digits of M. Then, since M and N consist of the same digits, they have [by 7.9 (a)] the same excess e. Thus we have

$$M = 9m + e, \qquad N = 9n + e,$$

and

$$M - N = 9(m - n).$$

7.9 (e). By 7.9 (d) the final product must be divisible by 9, whence, by 7.9 (a), the excess for the sum of the digits in the product must be 0.

7.9 (f). Replace 9 by $b - 1$.

7.10 (b). $x = 2.3696$.

7.10 (c). $x = 4.4934$.

7.11 (a). Find z such that $b/a = a/z$, then m such that $n/z = a/m$.

7.12 (a). Draw any circle Σ on the sphere and mark any three points A, B, C on its circumference. On a plane construct a triangle congruent to triangle ABC, find its circumcircle, and thus obtain the radius of Σ. Construct a right triangle having the radius of Σ as one leg and the polar chord of Σ as hypotenuse. It is now easy to find the diameter of the given sphere.

7.12 (b). If d is the diameter of the sphere and e the edge of the inscribed cube, then $e = (d \sqrt{3})/3$, whence e is one-third the altitude of an equilateral triangle of side $2d$.

7.12 (c). If d is the diameter of the sphere and e the edge of the inscribed regular tetrahedron, then $e = (d \sqrt{6})/3$, whence e is the hypotenuse of a right isosceles triangle with leg equal to the edge of the inscribed cube. See 7.12 (b).

8.1 (a). Let x, y, z denote the number of men, women, and children. Then we must have

$$6x + 4y + z = 200 \quad \text{and} \quad x + y + z = 100,$$

or $5x + 3y = 100$. It follows that y must be a multiple of 5, say $5n$. Then $x = 20 - 3n$ and $z = 80 - 2n$. One easily finds that the only permissible values for n are 1, 2, 3, 4, 5, 6. The solution given in Alcuin's collection corresponds to $n = 3$, namely 11 men, 15 women, 74 children.

8.1 (b). It is easily shown that each son must receive the same number of entirely empty flasks as full ones. There are many solutions.

8.1 (c). Let x denote the required number of leaps. Then $9x - 7x = 150$.

8.1 (d). Find two solutions. For other problems of this sort see Maurice Kraitchik, *Mathematical Recreations*, W. W. Norton & Company, Inc., N. Y., 1942, pp. 214–222.

8.1 (e). How about 5/27 to the mother, 15/27 to the son, and 7/27 to the daughter?

8.1 (f). Let the legs, hypotenuse, and area of the triangle be a, b, c, K, $a \geq b$. Then

$$a^2 + b^2 = c^2, \qquad ab = 2K.$$

Solving for a and b we find

$$a = [\sqrt{c^2 + 4K} + \sqrt{c^2 - 4K}]/2, \qquad b = [\sqrt{c^2 + 4K} - \sqrt{c^2 - 4K}]/2.$$

8.2 (b).   Use mathematical induction. Assume the relation true for $n = k$. Then

$$
\begin{aligned}
u_{k+2}u_k &= (u_{k+1} + u_k)u_k \\
&= u_{k+1}u_k + u_k{}^2 \\
&= u_{k+1}u_k + u_{k+1}u_{k-1} - (-1)^k \\
&= u_{k+1}(u_k + u_{k-1}) + (-1)^{k+1} \\
&= u_{k+1}{}^2 + (-1)^{k+1},
\end{aligned}
$$

and so forth. Or use the expression for u_n given in 8.2 (c).

8.2 (c).   Set $v_n = [(1 + \sqrt{5})^n - (1 - \sqrt{5})^n]/2^n\sqrt{5}$.   Show   that $v_n + v_{n+1} = v_{n+2}$ and that $v_1 = v_2 = 1$. Then $v_n = u_n$.

8.2 (d).   Use the expression for u_n given in 8.2 (c).

8.2 (e).   Use the relation given in 8.2 (b).

8.3 (a). A has 121/17 denare and B has 167/17 denare.

8.3 (b).   33 days. This may be solved as a problem in *variation*.

8.3 (c).   Let x represent the value of the estate and y the amount received by each son. Then the first son receives $1 + (x - 1)/7$, and the second receives

$$2 + \{x - [1 + (x - 1)/7] - 2\}/7.$$

Equating these we find $x = 36$, $y = 6$, and the number of sons was $36/6 = 6$.

8.4 (b).   The following is essentially Fibonacci's solution to the problem. Let s denote the original sum and $3x$ the total sum returned. Before each man received a third of the sum returned, the three men possessed $s/2 - x$, $s/3 - x$, $s/6 - x$. Since these are the sums they possessed before putting back $1/2$, $1/3$, $1/6$ of what they had first taken, the amounts first taken were $2(s/2 - x)$, $(3/2)(s/3 - x)$, $(6/5)(s/6 - x)$, and these amounts added together equal s. Therefore $7s = 47x$, and the problem is indeterminate. Fibonacci took $s = 47$ and $x = 7$. Then the sums taken by the men from the original pile are 33, 13, 1.

8.4 (c).   382 apples.

8.6 (a).   Following is essentially the solution given by Regiomontanus. We are given (see Figure 86) $p = b - c$, h, $q = m - n$. Now $b^2 - m^2 = h^2 = c^2 - n^2$, or $b^2 - c^2 = m^2 - n^2$, or $b + c = qa/p$. Therefore

$$b = (qa + p^2)/2p \quad \text{and} \quad m = (a + q)/2.$$

Substituting these expressions in the relation $b^2 - m^2 = h^2$ we obtain a quadratic in the unknown a.

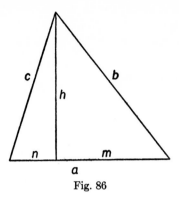

Fig. 86

8.6 (b). Following is essentially the solution given by Regiomontanus. Here we are given (see Figure 86) a, h, $k = c/b$. Set $2x = m - n$. Then

$$4n^2 = (a - 2x)^2, \qquad 4c^2 = 4h^2 + (a - 2x)^2,$$

$$4m^2 = (a + 2x)^2, \qquad 4b^2 = 4h^2 + (a + 2x)^2.$$

Then

$$k^2[4h^2 + (a + 2x)^2] = 4h^2 + (a - 2x)^2.$$

Solving this quadratic we obtain x, and thence b and c.

The triangle is easily constructed by using a circle of Apollonius. See Problem Study 6.8 (b).

8.6 (c). On AD produced (see Figure 87) take $DE = bc/a$, the fourth proportional to the given segments a, b, c. Then triangles DCE and BAC are similar and $CA/CE = a/c$. Thus C is located as the intersection of two loci, a circle of Apollonius and a circle with center D and radius c.

8.7 (a). \$29.

8.7 (b). 180/11 days.

8.7 (c). The price of each cask is 120 francs and the duty on a cask' is 10 francs.

8.7 (d). Suppose $a/c < b/d$. Then $ad < bc$, $ac + ad < ac + bc$, $a(c + d) < c(a + b)$, $a/c < (a + b)/(c + d)$, and so forth.

8.8 (a). Using standard notation we have

$$(rs)^2 = s(s - a)(s - b)(s - c),$$

or

$$16s^2 = s(s - 14)(6)(8),$$

and $s = 21$. The required sides are then $21 - 6 = 15$ and $21 - 8 = 13$. This is not Pacioli's method of solving the problem; his solution is needlessly involved.

8.9 **(b).** 463 7/23.

8.9 **(f).** Profits are proportional to the time the money is in the service of the company as well as to the amount.

8.9 **(g).** Over 19 per cent!

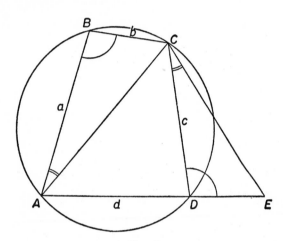

Fig. 87

8.10 **(b).** $H = (3ac - b^2)/9a^2$, $G = (2b^3 - 9abc + 27a^2d)/27a^3$.

8.10 **(d).** $x = 4$. The other two roots are imaginary.

8.11 **(a).** $(3 \pm \sqrt{5})/2$, $(-5 \pm \sqrt{21})/2$.

8.11 **(c).** $y^3 + 15y^2 + 36y = 450$, $y^6 - 6y^4 - 144y^2 = 2736$.

8.12 **(a).** $R\,q \lfloor \_\_ R\,c \lfloor \_\_ R\,q\,68\,p\,2\,\_\rfloor\,m\,R\,c \lfloor \_\_ R\,q\,68\,m\,2\,\_\rfloor\,\_\rfloor$.

8.12 **(b).** $\sqrt[3]{(4 + \sqrt{-11})} + \sqrt[3]{(4 - \sqrt{-11})}$.

8.12 **(c).** A cub $- B\,3$ in A quad $+ C$ plano 5 in A aequatur D solido 2.

8.13 **(b).** $\cos 5\theta = 16 \cos^5 \theta - 20 \cos^3 \theta + 5 \cos \theta$.

8.13 **(c).** $x = 243$.

8.13 **(d).** $x_2 = (r - qx - px^2 - x^3)/(3x^2 + 2px + q)$.

8.14 **(a).** 10.

8.14 **(b).** 28 beggars, $2.20.

8.14 **(c).** $92.

9.1 **(a).** See almost any text on college algebra or trigonometry.

9.1 **(b).** (1) Set $y = \log_b N$, $z = \log_a N$, $w = \log_a b$. Then $b^y = N$,

$a^z = N$, $a^w = b$, whence $a = b^{1/w}$, or $a^z = b^{z/w} = b^y$. Thus $y = z/w$.
(2) Set $y = \log_b N$ and $z = \log_N b$. Then $b^y = N$, $N^z = b$, whence
$N = b^{1/z} = b^y$. Thus $y = 1/z$. (3) Set $y = \log_b N$ and $z = \log_{1/n} (1/b)$.
Then $N^y = b$, $(1/N)^z = 1/b$, whence $N = b^{1/z} = b^{1/y}$. Thus $y = z$.

9.1 (c). $\log 4.26 = 1/2 + 1/8 + 1/256 + \cdots = 0.6294 \cdots$.

9.2 (b). $\cos c = \cos a \cos b$.

9.2 (c). (1) $A = 122° 39'$, $C = 83° 5'$, $b = 109° 22'$. (2) $A = 105° 36'$,
$B = 44° 0'$, $c = 78° 46'$.

9.5 (a). Acceleration is the increase in velocity during a unit period
of time.

9.6 (a). Open the compasses so that the given segment AA' stretches
between the 100 marks on the two simple scales of the compasses (see

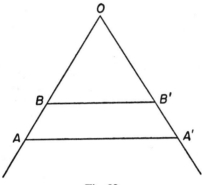

O

B   B'

A   A'

Fig. 88

Figure 88). Then the distance between the two 20 marks is one fifth of the
given segment. How does one solve this problem if the given segment is too
long to fit between the legs of the instrument?

9.6 (b). Open the compasses so that AA'/OA is the desired ratio of
scale. Then BB' is the new length to be associated with the old length OB.

9.6 (c). Connect a on one arm to b on the other. Through c on the
first arm draw a parallel to the line just drawn, to cut the other arm in the
sought fourth proportional.

9.6 (d). Open the compasses so that the distance between the 106
marks is equal to 150. Then the distance between the 100 marks represents
the amount of the investment one year ago. Perform this operation 5 times
to find the required amount.

9.7 (b). Show that $(HG)^2 = (HB)^2 = (BF)^2 - (HF)^2 = (HE)^2 - (HF)^2$,
and so forth.

9.7 (c). Two classes which can be placed in one-to-one correspondence are said to be *equivalent*, or to have the same *cardinality*. That is, it is possible to ascertain at a party of boys and girls, for example, that the count of boys is the same as that of the girls if each boy has one and only one partner among the girls, and vice versa. The difference between a finite and an infinite class is that an infinite class is equivalent to a part of itself.

9.8 (c). 1000 years.

9.8 (d). 25 A.U.

9.8 (f). 1 hour and 24 minutes.

9.10 (a). Choose for π' a plane parallel to the plane determined by point O and line l.

9.10 (c). Project line OU to infinity.

9.10 (d). Project line LMN to infinity and use the elementary fact that the joins of corresponding vertices of two similar and similarly situated triangles are concurrent.

9.10 (e). Choose a plane π' parallel to the minor axis of the ellipse and such that the angle θ between π' and the plane of the given ellipse is such that $\cos \theta = b/a$, where a and b are the semimajor and semiminor axes of the ellipse. Now project the ellipse orthogonally onto π'.

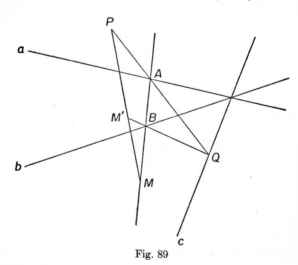

Fig. 89

9.10 (g). Let c be any line through the intersection of a and b (see Figure 89). Let PA cut c in Q and QB cut MP in M'.

9.11 (a). Let points 1 and 6 coincide so that the line 16 becomes the tangent to the conic at point 1.

9.11 (b).   Use 9.11 (a).

9.11 (c).   Let 1, 2, 3, 4 be the four points and 45 the tangent at $4 \equiv 5$, and let 12 cut 45 in P. Through 1 draw any line 16 cutting 34 in R, and then draw the Pascal line PR to cut 23 in Q. Then $5Q$ cuts 16 in a point 6 on the conic.

9.11 (d).   Take $1 \equiv 6$ and $3 \equiv 4$, and then take $2 \equiv 3$ and $5 \equiv 6$.

9.11 (e).   Take $1 \equiv 2$, $3 \equiv 4$, $5 \equiv 6$.

9.11 (f).   Use 9.11 (e).

9.13 (a).   This follows from the definition of the arithmetic triangle as given in Section 9-9.

9.13 (b).   By successive applications of 9.13 (a).

9.13 (c).   Use mathematical induction and 9.13 (a).

9.13 (d).   By 9.13 (c).

9.13 (e).   By 9.13 (a).

9.13 (f).   By 9.13 (e).

9.13 (g).   By 9.13 (c).

10.1 (a).

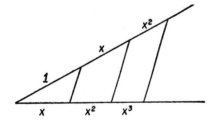

Fig. 90

10.1 (c).   By 10.1 (a), 10.1 (b), and 3.9 (c).

10.1 (d).   We have $r + s = g$ and $rs = h$.

10.1 (e).   We have $-r + s = -g$ and $-rs = -h$.

10.2 (a). $x^3 - 2ax^2 - a^2x + 2a^3 = axy$.

10.2 (b).   See 10.1 (c).

10.2 (c).   Consider the equations of L_1, L_2, L_3, L_4 in *normal*, or *perpendicular, form.* We then easily see that the equation of the locus is quadratic.

10.2 (d).   We find $x_2 - x_1 = m$.

10.4 (b). $r = (3a \sin \theta \cos \theta)/(\sin^3 \theta + \cos^3 \theta)$.

10.4 (c). $x = 3at/(1 + t^3)$, $y = 3at^2/(1 + t^3)$; loop $(0, \infty)$, lower arm $(-\infty, -1)$, upper arm $(-1, 0)$.

10.4 (d). $y = \pm x \sqrt{\dfrac{3 - x\sqrt{2}}{3x\sqrt{2} + 3}}$.

10.4 (e).   We find $h + m - k^2 = -2$, $k(m - h) = 8$, $mh = -3$. Eliminating m and h, by solving the first two relations for h and m in terms of k and then substituting in the third relation, we get

$$k^6 - 4k^4 + 16k^2 - 64 = 0,$$

a cubic in k^2.

10.5 (a).   (1) $\alpha = \beta$, (2) $\alpha + \beta = k$.

10.5 (b). $x = a(\cot\alpha - \cot\beta)/(\cot\alpha + \cot\beta)$, $y = a/(\cot\alpha + \cot\beta)$, $\alpha = \cot^{-1}[(a + x)/2y]$, $\beta = \cot^{-1}[(a - x)/2y]$.

10.5 (c).   (1) An ellipse, (2) a vertical straight line, (3) a straight line.

10.5 (d).   (1) $(x^2 + y^2)^2 = a^2(x^2 - y^2)$,   (2) $x^2 + y^2 + ax = a\sqrt{x^2 + y^2}$.

10.5 (f). $x = r\cos\phi\cos\theta$, $y = r\cos\phi\sin\theta$, $z = r\sin\phi$.

10.7 (a).   For the coordinates of a directed line segment of given length we may take the Cartesian coordinates of its initial point along with the angle the directed segment makes with the positive x-axis.

10.7 (b).   A line, not parallel to the xy-plane, is fixed by the x and y coordinates of the two points in which the line pierces the planes $z = 0$ and $z = 1$.

10.7 (c).   The straight line in hyperspace determined by the points $(x_1, \cdots, x_n)$ and $(y_1, \cdots, y_n)$ is the set of all points

$$((1 - t)x_1 + ty_1, \cdots, (1 - t)x_n + ty_n),$$

where t is an arbitrary real number.

10.7 (d).   Let d be the *distance* between the two points (see Section 10-3). Then we define the *direction cosines* of the line as the set of numbers

$$(y_1 - x_1)/d, \ (y_2 - x_2)/d, \cdots, \ (y_n - x_n)/d.$$

10.7 (e).   In 10.7 (c) limit t so that $0 \leqq t \leqq 1$.

10.8 (a). $\phi(n)$, for $n = 2, \cdots, 12$, is 1, 2, 2, 4, 2, 6, 4, 6, 4, 10, 4.

10.8 (b).   The only positive integers not exceeding p^α and not prime to p^α are the $p^{\alpha-1}$ multiples of p,

$$p, \ 2p, \cdots, \ p^{\alpha-1}p.$$

10.8 (e).   Let $n = ab$. Then, if $x^n + y^n = z^n$, we have $(x^a)^b + (y^a)^b = (z^a)^b$.

10.8 (f).   Suppose the point $(a/b, c/d)$, where a, b, c, d are integers, is on the curve. Then $(ad)^n + (bc)^n = (bd)^n$.

10.8 (g).   Consider the right triangle whose sides are given by

$$a = 2mn, \quad b = m^2 - n^2, \quad c = m^2 + n^2.$$

The area of this triangle is

$$A = \tfrac{1}{2}ab = mn(m^2 - n^2).$$

Taking $m = x^2$ and $n = y^2$, and setting $x^4 - y^4 = z^2$, we find

$$A = x^2 y^2 (x^4 - y^4) = x^2 y^2 z^2.$$

Therefore, if $x^4 - y^4 = z^2$ has a solution in positive integers x, y, z, there exists an integral sided right triangle whose area is a square number.

Finally, if $x^4 + y^4 = z^4$, then $z^4 - y^4 = (x^2)^2$.

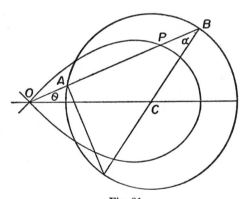

Fig. 91

10.10 (c).   For, by the definition of a cissoid (see Problem Study 4.4),

$$r = OP = AB.$$

By the law of sines applied to triangle OBC (see Figure 91),

$$(\sin \alpha)/(a\sqrt{2}/2) = (\sin \theta)/(a/2),$$

whence

$$r = AB = a \cos \alpha = a\sqrt{1 - 2 \sin^2 \theta},$$

and

$$r^2 = a^2 \cos 2\theta.$$

10.11 (b). Let the number chosen be x. Then

$$x = 3a' + a = 4b' + b = 5c' + c,$$

whence

$$(40a + 45b + 36c)/60 = 2(x - 3a')/3 + 3(x - 4b')/4 + 3(x - 5c')/5$$
$$= 2x - (2a' + 3b' + 3c') + x/60.$$

10.11 (c). In the general case, B ends up with $q(p + 1)$ counters.

10.12 (a). In the case of the ellipse consider a point on the curve as moving away from one focus and toward the other; in the case of the hyperbola consider a point on the curve as moving either away from or toward both foci. Now in the first case the sum of the focal radii of the moving point is constant, and in the second case the difference of the focal radii is constant.

10.13 (a). $\ln 2 = 0.69315$.

10.13 (b). $\ln 3 = 1.09861$.

10.13 (c). $\ln 4 = 2 \ln 2 = 1.3863$.

11.1 (a). Let M denote the given magnitude and m, taken less than M, any assigned magnitude of the same kind. By the axiom of Archimedes there exists an integer $n \geq 2$ such that $nm > M$. Since $n \geq 2$ it follows that $n/2 \leq n - 1$. Let M_1 be what remains after we subtract from M a part not less than its half. Then

$$M_1 \leq M/2 < (nm)/2 \leq (n - 1)m.$$

Continuing this process we finally get $M_{n-1} < m$.

11.1 (b). In Figure 92, $HA = HB < HD$. Therefore $\triangle HBD > \triangle HBA$, or $\triangle HKD > \frac{1}{2}(ABCD)$.

11.2 We have $(OM)(AO) = (OP)(AC)$. Summing we then find,

$$(\text{area of segment})(HK) = (\triangle AFC)(KC/3).$$

11.4 (a). Consider the triangular prism ABC–$A'B'C'$. Dissect the prism by the planes $B'AC$ and $B'A'C$.

11.4 (c). $V = 4a^3/3$.

11.4 (d). $V = \pi h^3/6$.

11.4 (e).   See 11.4 (d).

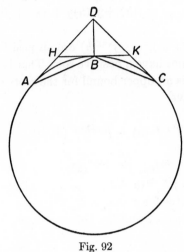

Fig. 92

11.5 (b).   Let O be any point in the mid-section and remove from the prismatoid the pyramids P_U and P_L having O as vertex and having the upper and lower bases, respectively, as bases. Then the volumes of P_U and P_L are given by $hU/6$ and $hL/6$. Now draw face diagonals, if necessary, so that all lateral faces of the prismatoid are triangles, and pass planes through O and the lateral edges, dividing the remaining piece of the prismatoid into a set of pyramids each having O as a vertex and a lateral triangular face of the prismatoid for opposite base. Show that the volume of one of these pyramids is $4hS/6$, where S is the area of the mid-section of the prismatoid included in the pyramid.

11.5 (c).   Any section, being a quadratic function of the distance from one base, is equal to the algebraic sum of a constant section area of a prism, a section area (proportional to the distance from the base) of a wedge, and a section area (proportional to the square of the distance from the base) of a pyramid. Thus the prismatoid is equal to the algebraic sum of the volumes of a parallelepiped, a wedge, and a pyramid. Now apply part (a).

11.5 (d).   Let $A(x) = ax^2 + bx + c$. Show that

$$V = \int_0^h A(x)\, dx = h[A(0) + 4A(h/2) + A(h)]/6.$$

11.6 (e).   Use mathematical induction.

11.8 (b).   Set $x = y + h$. Then, by 11.8 (a),

$$f(x) \equiv f(y + h) \equiv f(h) + f'(h)y + \cdots + f^{(n)}(h)y^n/n\ !.$$

If h is such that $f(h)$, $f'(h)$, $\cdots$, $f^{(n)}(h)$ are all positive, then the equation $f(y + h) = 0$ in y cannot have a positive root. That is, $f(x) = 0$ has no root greater than h, and h is an upper bound for the roots of $f(x)$.

11.8 (c).   We have

$$f^{(n-k)}(a + h) \equiv f^{(n-k)}(a) + f^{(n-k+1)}(a)h + \cdots + f^{(n)}(a)h^k/k\ !,$$

which shows that if $f^{(n-k)}(a)$, $f^{(n-k+1)}(a)$, $\cdots$, $f^{(n)}(a)$ are all positive, and h is also positive, then $f^{(n-k)}(a + h)$ must be positive. Similarly, the other functions are also positive for $x = a + h$.

11.8 (d).   The greatest root lies between 3 and 4.

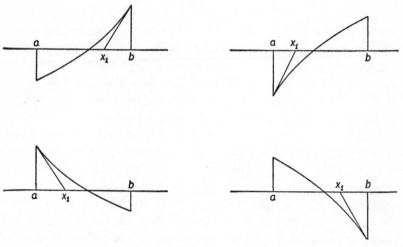

Fig. 93

11.9 (a).   Consider the four cases illustrated in Figure 93.

11.9 (b).   2.0945514, correct to seven places.

11.9 (c).   4.4934.

Index